W9-ABM-717

RECENT MATHEMATICAL DEVELOPMENTS IN CONTROL

RECENT MATHEMATICAL DEVELOPMENTS IN CONTROL

Proceedings of a conference held at the University of Bath in September, 1972, organised by the Institute of Mathematics and its Applications in association with the Institute of Physics, the Institution of Mechanical Engineers, the Institution of Electronic and Radio Engineers and the Institution of Electrical Engineers.

Edited by

D. J. BELL

Department of Mathematics and Control Systems Centre
The University of Manchester Institute of Science and Technology
Manchester, England.

1973

ACADEMIC PRESS · LONDON AND NEW YORK

A Subsidiary of Harcourt Brace Jovanovich, Publishers

ACADEMIC PRESS INC. (LONDON) LTD.
24–28 Oval Road,
London NW1

United States edition published by
ACADEMIC PRESS INC.
111 Fifth Avenue
New York, New York 10003

Library of Congress Catalog Card Number: 73–7031
ISBN: 0–12–085050–8

PRINTED IN GREAT BRITAIN BY
ROYSTAN PRINTERS LIMITED
Spencer Court, 7 Chalcot Road
London NW1

Contributors

J. C. ALLWRIGHT; *Department of Computing and Control, Imperial College of Science and Technology, Exhibition Road, London, SW7 2BT, England.*

S. BARNETT; *School of Mathematics, University of Bradford, Bradford, Yorkshire, England.*

J. M. C. CLARK; *Department of Computing and Control, Imperial College of Science and Technology, Exhibition Road, London, SW7 2BT, England.*

W. D. COLLINS; *Department of Mathematics, University of Strathclyde, Glasgow G1 1XH, Scotland.*

P. A. COOK; *Department of Mathematics, Loughborough University of Technology, Loughborough, Leicestershire, England.*

R. F. CURTAIN; *Control Theory Centre, University of Warwick, Coventry CV4 7AL, England.*

M. J. DAVIES; *Department of Applied Mathematics, University College of Wales, Aberystwyth, Wales.*

M. H. A. DAVIS; *Department of Computing and Control, Imperial College of Science and Technology, Exhibition Road, London, SW7 2BT, England.*

M. J. DENHAM; *Department of Computing and Control, Imperial College of Science and Technology, Exhibition Road, London SW7 2BT, England.*

P. J. FLEMING; *Department of Engineering Mathematics, The Queen's University, Belfast BT9 5AH, N. Ireland.*

E. A. FREEMAN; *Sunderland Polytechnic, Sunderland, England. (Present Address: Department of Engineering Science, University of Oxford, Parks Road, Oxford OX1 3PJ, England,)*

P. E. GILL; *Division of Numerical Analysis and Computing, National Physical Laboratory, Teddington, Middlesex, England.*

M. J. GREGSON; *School of Mathematics, Computing and Statistics, City of Leicester Polytechnic, Leicester LE1 9BH, England.*

C. J. HARRIS; *Department of Electronic Engineering, The University of Hull, England. (Present address: Control Systems Centre, U.M.I.S.T., Sackville Street, Manchester M60 1QD, England.)*

B. L. JOFFE; *Department of Chemical Engineering and Chemical Technology, Imperial College of Science and Technology, Exhibition Road, London SW7 2BT, England. (Present address: Computer Sciences S.A. Limited, 51 Juta Street, Braamfontein, Johannesburg, S. Africa).*

SIR JAMES LIGHTHILL, F.R.S.; *Department of Applied Mathematics and Theoretical Physics, Silver Street, Cambridge CB3 9EW, England.*

M. A. MAKHLOUF; *Texas Instruments Incorporated, Petroleum Exploration Division, P.O. Box 5621, Dallas, Texas 75222, U.S.A.*

M. J. E. MAYHEW; *Lanchester Polytechnic, Rugby, Warwickshire, England.*

A. MEES; *Department of Applied Mathematics and Theoretical Physics, Silver Street, Cambridge CB3 9EW, England.*

S. MIRICĂ; *Control Theory Centre, University of Warwick, Coventry CV4 7AL, England. (Present address: University of Bucharest, Bucharest, Romania.)*

W. MURRAY; *Division of Numerical Analysis and Computing, National Physical Laboratory, Teddington, Middlesex, England.*

M. M. NEWMANN; *Department of Engineering Mathematics, The Queen's University, Belfast BT9 5AH, N. Ireland.*

I. S. PACE; *School of Mathematics, University of Bradford, Bradford, Yorkshire, England.*

P. C. PARKS; *Control Theory Centre, University of Warwick, Coventry CV4 7AL, England.*

A. J. PRITCHARD; *Control Theory Centre, University of Warwick, Coventry CV4 7AL, England.*

H. H. ROSENBROCK; *Control Systems Centre, U.M.I.S.T., Sackville Street, Manchester M60 1QD, England.*

D. A. SÁNCHEZ; *Department of Mathematics, University of California, Los Angeles, California 90024, U.S.A.*

R. W. H. SARGENT; *Department of Chemical Engineering and Chemical Technology, Imperial College of Science and Technology, Exhibition Road, London SW7 2BT, England.*

B. SHANE; *School of Mathematics, University of Bradford, Bradford, Yorkshire, England.*

M. G. SINGH; *Control Engineering Group, Department of Engineering, University of Cambridge, Mill Lane, Cambridge CB2 1RX, England.*

C. STOREY; *Department of Mathematics, Loughborough University of Technology, Loughborough, Leicestershire, England.*

T. SUBBA RAO; *Department of Mathematics, U.M.I.S.T., Sackville Street, Manchester M60 1QD, England.*

H. TONG; *Department of Mathematics, U.M.I.S.T., Sackville Street, Manchester M60 1QD, England.*

J. H. WESTCOTT; *Department of Computing and Control, Imperial College of Science and Technology, Exhibition Road, London SW7 2BT, England.*

Preface

This book contains the proceedings of a Conference held from 5th–7th September, 1972 at the University of Bath, Somerset, England. The organizing committee for the Conference had two major aims in mind. First, to make known new developments in the theory and practice of automatic control in five main areas of research. These areas are stability, optimal control, filtering theory, distributed parameter systems and algebraic systems theory. Second, to help spread the message, particularly to young mathematicians in the United Kingdom, that Control Engineering is a worthwhile and stimulating field in which to apply Mathematics. It is a topic which should now take its place alongside those more traditional areas of Applied Mathematics which have figured so prominently in the syllabuses of University Mathematics Departments for so long. It is to be hoped that these proceedings will in some way help to stimulate those who teach Mathematics in establishments of higher education to include at least some introductory lectures on Control Theory.

The Conference programme was formulated by a committee consisting of Professor Diprose (Chairman), Dr. G. F. Bryant, Professor J. L. Douce, Dr. T. Fuller, Dr. E. T. Goodwin, Mr. W. G. Sherman, Professor C. L. Storey and myself. Special thanks are due to the chairman who organized the arrangements at Bath during the Conference and who also arranged for the lectures and discussions to be recorded.

The organizing committee arranged for invited speakers to present papers at the five sessions. These invited speakers were Professor Sir James Lighthill and Professor C. L. Storey (Stability), Professor J. H. Westcott and Professor L. Markus (Optimal Control), Dr. J. M. C. Clark (Filtering Theory), Dr. P. C. Parks (Distributed Parameter Systems), Professor H. H. Rosenbrock and Dr. S. Barnett (Algebraic Systems Theory). It was a disappointment that Professor Rosenbrock and Professor Markus were unable to be present at the Conference due to other commitments in Holland and America respectively.

The papers presented in this book are published in the order in which they were read at the Conference. Only an abstract of the paper read by Professor Sanchez appears in these proceedings. His contribution has already been published in another place (albeit in Spanish) and permission was not granted for it to appear again (even in English). On the other hand, there is one paper included in this book which was not presented at the Conference. This is the paper by Dr. P. A. Cook of the University of Technology, Loughborough. The paper describes recent work closely related to that described in the paper by Professor Rosenbrock and it was felt that such related research should, as far as possible, be kept together in one place.

The Bath Conference followed the fifth International Federation of Automatic Control (IFAC) Congress held in Paris, France during June 1972. Some papers reflect the work described at that earlier Conference and are thus of particular value to those who were unable to visit Paris.

No attempt has been made to standardize notation throughout this book. Inevitably notation becomes standardized within a particular area of automatic control but may well differ from that used in another area. For example, the notation for the transpose operation on a matrix is denoted by superscript T or prime. Both will be found in the following pages. However, whenever it has seemed necessary, an explanation of notation has been included in the text.

It is to be hoped that the list of references at the end of each paper will act as a guide to further reading for newcomers in the field. Similarly the discussions which took place after most lectures should contribute to the overall usefulness of the proceedings.

D. J. BELL

U.M.I.S.T., Manchester,
March 1973.

Acknowledgements

My thanks are due to many for help in the preparation of this book. In particular I must first thank the Institute of Mathematics and its Applications for assistance with the secretarial work and much of the Conference organization. Secondly, my thanks to Academic Press for the efficient and very professional way in which they produced the volume which is now before you. Finally I acknowledge with gratitude the unstinting help received from the authors of the papers and contributors to the discussions during the draft and proof stages of this book.

Contents

Part 1 Stability of Non-Linear Systems

Part II Optimal Control

Part III Filtering Theory

Part IV Control of Systems Governed by Partial Differential Equations

Part V Algebraic Systems Theory

1. Stability of Nonlinear Feedback Systems

SIR JAMES LIGHTHILL, F.R.S. AND A. MEES,

Department of Applied Mathematics and Theoretical Physics, University of Cambridge, England

LIGHTHILL. We are going to present this paper jointly. It is going to be a survey paper but one in which individual researches by Alistair Mees will be incorporated. The paper is concerned with nonlinear feedback systems. We are shown one here in Fig. 1.

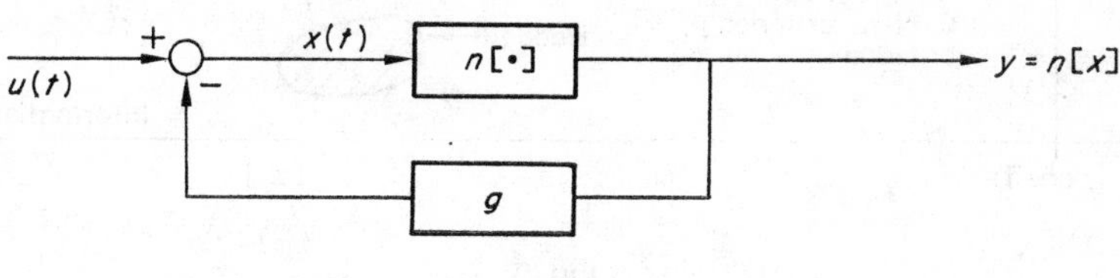

FIG. 1.

n stands for a nonlinear function and many of you will know how in mechanical or electrical or chemical problems there are types of nonlinear functions which may appear in the process and they may or may not include hysteresis: that is, they may represent one-valued functions *or* hysteresis loops. We show the system in Fig. 1 controlled by a linear feedback loop so that g is a linear functional involving perhaps time delays. Of course, the process may also involve inertias, inductances and time delays generally but, provided they are in series with the nonlinear function, then they can all be combined into a single linear functional, and the same type of equation

$$(gn + 1)\, x = u$$

holds provided x always represents the input into the nonlinear part of the system. We want to study methods of analysing the stability of these systems.

There are an enormous number of methods; we confine ourselves to practical ones, that is, those of practical use to designers. Generally speaking it is those which involve graphical output that are of greatest practical use. In the alliance between the human brain and the computer there is a lot to be said for making the computer produce its output in graphical form because this is easily assimilated by the human brain which has unequalled ability in pattern recognition. Also, graphical output is particularly suitable for making modifications, for example, seeing what the effect of changes in the design of the systems will be. Now how can we classify the different kinds of system that exist for analysing stability?

MEES. The methods we are going to discuss can be summarised on the following graph (Fig. 2).

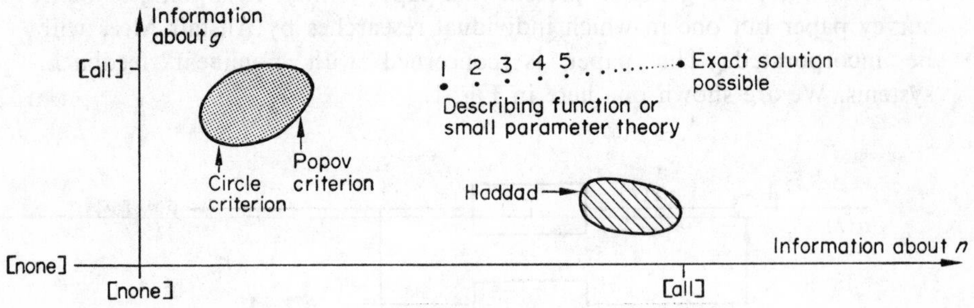

FIG. 2.

On the horizontal axis we measure the amount of information about the nonlinear part and on the vertical axis we measure the amount of information about the linear part. Useful methods, like the circle criterion, aren't at the top right of this graph at all, and don't try to hold on to all the available information. What I mean by this is that, while it's true that knowing nothing about the system means you can say nothing, it's also true that if you know everything about the system (and you refuse to temporarily sweep anything under the carpet) then you can probably still say nothing useful unless you can get an exact solution.

LIGHTHILL. In general most systems are too complicated to do that.

MEES. Yes, exactly. So the most successful methods usually try to keep a little bit of information about some aspect of the system and quite a lot about some other aspect, and see what can be said from that. For example, in Fig. 2 the shaded area is supposed to represent where the circle criterion would lie in the information space. It doesn't remember very much about the nonlinearity but it has a lot of information about the linear part through the Nyquist diagram. The Popov criterion is rather similar. Another method we are going to describe, due to Haddad, does a sort of conjugate of this. It keeps very little information about the linear part—just two numbers—but it keeps the whole of the graph for the nonlinear part. Then again, in Fig. 2, the dots are supposed to represent successive approximation methods such as the describing function method. If you could work your way from the first approximation, represented by the first dot, through an infinite number of dots, then you would get an exact solution. So, in general, the further you get from the axis the more work there is to do but the more information you receive, which isn't all that surprising.

Perhaps we could now talk about some representative methods.

LIGHTHILL. Let us remind you about the circle criterion. This is the most natural possible extension of the familiar Nyquist diagram for analysing a linear system. We take the case where the nonlinear function satisfies a bound criterion that $n[x]/x$ lies between the bounds α and β as in Fig. 3.

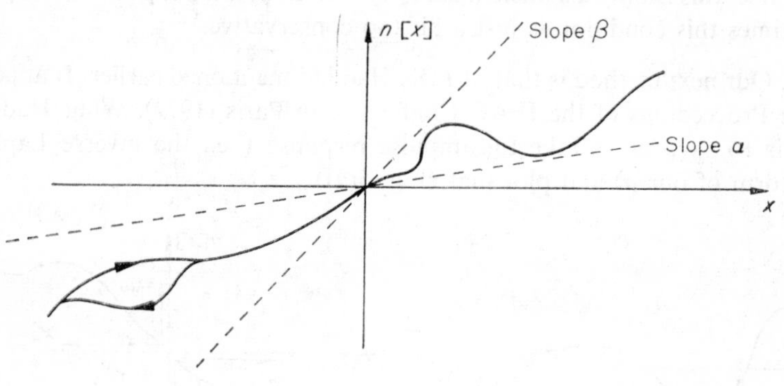

FIG. 3.

It is quite possible that $n[x]$ includes some sort of hysteresis loop but provided the whole loop lies between the lines with slopes α and β then the circle criterion applies. Instead of the ordinary criterion that the system is stable provided it does not encircle the point -1, the circle criterion, which is most easily derived from a contraction mapping principle, merely requires that the Nyquist diagram (i.e., the locus of the linear part of the system written as a frequency response in the complex plane, so that this is the locus of $g(i\omega)$ in the complex plane) must not encircle, or even intersect, a circle whose diameter stretches between $-1/\alpha$ and $-1/\beta$. Thus in Fig. 4 below, provided the dashed line does not intersect or embrace the shaded circle, one has a stable system.

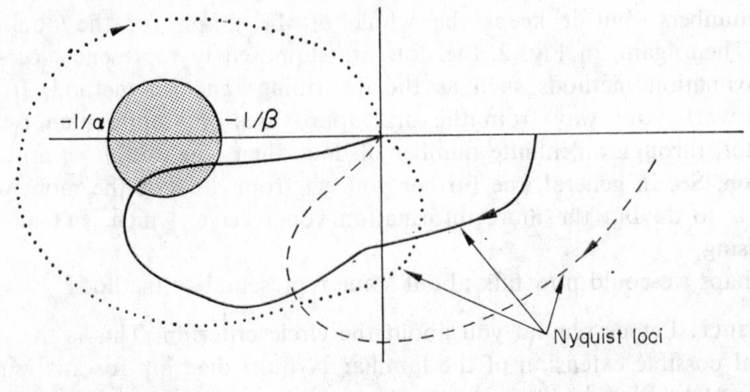

FIG. 4.

Of course, this is only a sufficient stability condition and, as you can imagine, sometimes this condition is just a bit too conservative.

MEES. Our next method is that of E.K. Haddad mentioned earlier. It appears in the Proceedings of the IFAC Conference in Paris (1972). What Haddad does is to say, let us take the impulse response (i.e., the inverse Laplace transform of our g) and plot that (Fig. 5(a)).

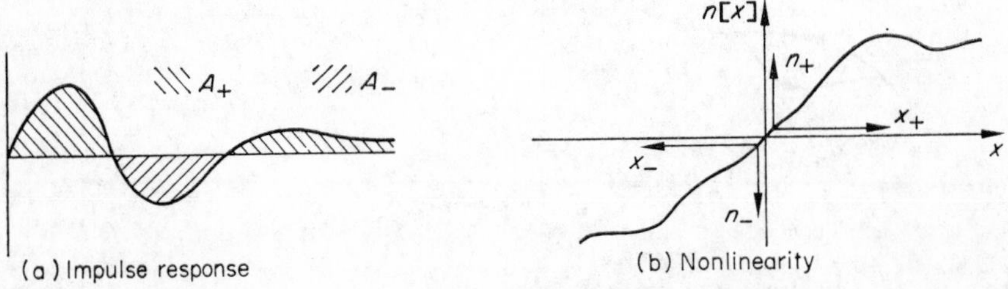

FIG. 5.

Work out the area above the axis (A_+) and the area below the axis (A_-). Now take the nonlinearity and just for convenience, split that up into parts of the graph which lie in different sectors. In Fig. 5(b) a nonlinearity that lies in the first and third quadrants has been taken; n_+ is the part that lies in the first quadrant and n_- the part that lies in the third quadrant. Observe that n_-

has been made a positive function of a positive variable x_- which is just the negative of x. Now we take a couple of loci which are defined by the equations

$$L_1: x_+ = A_- n_+(x_+) + A_+ n_-(x_-)$$
$$L_2: x_- = A_+ n_+(x_+) + A_- n_-(x_-)$$

and we plot these in the (x_+, x_-)—plane (Fig. 6).

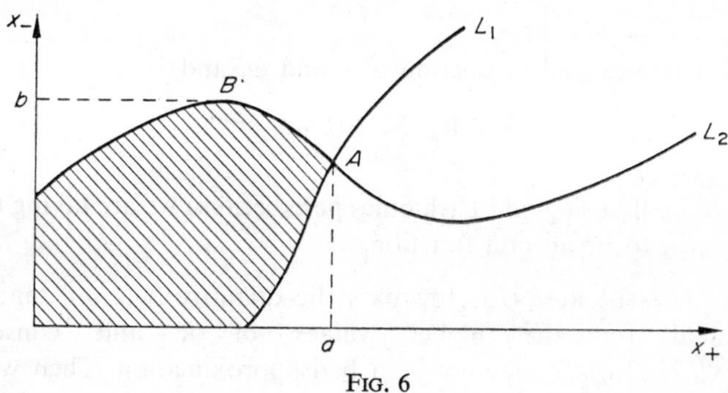

FIG. 6

Haddad says that if there is no area contained between these two loci then the system is stable. If, as in Fig. 6, an area is contained there, then the system is not known to be stable. Once more, this is just a sufficient condition so you cannot say it is unstable. However, what you can say is that x_+, the positive value of x, can never get further to the right than the point A and x_- can never get further up than the point B. x must therefore lie between $-b$ and a. If the system were completely unstable then the area enclosed by L_1 and L_2 would be infinite and a or b (or both) would be infinite.

LIGHTHILL. Well, these two methods we have described are extreme methods where it is particularly important to get as much information about the linear part of the system or the nonlinear part respectively, as possible. But there is a very well-known method which takes an intermediate course and that is the describing function method. This may be particularly valuable for a large class of control systems because basically a control system tends to be a low-pass filter; you are trying to design the system so that it responds to variations in the input which take place on a reasonable time scale, but it does not respond to noise, which takes place at high frequency. So you design your system, especially the linear part of your system, in the nature of a low-pass filter. For that reason the describing function method, which works on the basis of the so-called equivalent complex gain in response to a harmonic input, works quite well. The idea is that we are going to try to find a possible oscillation with a given amplitude a. This will be something like a limit cycle with given amplitude a and frequency ω and the describing function method works by saying that because the input is taken to be a periodic function, the output from the nonlinear part of the system will also be periodic and so can be represented by some sort of Fourier series. Thus,

let
$$x = a\cos\omega t = \mathrm{Re}\, ae^{i\omega t}$$
$$nx = \mathrm{Re}\sum_{k\ \mathrm{odd}} c_k e^{ik\omega t}$$

where c_k is complex and a function of a and ω, and

$$gnx = \mathrm{Re}\sum_{k\ \mathrm{odd}} G(ki\omega)c_k e^{ik\omega t}.$$

MEES. We are only using odd harmonics here because we are taking the nonlinear part just to be an odd function.

LIGHTHILL. Now because G is low-pass the quantity $|G(ki\omega)|$ tends to be rather small for the higher values of k and consequently $gnx \simeq \mathrm{Re}\, G(i\omega)c_1(a)e^{i\omega t}$ may not be a bad approximation. Then what that really means is that the nonlinear part of the system involves a nonlinear response function $N(a) = c_1(a)/a$:

$$gnx = -x \Rightarrow GNae^{i\omega t} \simeq -ae^{i\omega t}$$

i.e.
$$[G(i\omega)N(a) + 1]a = 0$$

i.e.
$$G(i\omega) = -1/N(a).$$

This enables one then to analyse a system using this standard criterion. In order to find a limit cycle the frequency ω and amplitude a must satisfy this criterion.

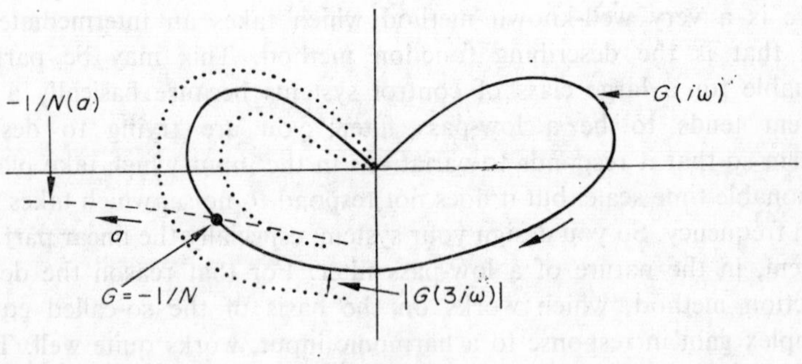

FIG. 7.

This case enables us again to use a Nyquist diagram as shown in Fig. 7. Here $G(i\omega)$ is the Nyquist locus for the linear part of the system and the limit cycle will be where that intersects the dashed curve which is the locus of $-1/N(a)$.

In Fig. 7 we have shown the case in which the system is unstable at zero amplitude and then as a increases an intersection is obtained which will be a limit cycle. The form of the instability will be a hunting with definite amplitude given by that point of intersection. So this, as it were, limits the instability of that particular system—it will only be an instability involving oscillations at that amplitude. But if, instead, the diagram was one where a was increasing in the opposite direction, where for small amplitude one had stability, there would nevertheless be a limit cycle but it would be an unstable one—what you might call a threshold of excitation for instability. But if the dashed curve were to cut the Nyquist locus again we would find a stable limit cycle, the oscillation that would ultimately take place if you once had noise in the input which exceeded that threshold.

Some idea of how good this approximation is can be obtained by drawing in bounds corresponding to $|G(3i\omega)|$ because it can easily be shown that the error in this method depends on the magnitude of $G(3i\omega)$. Thus, when these bands (Fig. 7) are drawn around the Nyquist locus, provided it isn't making any difference to the behaviour of the intersection, there is reasonable assurance that this method is giving fair accuracy.

MEES. These bands aren't giving us a genuine measure of the error, of course, but if the band is very wide then it is very likely that the error will be large. Incidentally, the dashed line in Fig. 7, the inverse describing function locus, can be shown to be inside the circle of the circle criterion (the shaded disc in Fig. 4) which is really a good thing because if it didn't, then the two would contradict one another rather more often than they do already, which is quite often enough. The Nyquist locus intersecting the describing function locus corresponds to the Nyquist locus entering the circle and indeed the inverse describing function locus is inside the circle. This was proved to me by Dr. Cook at Loughborough [1]. I always thought it must be true but couldn't prove it for a very general case.

LIGHTHILL. Let's take an example now.

MEES. I have an example here which is quite a simple feedback system and is shown in Fig. 8.

We are going to find that the circle criterion, Haddad's method, the describing function method and the Popov criterion (this latter we have

1. P. A. Cook. A note on the describing function for a sector nonlinearity *Proc. I.E.E.* **120** (1) (1973) p. 143

not described) are going to disagree about what actually happens. The circle criterion is going to say that perhaps the system is unstable although it can't be sure. Haddad will say something similar except that if it is unstable then the system response is bounded. The describing function is going to think that the system is stable and so is the Popov criterion.

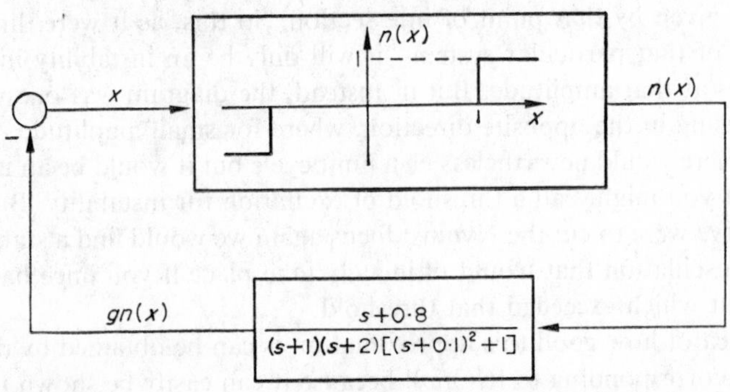

FIG. 8.

In this particular example the circle in the circle criterion degenerates into a half-plane: the shaded disc in Fig. 4 is now everything to the left of the vertical dashed line in Fig. 9.

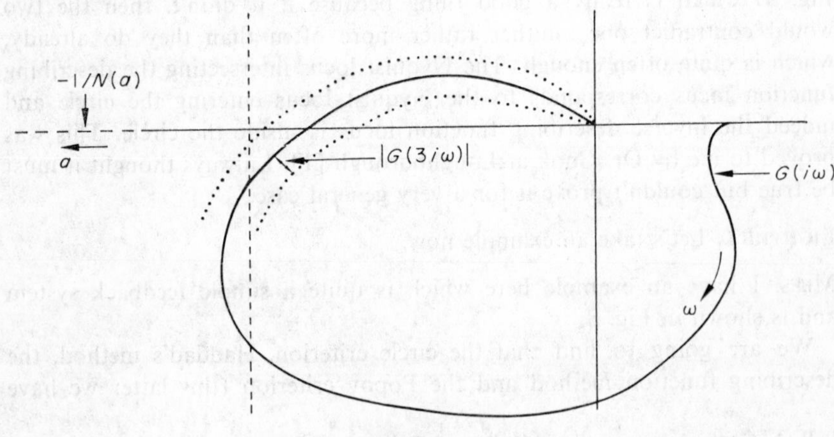

FIG. 9.

LIGHTHILL. It is a very bad case for the circle criterion.

MEES. Yes, we are being a little bit hard on it. However, the Nyquist locus is shown in Fig. 9, and it passes inside the degenerate circle so the circle criterion can't say very much. The other dashed line shows the describing function locus; it goes to $-\infty$ along the axis and as you can see from the figure the solid line doesn't intersect it. Even when we draw in the $3i\omega$ bands as shown it steers pretty well clear so the describing function says the system is stable and seems reasonably confident about it as well. Now Haddad's idea; the impulse response is shown in Fig. 10(a) and we find that the area above the axis is about 1·5 and the lower one is about 1·1.

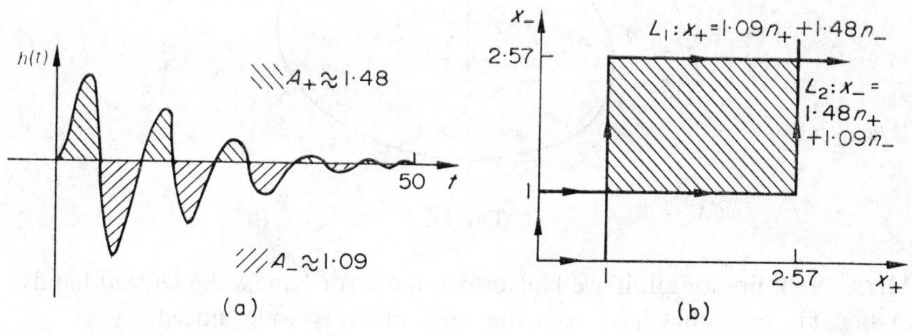

FIG. 10.

Fig. 10(b) shows the two loci L_1 and L_2 and sure enough there is an area contained between them. However, the area is bounded so Haddad says the worst that can happen is that there might be some sort of instability, but x can never get greater than 2·57 or less than -2·57. Haddad's criterion is therefore doing somewhat better in this case than the circle criterion. We haven't described the Popov criterion nor do we propose to, but it is a proper sufficient condition and it says that the system in the present example is stable.

LIGHTHILL. It is really a case where you had to have quite a lot of information about that nonlinear function and you also had quite a bit of information about the linear part of the function. The describing function was really the ideal one for analysing that. Now you can find cases where the describing function doesn't work. Jan Willems showed us a case (Fig. 11) where the nonlinear function is x^3 so the describing function locus is the whole of the negative axis. You have an ambiguous situation here if a simple describing function is used; it is not clear whether you will have stability or not, whether there is a limit cycle or not. The question is how can one do better when the describing function method gives an ambiguous answer? How can you improve it?

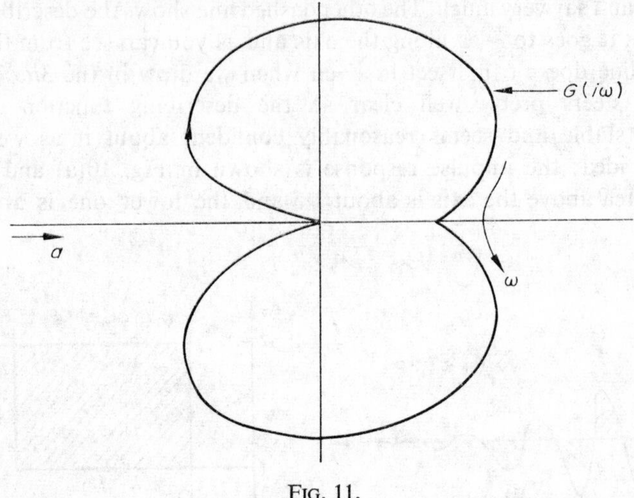

FIG. 11.

MEES. Well, first of all, if we had drawn the error bands, the $G(3i\omega)$ bands, in Fig. 11, we would have seen that they are very wide indeed.

LIGHTHILL. You've got to take the third harmonics into consideration.

MEES. Yes, certainly. We would hope that by taking the first and third harmonics and balancing both of these then things might work a little bit better. In general, we might hope that however bad the system was, if it limit cycles then by taking enough harmonics we would be able to find the limit cycle. Not only that; having found the limit cycle, by taking more harmonics we could achieve as much accuracy as we like.

LIGHTHILL. This is Alistair's own way of using the describing function method with several harmonic terms.

MEES. Let's try an input to the nonlinearity which has m harmonics so that we have this sum

$$x = \operatorname{Re} \sum_{k=1}^{m} a_k e^{ik\omega t}$$

where all the a_k can be complex. I shall assume that the nonlinear part is odd for convenience. Write

$$\mathbf{a}^m = (a_1, a_3, a_5, \ldots, a_m)^T.$$

If we put such an x into the nonlinearity then we get out an infinite number of harmonics:

$$nx = \text{Re} \sum_{k=1}^{\infty} c_k e^{ik\omega t}.$$

Each coefficient c_k in the Fourier series for the output will be a function of $\mathbf{a}^m$, i.e. a function of the first m harmonics. I'm going to suggest that we write these as column vectors. We write $\mathbf{c}(\mathbf{a}^1)$ as the first column. Beside that write $\mathbf{c}(\mathbf{a}^3)$: in other words we write the output vector when we try a first and third harmonic at the output. There is no point in repeating what has already been done so we just subtract off what was worked out the first time to give $\mathbf{c}(\mathbf{a}^3) - \mathbf{c}(\mathbf{a}^1)$. When we try and put in harmonics 1, 3, 5 we get $\mathbf{c}(\mathbf{a}^5)$ and subtract off what we already have, and so on. Thus we form the difference matrix

$$[\mathbf{c}(\mathbf{a}^1), \mathbf{c}(\mathbf{a}^3) - \mathbf{c}(\mathbf{a}^1), \mathbf{c}(\mathbf{a}^5) - \mathbf{c}(\mathbf{a}^3) \dots].$$

If Fourier series are any good at all, these columns will get smaller and smaller as we move off to the right. It looks as if we are some way to getting a successful approximation method.

To make this look a little more like the describing function method already discussed we divide the jth column of the difference matrix by a_j and define a matrix $\mathbf{N}$ by

$$N_{kj} = \{c_k(\mathbf{a}^j) - c_k(\mathbf{a}^{j-2})\}/a_j.$$

This definition has been arranged so that, for any value of m at all, $\mathbf{N}^m(\mathbf{a}^m)\mathbf{a}^m = \mathbf{c}^m(\mathbf{a}^m)$ where $\mathbf{N}$, $\mathbf{c}$ are matrices and vectors now instead of scalars.

LIGHTHILL. This is your describing function matrix and each column of it tells the effect of adding one more harmonic term and then dividing by the coefficient of that term to see what the response to adding that harmonic is.

MEES. That's right. The nice thing about it is that you only have to take another row and another column when you wish to take an extra harmonic into account. You don't need to work out everything again.

LIGHTHILL. Everything is worked out up to a certain level and then you add another column to the matrix.

MEES. Yes. It is an attempt to make the 1, 5 element the effect of the fifth harmonic of the input on the first harmonic of the output. It isn't quite that but it's as near as you seem able to manage.

I'm going to let

$$\mathbf{G} = \begin{pmatrix} G(i\omega) & 0 & --- \\ 0 & G(3i\omega) & \\ | & & \ddots \\ | & & \end{pmatrix}.$$

We have system equations which are

$$gnx = -x,$$

just to close the loop. Plugging in the above expressions gives

$$\mathbf{GNa} \simeq -\mathbf{a}.$$

$$[\text{cf. } G_1 N_{11} a_1 \simeq -a_1].$$

I've written "approximately equals" because we're only going up to order m. If we go to infinity then I'm claiming it would be exactly true.

LIGHTHILL The reason why **G** is a diagonal matrix is that g is a purely linear part of the system and so different harmonics don't interact as far as the G's are concerned.

MEES. The top left element of the describing function matrix is the describing function as it is normally thought of. Our equation for the system is

$$(\mathbf{GN} + 1)\mathbf{a} = 0.$$

We want to see what happens when we truncate this. Let the bit that we are going to ignore in **G** be called **H** and the bit we are going to ignore in **N** be called **W**. For **a**, the bit left over is $\boldsymbol{\alpha}$. Thus, let

$$\mathbf{G} = \begin{pmatrix} \mathbf{G}^m & 0 \\ 0 & \mathbf{H} \end{pmatrix}, \quad \mathbf{N} = \begin{pmatrix} \mathbf{N}^m & \mathbf{W} \\ \mathbf{V} & \mathbf{U} \end{pmatrix}, \quad \mathbf{a} = \begin{pmatrix} \mathbf{a} \\ \boldsymbol{\alpha} \end{pmatrix}.$$

Now use these to write out our system equation again. It turns out to be the two equations

$$(\mathbf{G}^m\mathbf{N}^m + 1)\mathbf{a}^m + \mathbf{G}^m\mathbf{W}\boldsymbol{\alpha} = 0$$

$$\mathbf{H}\mathbf{V}\mathbf{a}^m + (\mathbf{HU} + 1)\boldsymbol{\alpha} = 0.$$

The first equation consists of the mth order describing function terms plus a correction term. We would like to think that $\mathbf{G}^m\mathbf{W}\boldsymbol{\alpha}$ is rather small: that any solutions obtained by neglecting this term are reasonably good approximations to the true limit cycle. In fact $\mathbf{G}^m\mathbf{W}\boldsymbol{\alpha}$ is small of $\boldsymbol{\alpha}$ is small, the bit left over from **a**. The second of the above equations may be written as

$$\boldsymbol{\alpha} = -\,\mathbf{H}(\mathbf{V}\mathbf{a}^m + \mathbf{U}\boldsymbol{\alpha}).$$

The fact that α occurs on both sides of this equation doesn't matter because H is small, so we may do a contraction mapping, just a successive approximation method, and find a value of α. It will be governed by the size of $\mathbf{H}$ so that if $\mathbf{H}$ is small enough (i.e. if we've taken enough harmonics that the bit left over from the filter is low-pass enough), then α will be small. You can use this approach to prove convergence.

LIGHTHILL. That's what you have done in the recent number of the *J. Inst. Maths Applics.*

MEES. Yes, and it can also be used as a practical method because most people, particularly in this sort of case, would not want to work out error bounds. It's rather easy to plug this into a computer, e.g. suppose one has a first order describing function approximation to a limit cycle obtained by a graphical method. Put this into a computer and work out the third or fifth harmonic or whatever. If it looks as though it is converging then we know that the method is working properly.

LIGHTHILL. How easy is it to actually carry out the describing function matrix method with more than one harmonic?

MEES. It doesn't take very long. Our computer at Cambridge is not very fast and it takes perhaps a couple of seconds. On Jan Willem's example I went from third order to seventh in about two seconds with a very inefficient programme on a slow machine.

The describing function matrix method appears in *J. Inst. Maths Applics.* [2] and the stability question mentioned by Professor Lighthill earlier (i.e. whether the limit cycle is stable) is answered in a paper which is to appear in the same journal [3].

LIGHTHILL. Well, all that is what one can say about various methods for single-loop systems. But of course, an enormous number of practical control systems are multi-loop systems. So the question is whether one can extend any of these methods. Of course, the concept of graphical methods is essentially harder with multi-loop systems because in the single-loop complex plane you have all the simple topology of whether or not a curve encloses a point or another curve. Once you have more than one complex variable things tend to be a bit difficult. But there is a very interesting paper [4] by Falb, Freedman and Zames which does extend the circle criterion rather effectively in an interesting group of cases.

Suppose one has a nonlinear function which now is a vector function

2. MEES, A. I. The describing function matrix. *J. Inst. Maths. Applics.* **10,** (1972) 49–67.
3. MEES, A. I. Limit cycle stability. To appear in *J. Inst. Maths. Applics.*
4. FALB, P. L., FREEDMAN, M. I., ZAMES, G. Input-output stability—a general viewpoint. *Proc. IFAC* (*Warsaw*) (1969) §4.1.

of a vector. That is to say, one has a multi-loop system involving a number of nonlinear functions of a number of variables. Think of this as a vector function of a vector and suppose that this function satisfies the norm condition in Fig. 12, which is almost the same as the condition on $|n(x)|$ in the scalar case.

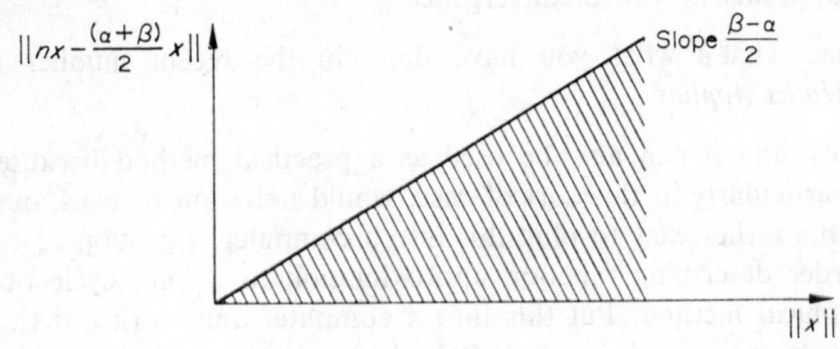

FIG. 12.

Then the paper referred to above, [4], proves by an extension of the same contraction mapping principle used originally to prove the circle criterion, the result illustrated in Fig. 13.

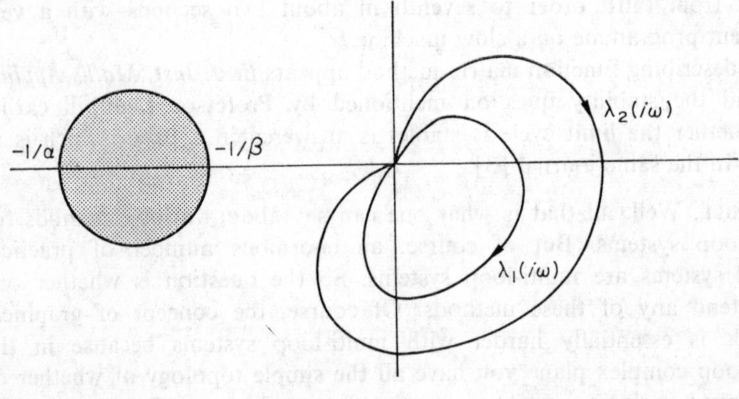

FIG. 13.

Here $\lambda_1, \lambda_2, \ldots$ are the different eigenvalues of the linear part of the system with a vector input. Unfortunately, we have to restrict this linear part to be normal, i.e. it is a matrix which commutes with its adjoint.

If you can draw these eigenvalues in an ordinary sort of Nyquist diagram then, provided none of them intersect or embrace the circle, the system is stable.

That is really the only result of general type we can find which would apply to a general multi-loop system. But there are some other criteria which can apply in what Professor Rosenbrock calls diagonally dominant systems, aren't there?

MEES. Yes. Professor Rosenbrock will be justifying diagonal dominance later.† We just state that such systems are very important. Figure 14 represents a system with a nonlinear part which can be described by a diagonal matrix.

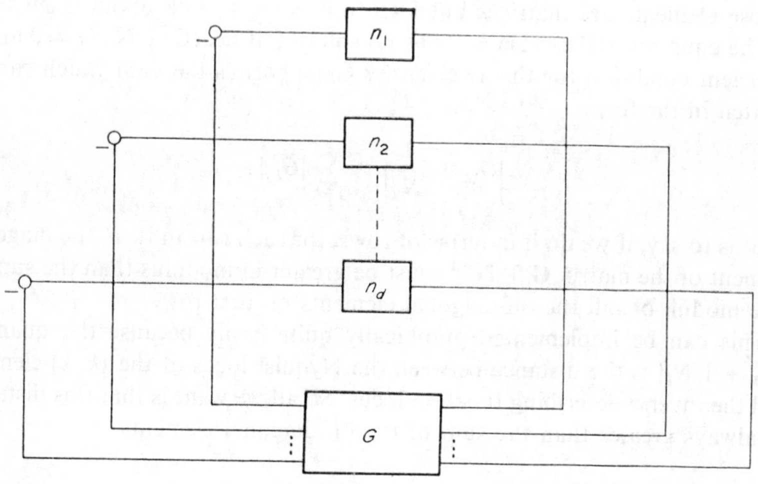

FIG. 14.

Each one of the nonlinearities is single input–single output but they are all incorporated into a large transfer function matrix. We should like to try to apply the describing function method again. I'm only going to talk about the first order graphical-type method although it will look very similar to our more general *m*th order method we did earlier.

LIGHTHILL. So the nonlinearity is diagonal and the linear part is diagonally dominant in the sense that diagonal terms dominate.

MEES. Yes, we are going to find that this is necessary to show stability. Now

$$(\mathbf{GN} + 1)\mathbf{a} = 0$$

is the first order describing function equation for the system, where $\mathbf{N}$ is a

† See pp. 345–365 of these Proceedings.

diagonal matrix with describing functions down the diagonal. This is the equation, according to the first-order describing function, for this system to have a limit cycle.

LIGHTHILL. Now we have a different kind of describing function matrix. In your method you have several harmonics, a matrix corresponding to the response of all those different harmonics. Now you are going back to the single frequency case but you've got a matrix because it's a multi-loop system.

MEES. Yes. We can put them together of course. You can have matrices. whose elements are matrices but I am not going to talk about them here.

The equation $(\mathbf{GN} + 1)\mathbf{a} = 0$ has no solution if $\det(\mathbf{G} + \mathbf{N}^{-1}) \neq 0$ and a sufficient condition for this is given by Gershgorin's theorem which can be written in the form

$$\left|G_{kk} + \frac{1}{N_k}\right| > \sum_{j \neq k} |G_{kj}|.$$

That is to say, if we do it in terms of rows, for each row in turn, the diagonal element of the matrix $\mathbf{G} + \mathbf{N}^{-1}$ must be greater in modulus than the sum of (the moduli of) all the off-diagonal elements on that row.

This can be implemented graphically quite easily because the quantity $|G_{kk} + 1/N_k|$ is the distance between the Nyquist locus of the (k, k)-element and the inverse describing function locus. So all we want is that this distance be always greater than the sum of the off-diagonal elements.

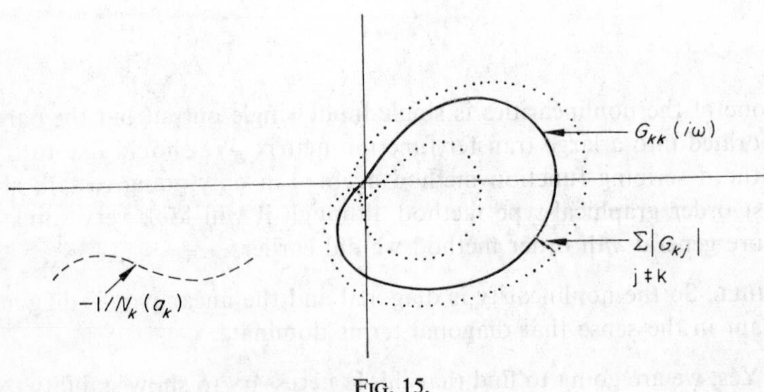

FIG. 15.

In Fig. 15 the dotted line is the edge of a band which has a width equal to the sum of the off-diagonal elements at each value of ω. If the describing function locus does not get inside the band then there cannot be a first order describing function solution. As usual, because a describing function is an approximation method, this could be wrong. On the other hand, I think it is likely to be right in a great many cases.

We can carry out the same procedure for a non-diagonal **n** and the same sort of thing happens. In this case the *N*-locus has a band around it too and the two bands must keep away from one another. Professor Rosenbrock has produced a circle criterion for diagonally dominant matrices which is rather similar. There is the same Nyquist locus and the same band of off-diagonal elements and the latter has to keep clear of the circle. By what has been said earlier, the describing function locus lies inside the circle (Fig. 16) so again they do not contradict one another.

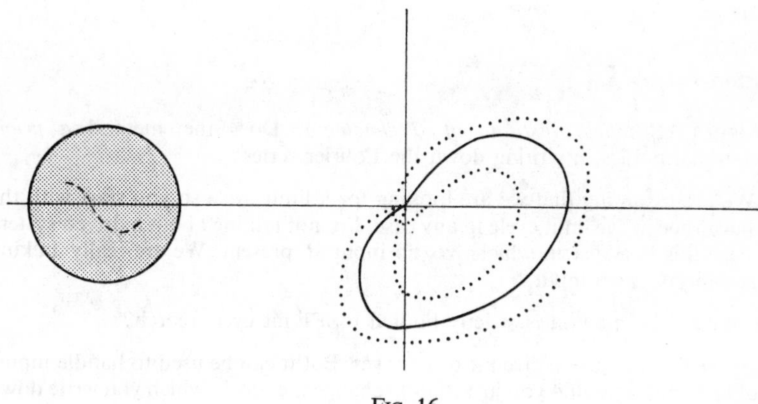

FIG. 16.

To learn more about diagonal dominance one should refer to the many papers by Professor Rosenbrock in the Proceedings of the Institute of Electrical Engineers. The circle criterion of Professor Rosenbrock will be described in a later paper† at this Conference and what I've said about the describing function is to appear in *Proceedings IEE* [5].

LIGHTHILL. I think we can conclude from all this that we have a pretty good spectrum of methods in the single-loop cases. I don't think any one of those methods offers a cure-all; one has always to match the methods to the type of problem that you've got. But for the different types of problem you have a fair range of methods and you can bring out the one that is appropriate. It's pretty clear that more work is needed on

† See pp. 345–365 of these Proceedings.

5. MEES. A, I., Describing functions, circle criteria and multiple-loop feedback systems. *Proc. I.E.E.* **120** (1) (1973) 126.

multivariable systems and that is where it is well worth pushing a good deal beyond these methods; getting better error bands and particularly being able to deal with more general multi-loop systems than those which are strictly diagonally dominant.

I think what we have tried to put across to you above all is the merit of these graphical methods, with curves intersecting curves and bands round the curves. After all, when one is analysing a feedback system of this kind one has other things to assure besides stability. The same diagrams are very good for assuring quality of control systems. By gradually making changes to the system one can simultaneously investigate graphically the offset and other features of importance for quality purposes and assure stability within the degree of accuracy given by the type of methods we've tried to describe.

Discussion

CHRISTOPHER (*Cranfield Institute of Technology*). Does the method *a priori* exclude subharmonics in writing down the Fourier series?

MEES. We are assuming that we are looking for a limit cycle so you start with the lowest harmonic in the limit cycle in any case. I'm not talking about a driven system in this case this is a system which has no input at present. We are really talking about stability to zero input.

CHRISTOPHER. The method is strictly limited to a limit cycle search?

MEES. In the form that I've given it to you, yes. But it can be used to handle inputs and subharmonics as well if you just slightly change the way in which you write down the equation.

CHRISTOPHER. How do you decide which is the lowest subharmonic?

MEES. Well, this is the sort of thing that graphical methods aren't so good at. You seem to have to try each subharmonic in turn: 1st, then 3rd, then 5th and so on. Very recently I've been doing some work on subharmonic prediction, and it's going quite well. For example, I have a graphical method for seeing if there's a subharmonic solution of any given order.

FREEMAN (*Sunderland Polytechnic*). On the question of convergence I wasn't quite sure whether your method gives an *a priori* estimate of the convergence rate or if one must calculate the rate as one proceeds with the expansion of the describing function matrix.

MEES. For practical calculations you examine the convergence in an *a posteriori* sense. At the end of the work, when you have some numbers out, it's a lot easier to get error bounds where you want them. However, you're probably not interested in them at that stage when it looks as though it's convergent.

For proving convergence we can show *a priori* that under certain conditions, which aren't too restrictive, the method will always work if you take a large enough number of harmonics to start off with.

LIGHTHILL. In your paper you do in fact give a convergence proof for a pretty general situation, don't you?

MEES. Yes.

PARKS (*University of Warwick*). Could the authors clarify the relationships between their work and the dual input describing function that was all the rage a few years ago?

MEES. Our method is a generalization of that type of idea. Dual input describing function is used in two different senses: sometimes it means the describing function with first and third harmonic, sometimes a constant term and a first harmonic. But all of this comes in as special cases.

LIGHTHILL. I think the valuable new feature is getting a describing function matrix which enables the formalism of the single-input describing function to be taken over to any number of harmonics. Since computers handle matrices with great facility it is very convenient for computational analysis.

MEES. And as with the describing function, you work it out once for a nonlinearity and then apply it to any system containing that nonlinearity. You don't have to work it out afresh everytime.

REVINGTON (*C.E.G.B.*). I wonder if Mr. Mees is familiar with the Bogoliubov–Krylov–Mitropolsky method. How does it fit in with your work?

MEES. They approach the problem from a slightly different view-point (slowly varying amplitude and phase) but it amounts to the same thing in the first order case. I'm not quite sure whether it's exactly the same in the higher order cases but it presumably gets to the same result in the end.

LIGHTHILL. In the single-input case the mathematics is the same but I think the reasoning is different. Applying the describing function to a control system is very much using the fact that you have designed your system to be a low-pass filter. They are analysing systems which are not specifically control systems and so it is more uncertain whether it's a good approximation to ignore higher harmonics. But in the single-input case it does come out the same, possibly because those authors have in mind resonant systems, where the response near resonance greatly exceeds the response at higher frequencies.

MEES. The Russians have done a lot of work on this. They call it the method of harmonic balance which is more descriptive than describing function. The point is, this is not a new idea I've used in my work; it's a different approach, different formalism, which happens to ease the work involved.

LIGHTHILL. It makes it practical to include more harmonics than it would normally be feasible to do.

2. Stability and Controllability

C. STOREY

Department of Mathematics,
University of Technology, Loughborough, England

1. Introduction

In this lecture I would like to compare the development of the ideas of controllability and stability beginning with well-known results and leading up to the concept of "final stability" which it is hoped will allow Liapunov-like theorems to be used to discuss the controllability of non-linear systems. I have made no attempt to deal exhaustively with controllability and stability and give only representative results and definitions from each field.

2. Controllability

Consider a control system defined by the set of differential equations (using the usual notation)

$$\dot{x} = f(x, u) \tag{1}$$

where

$$x \in E^n, \qquad u \in E^m.$$

Then, roughly speaking, the system (1) is said to be completely controllable to the origin (or to any point x_0) if every state can be transferred to the origin (to x_0) by some admissible control $u(t)$ in a finite time. There are many refinements of this idea (see, for example, [1] and [2]).

When (1) is a linear time-invariant system

$$\dot{x} = Ax + Bu \tag{2}$$

then the theory of controllability is almost complete. Thus (2) is completely controllable if and only if

$$\operatorname{rank}(B, AB, A^2B, \ldots, A^{n-1}B) = n.$$

Notice that this is a purely algebraic condition concerning the matrices A and B.

For a linear system the reachable set (i.e. the set of points attainable from a given point with all admissible controls) has the attractive geometric properties of compactness and convexity. It is these properties that allow controllability theory to be developed with relative ease for linear systems (see [3]). Of course, the time-varying linear case

$$\dot{x} = A(t)x + B(t)u \tag{3}$$

is more difficult to treat than (2). However, a typical result due to Kalman is that if the controllability matrix

$$W(t_0, t) = \int_{t_0}^{t} \Phi(t_0, s)B(s)B'(s)\Phi'(t_0, s)\, ds \tag{4}$$

has maximal rank for some t_0 and every t then (3) is completely controllable. In (4) Φ is the transition matrix of

$$\dot{x} = A(t)x.$$

Turning now to the general non-linear system (1) there are some "linearization" or "local" results [3]. For example, suppose that $f(0, 0) = 0$ and the class of admissible controls satisfies $\|u(t)\| \leqslant \varepsilon$ for some positive ε. Then provided

$$\text{rank}\,(B, AB, A^2B, \ldots, A^{n-1}B) = n$$

where

$$A = \left(\frac{\partial f_i}{\partial x_j}(0, 0)\right)$$

$$B = \left(\frac{\partial f_i}{\partial u_j}(0, 0)\right)$$

the domain of controllability contains an open neighbourhood of the origin. The size of this domain is not known but if the above rank condition holds then such a domain exists. Roughly speaking the controllability of the first approximation to (1) (with respect to both state and control) implies that there is an open neighbourhood of the origin in which every point is controllable to the origin for the original system (1).

There has been a generalization of this result to the case where although x is small u may not necessarily be so [4]. This could be important in practice since it is not always convenient to consider small controls.

Recent mathematical research on controllability has tended towards generalization and abstraction. A few areas in which work is progressing are given below:

(i) Controllability for abstract linear dynamical systems.

(ii) Extension to discrete systems (including analogues of the above local results) and to systems of functional differential equations (e.g. systems with time lags).

(iii) Systems in non-linear state spaces.

(iv) The general geometry of attainable sets.

The literature on the above work is extensive and rather than list specific references I refer the interested reader to recent issues of *Journal of Differential Equations* and *SIAM Journal of Control.* In particular *SIAM J. Control* (1972), **10**, No. 2, is completely devoted to recent theoretical advances.

3. Liapunov Stability

There has been a somewhat parallel development in stability theory. The origin of the n-vector differential system

$$\dot{x} = f(x), \qquad f(0) = 0 \tag{5}$$

is said to be stable (in the sense of Liapunov) if given a region $||x|| < R_1$ about the origin a second region $||x|| < R_2$ exists such that any trajectory starting inside the second region will never leave the first no matter how large t becomes. If the origin of (5) is stable and, in addition, a region $||x|| < R_3$ can be found such that all trajectories with initial points in this region tend to the origin as t tends to infinity then the origin is said to be asymptotically stable. (These geometric definitions can easily be couched in an analytical manner). Many extensions and refinements of these fundamental definitions exist (see, for example, [5]–[11], [13] and [14]).

As was the case with controllability much is known when the system is linear. Thus for the time invariant system

$$\dot{x} = Ax$$

the origin is asymptotically stable if and only if the solution P of the matrix equation

$$A'P + PA = -Q$$

is positive definite for any positive definite, symmetric Q.

The origin of the linear time-varying system

$$\dot{x} = A(t)x \tag{6}$$

is asymptotically stable if $\|\Phi(t)\| \to 0$ as $t \to \infty$ where $\|.\|$ is some suitable norm and $\Phi(t)$ is a transition matrix for (6).

This theorem and the controllability theorem given above say little about how the relevant properties depend upon the elements of the matrix $A(t)$ and are, in general, numerical in nature only. As before, the general non-linear form of (5) is more difficult to deal with but the following important result concerning the first approximation is valid.

Let

$$\dot{x} = Ax$$

be the first approximation to (5) so that (5) can be written

$$\dot{x} = Ax + g(x), \qquad g(0) = 0$$

where $g(x)$ is analytic and of degree greater than or equal to 2. Then if the stability behaviour of the matrix A is significant the origin of (5) and its first approximation have equivalent stability behaviour. The stability behaviour of A is significant if its eigenvalues either have all their real parts negative or at least one of them has a positive real part. In the former case (5) and its linear approximation would both be asymptotically stable and in the latter case both would be unstable.

Notice that just as in the case of the local controllability result no idea of the extent of the region of asymptotic stability is given for the non-linear system. (If a linear system is asymptotically stable at all, it is so everywhere). It is possible to extend this result of linearization somewhat (see [5], [9] and [10]). If the stability behaviour of the matrix A is not significant then the stability behaviour of (5) cannot be inferred from that of its first approximation and the terms in $g(x)$ must be taken into account. (For this "critical" case see [5], [11] and [12]).

For stability theory there are a series of powerful results due to Liapunov to enable the general non-linear case to be dealt with. These results involve the sign-definiteness of certain scalar functions known as Liapunov functions. A typical such theorem is the following. The origin of (5) is asymptotically stable provided a positive definite function $V(x)$ can be found such that its derivative with respect to (5), i.e., $\dot{V} = \nabla V.f$, is negative definite. When such a Liapunov function can be found the stability of (5) is determined without any need to solve the system equations. The definitions of stability and the use of Liapunov functions can, of course, with suitable care, be extended to the non-autonomous system $\dot{x} = f(x, t)$.

(For extensions see [5]–[11], [13] and [14].) Current research on stability theory includes:

(i) The construction of Liapunov functions.

(ii) Usual abstract generalizations.

(iii) Partial stability (i.e. stability involving only some of the components of the system).

(iv) Stability of stochastic systems.

(v) Comparison theory and the use of differential inequalities.

(vi) Vector Liapunov functions.

(vii) Liapunov functionals and the stability of partial differential equations.

(Again, see recent literature, I.F.A.C. proceedings, etc.).

Unfortunately there are a number of difficulties associated with the Liapunov concept of stability.

(i) It deals with an infinite time interval when obviously a finite time interval is more important in practice.

(ii) In the definition the number R_2 depends on R_1, so that for R_1 given, R_2 may be very small and the system, although stable according to the definition, would not be so in a practical sense.

(iii) Perturbations are not taken into account.

The above disadvantages have led to generalizations of Liapunov stability three of which are now discussed.

4. Final-time Stability

The first generalization is known as practical stability and is due to La Salle [14]. The vector system equation is

$$\dot{x} = f(x, t) + p(x, t) \tag{7}$$

where $p(x, t)$ is a perturbation. Suppose that a set Q_0 of initial states, a set $Q \supset Q_0$ of acceptable responses, and a positive number δ are given. Then the system (7) is practically stable if for each perturbation $p(x, t)$ satisfying $\|p(x, t)\| < \delta$ (for all $x \in R^n$ and $t \geqslant 0$), each $x_0 \in Q_0$ and each $t_0 \geqslant 0$, the solution $x(t, x_0, t_0) \in Q$ for all $t \geqslant t_0$.

Originally practical stability was still concerned with an infinite time interval and a second generalization was introduced by Weiss and Infante [15] (among others) to deal with a finite time interval. Let the three sets α, β and γ be defined as follows:

$$\alpha: ||x|| < a, \qquad \beta: ||x|| < b, \qquad \gamma: ||x|| < c$$

with $c < a \leqslant b$ and let J be the finite time interval $[t_0, t_0 + T)$. Then the system

$$\dot{x} = f(x, t)$$

is said to be stable with respect to α, β, J if, for any trajectory $x(t)$, the condition $x(t_0) \in \alpha$ implies that $x(t) \in \beta$ for all $t \in J$. If, in addition, any trajectory $x(t)$, with $x(t_0) \in \alpha$, is such that $x(t) \in \gamma$ for all $t \in J$, $t \geqslant t_1$, for some $t_1 \in J$ (which may depend on the particular trajectory), then the system is contractively stable with respect to α, β, γ and J. Liapunov-like theory may be extended to finite-time theory and its generalizations.

The third generalization is a natural extension of finite time stability to the situation where α is not necessarily a sub-set of β[16]. That is to say the general behaviour of trajectories of a system of differential equations with respect to two arbitrary sets α and β and a finite time interval J is considered. Thus the system

$$\dot{x} = f(x, u)$$

is said to be semi-finally stable with respect to the sets $[\alpha, \beta, U, t_1 \in J]$ if for any trajectory $x(t)$ with $u \in U$ (a subset of the admissible controls) the condition $x(t_0) \in \alpha$ implies $x(t_1) \in \beta$.

The system is finally stable with respect to $[\alpha, \beta, U, t_1 \in J]$ if it is semi-finally stable with respect to the sets $[\alpha, \beta, U, t_2 \in J]$ for all $t_2 \in J$ with $t_2 \geqslant t_1$. Geometrically speaking semi-final stability occurs when all trajectories starting in α are in β at time t_1 and final stability occurs if once the trajectories are in β at time t_1, they remain in β for the rest of the interval J. (Many refinements of these definitions and extensions to dynamical systems and discrete systems can be found in [17]).

Final-time stability can now be related to controllability in an obvious manner. For example, if α is a point x_0 and β the set $||x|| < \varepsilon$ then x_0 is "approximately" controllable to the origin if it is semi-finally stable with respect to $[\alpha, \beta, U, t_1 \in J]$. If the system is finally-stable with respect to the above sets then there exists at least one admissible control which will drive the trajectory(ies) starting at x_0 into an ε-neighbourhood of the origin and keep it in this neighbourhood for the rest of the time in J.

The interest, of course, in this connection between final-stability and controllability is in the hope that Liapunov-like theory can be used to deal with controllability of non-linear systems. Sufficient (and in some cases necessary) conditions can be given for the various kinds of final-stability in terms of Liapunov theory [16], [17]. Comparison theory [13] is used to avoid undue restrictions on the functions employed. The following is a typical theorem.

The system $\dot{x} = f(x, u)$ *is semi-finally stable with respect to the sets* $[\alpha, \beta, U, t_1 \in J]$ *if there exist two functions* $V(x,t)$ *and* $w(t,r)$ *such that*:

(i)
$$\left.\begin{aligned} V_M{}^{\alpha}(t_0) &= \sup_{x \in \alpha} V(t_0) \\ V_m{}^{\beta^c}(t_1) &= \inf_{x \in \beta^c} V(t_1) \end{aligned}\right\} \textit{are finite.}$$

(ii)
$$\dot{V}(x,t) \leqslant w(t, V(x,t))$$

(iii) *The maximal solution* $r_M(t)$ *of the scalar equation*

$$\dot{r} = w(t, r)$$

with $r_M(t_0) = V_M{}^{\alpha}(t_0)$, *is such that*

$$r_M(t_1) < V_m{}^{\beta^c}(t_1)$$

In (ii) *above the derivative* $\dot{V}$ *is given by*

$$\dot{V} = \frac{\partial V}{\partial t} + \nabla V \,.\, f(x,u)$$

but the theory holds when the solutions of the system $\dot{x} = f(x, u)$ *are not necessarily unique and a generalized derivative is taken* [16], [17].

In the above theorem the set β cannot be reduced to a point but this difficulty can be avoided by imposing extra conditions on $V(x,t)$. Gershwin and Jacobson [18] have used such modified theorems to reproduce the results for controllability of linear systems. They also give some non-linear examples and deal with optimal control problems. Other applications can be found in [16], [17].

In conclusion I give a simple illustrative example [16]. Consider the scalar differential equation

$$\dot{x} = u$$

and let

$$\alpha \equiv x_0, \qquad \beta \equiv ||x|| < \varepsilon, \qquad \varepsilon > 0,$$

and suppose that a control $u(t)$ is required to drive a trajectory from α to β at time $t = t_1$.

Choose

$$V = x^2;$$

then

$$V_M^{\alpha}(t_0) = x_0^2, \qquad V_m^{\beta^c}(t_1) = \varepsilon^2.$$

Let

$$u = \lambda(t)x$$

with

$$\int_{t_0}^{t_1} \lambda(t)dt < \log_e \frac{\varepsilon}{\|x_0\|} \tag{8}$$

to give

$$\dot{V} = 2\lambda(t)x^2 = 2\lambda(t)V.$$

The comparison equation is

$$\dot{r} = 2\lambda(t)r, \qquad r(t_0) = x_0^2$$

so that

$$r_M(t) = x_0^2 \exp\left\{2\int_{t_0}^{t_1} \lambda(s)\,ds\right\}$$

and

$$r_M(t_1) < \varepsilon^2 = V_m^{\beta^c}(t_1).$$

Hence any $u = \lambda(t)x$ satisfying (8) is an adequate control. Notice that this control will drive the trajectory as close as is wished to the origin though not to the origin itself. To actually get to the origin with this type of control would require the control to be unbounded. This illustrates the point that with a linear system if it is required to drive a state to the origin in finite time it is not possible to use a feedback control (i.e. proportional to the state). This is because with such a control the origin of the system would be a critical point and, therefore, impossible to reach in a finite time.

References

1. Rosenbrock, H. H. "State-space and Multi-variable Theory". Nelson, London (1970).
2. Kalman, R. E., Falb, P. L. and Arbib, M. A. "Topics in Mathematical System Theory". McGraw-Hill, New York (1969).
3. Lee, E. B. and Markus, L. "Foundations of Optimal Control Theory". John Wiley and Sons (1967).
4. Yorke, J. A. *S.I.A.M. J. Control* **10** (2) (1972), 334–338.

5. Hahn, W. "Theory and Application of Liapunov's Direct Method". Prentice-Hall (1963).
6. Willems, J. L. "Stability Theory of Dynamical Systems". Nelson (1970).
7. Krasovskii, N. N. "Stability of Motion". Stanford University Press (1963).
8. Barbasin, E. A. "Introduction to the Theory of Stability". Wolters-Noordholt (1970).
9. Lehnigk, S. H. "Stability Theorems for Linear Motions with an Introduction to Liapunov's Direct Method". Prentice-Hall (1966).
10. Barnett, S. and Storey, C. "Matrix methods in Stability Theory". Nelson (1970).
11. Zubov, V. I. "Methods of A. M. Liapunov and their Application". Noordhoff (1964).
12. Liapunov, A. M. "Stability of Motion". Academic Press, London and New York (1966).
13. Lakshmikantham, V. and Leela, S. "Differential and Integral Inequalities", Vol. 1. Academic Press, London and New York (1969).
14. La Salle, J. P. and Lefschetz, S. "Stability by Liapunov's Direct Method with Applications". Academic Press, London and New York (1961).
15. Weiss, L. and Infante, E. F. *Proc. Nat. Acad. Sci., U.S.A.* **54** (1965), 44–48.
16. Ben Lashiher, A. M. and Storey, C. *J. Inst. Maths. Applics.* **9**, (1972), 397–410.
17. Ben Lashiher, A. M. "Stability over a Finite-time Interval" *Ph.D Thesis, Loughborough University of Technology*, (1970).
18. Gershwin, S. B. and Jacobson, D. H. *Technical Report No.* 592, *Division of Engineering and Applied Physics, Harvard University*, (1969).

Discussion

CHRISTOPHER (Cranfield Institute of Technology). Professor Storey's approach indicates that the search for a Liapunov function is a rather useless occupation. One might assume that the application of Liapunov-type techniques to the present problem is, therefore, inconsistent.

STOREY. One mitigating factor is that comparison theory is used rather than the original Liapunov approach. The advantage of the former method is that the usual conditions on the V function are not required. V need not be positive definite nor need its derivative be negative definite. Provided that the conditions on the comparison equation are satisfied then stability can be shown. The stability behaviour of the vector system is determined from a simple scalar system.

Whether anything useful can be done in practice with this alternative approach to controllability is another matter. We are currently looking at the possibilities and the technique appears quite promising.

CHRISTOPHER. Do you think that discarding the search for a Liapunov function in favour of solving the Zubov partial differential equation is something that is worth following up?

STOREY. Having to resort to hybrid computing techniques† to solve the Zubov partial differential equation directly indicates the lack of progress in analytical methods. If it is possible to solve the Zubov equation analytically then the exact domain of asymptotic stability can be obtained. On the other hand approximation

† See paper by Troch, IFAC, Paris, 1972.

by series does not converge uniformly so, for example, a second order Liapunov function might give a greater area than a twentieth order. This lack of uniformity in the convergence is very unsatisfactory.

CHRISTOPHER. Is there any point in trying to obtain a Liapunov function of the Zubov partial differential equation in order to get properties of the Liapunov function?

STOREY. I don't know. I wouldn't have thought that the Zubov equation bears any relation. One is merely setting $\dot{V}$ equal to some convenient negative definite function.

PARKS (University of Warwick). Could you explain a little on this maximal solution $r_M(t)$ which you used in your theorem?

STOREY. A good account of comparison theory and maximal and minimal solutions can be found in Ref. 13 of my lecture.

MEES (University of Cambridge). Are there cases when the system is unstable but the instability is not too important; perhaps you have a control on the system and it makes the system unstable but in fact the instability does not dissipate much power and you needn't worry about it? I thought perhaps in the analysis you gave the fact that you are not trying to control to a point would be useful in cases like this.

STOREY. The whole point of practical stability is to allow this, because a region of initial conditions and an acceptable region of disturbances are both given in advance. For example, a rocket which was oscillating slightly could be practically stable but not stable in the Liapunov sense.

3. Statistical Analysis of Multi-nonlinear Continuous Feedback Systems by Functional Series Expansions

C. J. HARRIS†

Department of Electronic Engineering, The University of Hull, England.

1. Introduction

N. Wiener [7] showed that if the realisation of a non-linear function of a stochastic process is square or Lebesgue integrable, then it can be represented exactly by a series expansion in orthogonal polynomial functions. Wiener's results were confined to open-loop non-linear systems with stochastic perturbations derived from some linear operation on white noise. Later, Van Trees [6] and Shen and Rosenberg [5] made use of the Volterra series representation, which has a natural link with the system impulse response, in the analysis of stochastic non-linear feedback systems. However, such a representation requires the analyticity of the functionals describing the system non-linear elements, and consequently excludes from investigation systems with piecewise linear elements such as Limiters, Relays, Hysteresis etc. This problem was overcome by the introduction of Hermite polynomial functional expansions for non-linear stochastic systems by Barrett [1] and of Chebyschev polynomials, for which the intervals of Lebesgue integrability is satisfied by the physical characteristics of almost all of the single-valued and multi-valued non-linearities met with in feedback control systems.

In the following an analytic method is given for the exact statistical representation of non-linear feedback systems with stochastic inputs. This series representation is shown to require the evaluation of an infinite number of kernels; in consequence for practical usage the exact statistical model is replaced by a least mean squares linearised model of two parallel components: one of which relates to the transfer of the random process through the system and the other describes the non-random or steady-state periodic behaviour of the feedback system. These linearised models are similar to those obtained by the technique of statistical linearisation due to

† Present address: Control Systems Centre, UMIST, Manchester, England.

Kazakov, except that these models are derived from closed-loop considerations.

2. Least squares approximation by orthogonal polynomial functionals

2.1 *Definitions*

The function $f(x, t)$, real or complex, is square integrable in the interval (a, b) (or $f \in L_2(a, b)$)† if the integral $\int_a^b |f(x)|^2 \, dx$ exists and is finite; special classes of function(al)s that have finite mean square are analytic functionals (including linear and polynomial functions) and certain discontinuous functions.

The scalar or inner product

$$\langle \psi_1 \psi_2 \rangle \equiv \int_a^b w(x) \psi_1(x) \psi_2{}^*(x) dx \tag{1}$$

is defined for all functions $\psi(x)$ which are Lebesgue integrable in (a, b), where $w(x)$ is a non-negative weighting function.

The system of functions $\psi_n(x) \in L_2(a, b)$ are said to be orthogonal in (a, b) if, and only if

$$\langle \psi_i \psi_j \rangle = h_i \delta_{ij} \tag{2}$$

If in addition $\|\psi_n\| = \langle \psi_n \psi_n \rangle^{\frac{1}{2}} = 1$, then the system is orthonormal and $h_i = 1$ for all i. (An orthogonal system $\psi_i(x)$ may be normalised by replacing $\psi_i(x)$ by $\psi_i(x) \langle \psi_i \psi_i \rangle^{-\frac{1}{2}}$).

A finite or infinite arbitrary sequence of functionals $\{\beta_i(x)\}$ in which any subsequence is linearly independent can be orthonormalised with respect to the scalar product equation (1) by the formulation of suitable linear combinations, such as the Gram–Schmidt transformations,

$$\psi_j(x) = \beta_j(x) - \sum_{i=1}^{j-1} \frac{\langle \psi_i \beta_i \rangle}{\langle \psi_i \psi_i \rangle} \psi_i(x), \qquad (j = 2, 3, \ldots, n) \tag{3}$$

which forms an orthogonal set and the unit vectors $\psi_j = \psi_j(x) \langle \psi_j \psi_j \rangle^{-\frac{1}{2}}$ form an orthonormal set.

A generalised choice for the linearly independent sequence $\{\beta_i(x)\}$ is the polynomial integral functionals [7]

$$\beta_0(x) = \int_0^t b_0(u) \, du, \qquad \beta_1(x) = \int_0^t b_1(u) x(u) \, du,$$

$$\ldots \beta_n(x) = \int_0^t \ldots (n) \ldots \int_0^t b_n(u_n) x(t - u_n) \, du_n \tag{4}$$

† The notation $L_2(c)$ is used for functions square integrable in the Lebesgue sense.

where

$$b_n(u_n) = b_n(u_1, u_2, \ldots u_n) \in L_2(0, t),$$

$$x(t - u_n) \triangleq \prod_i^n x(t - u_i), \qquad du_n = \prod_i^n du_i$$

and

$$\int_0^t \ldots (n) \ldots \int_0^t b_n(u_n)\, du_n < \infty$$

for all n and t.

If an infinite orthonormal set of functionals $\{\psi_i(x)\}$ exist such that any Lebesgue integrable functional $f(x)$ can be expressed by a finite linear combination of functionals $\{\psi_i\}$ with any prescribed accuracy, then the set $\{\psi_i\}$ is said to form a complete orthonormal set.

2.2 *Theorems*

THEOREM 1. *If $f(x) \in L_2(a, b)$ is any functional for which the Lebesgue integral $\int_a^b w(x)\,|f(x)|^2\, dx$ exists and is finite, then for a complete orthonormal system of polynomial functionals $\{\psi_i(x)\}$ in $L_2(a, b)$*

$$\lim_{N\to\infty} \int_a^b w(x) \left| f(x) - \sum_{k=0}^{N} A_k \psi_k(x) \right|^2 dx \to 0 \tag{5}$$

where

$$A_k = \int_a^b f(x) w(x) \psi_k(x)\, dx$$

This theorem is a generalisation of that of Cameron and Martin [2], who considered a complete orthonormal set based on Hermite polynomials and a weighting function $w(x)$ which is a Gaussian probability density function for an input stochastic process $x(t)$ which is completely random.

A corollary to the scalar orthogonal expansion representation of a non-linear functional is that which includes orthogonal sequences of complex functions, allowing the orthogonal series expansion of multi-valued non-linearities. One such obvious representation is the Fourier series expansion for which a complete orthonormal sequence of polynomial functions $J_k(x) \in L_2(-a, a)$ is defined such that

$$\lim_{N\to\infty} \int_{-a}^{a} \left| f(x) - \sum_{k=-N}^{N} A_k J_k(x) \right|^2 w(x)\, dx \to 0 \tag{6}$$

for

$$f(x) \in L_2(-a, a) \qquad \text{and} \qquad \int_{-a}^{a} f^2(x) w(x)\, dx < \infty,$$

where

$$J_k(x) = \exp\left\{jk\cos^{-1}\left(\frac{x}{a}\right)\right\}$$

$$= \cos\left\{k\cos^{-1}\left(\frac{x}{a}\right)\right\} + j\sin\left\{k\cos^{-1}\left(\frac{x}{a}\right)\right\} \tag{7}$$

$$A_k = \tfrac{1}{2}(a_k + j\,\mathrm{sgn}(k)b_k) \tag{8}$$

$$a_k = \frac{1}{h_k}\int_{-a}^{a} f(x)\,\mathrm{Re}\,(J_k(x))w(x)\,dx \tag{9}$$

$$b_k = \frac{1}{h_k}\int_{-a}^{a} f(x)\,\mathrm{Im}\,(J_k(x))w(x)\,dx \tag{10}$$

Taking only the real part of $J_k(x)$ gives the Chebyschev polynomial of the first kind with $w(x) = (1-x^2)^{-\frac{1}{2}}$, $h_k = \pi/2$ for $k \neq 0$ and $h_k = \pi$ for $k = 0$.

For $\mathbf{x} = \{x_1, x_2, \ldots x_n\}$ a n-dimensional vector, Theorem 1 can be extended to functionals $f_r(\mathbf{x}) \in L_2(\mathbf{c})$ where $\mathbf{c} = \{c_1, c_2, \ldots c_n\}$ which satisfy

$$\int_{c_1} \ldots (n) \ldots \int_{c_n} |f_r(x)|^2 w(x_1)\,dx_1 \ldots w(x_n)\,dx_n < \infty$$

THEOREM 2. *For a complete orthonormal system of polynomial functionals* $\{\psi_k(\mathbf{x})\} \in L_2(\mathbf{c})$,

$$\psi_k(\mathbf{x}) = \prod_{j=1}^{n} \psi_{k_j}{}^{(j)}(\mathbf{x}_j), \quad (k_j = 0, 1, \ldots)$$

$$\lim_{N\to\infty}\left\{\int_{c_1} \ldots (n) \ldots \int_{c_n} |f_r(\mathbf{x}) - \sum_{k_1,\ldots k_n=0}^{N} A^{(r)}_{k_1,\ldots k_n} \prod_{j=1}^{n} \psi_{k_j}{}^{(j)}(x_j)|^2 \times w(x_1)\,dx_1 \ldots w(x_n)\,dx_n\right\} \to 0 \tag{11}$$

where

$$A^{(r)}_{k_1,\ldots k_n} = \int_{c_1} \ldots (n) \ldots \int_{c_n} f_r(\mathbf{x}) \prod_{j=1}^{n} \psi_{k_j}{}^{(j)}(x_j) \prod_{i=1}^{n} w(x_i)\,dx_i. \tag{12}$$

For a rigorous proof of these theorems see E. W. Cheney [3]; a simplified proof results from finding the minimum of Eqns (5), (11) with respect to the polynomial coefficients $A^{(r)}_{k_1, \dots k_n}$ and by utilising the orthogonality condition of Eqn (2).

3. Orthogonal Polynomial Representation of Feedback Multi-nonlinear Systems

3.1. *Exact stochastic models of feedback systems*

Consider the general multi-nonlinear closed-loop system shown in Fig. 1, in which the linear transfer functions $G(p)$, $H(p)$ have denominator polynomials which are Hurwitz with no zero roots, and the system input $r(t)$ is a zero mean stochastic process derived from some linear operation on white noise.

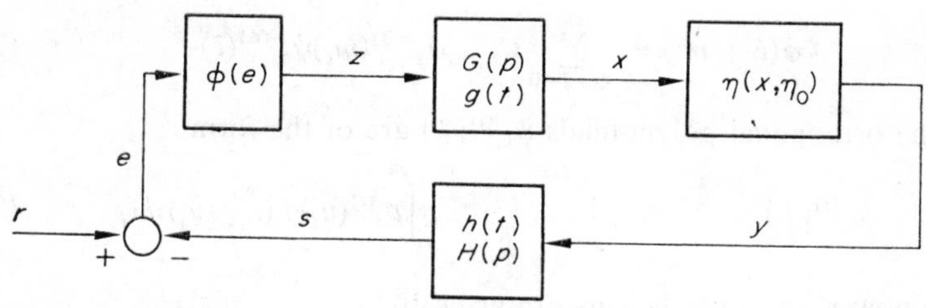

FIG. 1.

Assume that the feedback system is stable or at least that there exists an asymptotically stable set in state space which contains all stable limit cycles and sets of equilibrium. A necessary condition of stability is that the single-valued and multi-valued non-linearities $\phi(e)$, $\eta(x, \eta_0)$ must respectively satisfy the sector inequalities

$$e^2K_1 \geqslant e\phi(e) \geqslant 0 \tag{13}$$

$$x^2K_2 \geqslant x\eta(x, \eta_0) \geqslant 0 \tag{14}$$

for constants $(K_1, K_2) > 0$

The system loop equations are

$$x(t) = \int_0^t g(t-\tau)\phi(e)_\tau \, d\tau \tag{15}$$

$$e(t) = r(t) - \int_0^t h(t-\tau)\eta(x, \eta_0)_\tau \, d\tau. \tag{16}$$

Expanding the random inputs to each non-linearity as the sum of a centralised random component and mean or non-random function:

$$x(t) = \tilde{x}(t) + m_x(t) \tag{17}$$

$$e(t) = \tilde{e}(t) + m_e(t). \tag{18}$$

For oscillatory modes within the loop the non-random functions $m_x(t)$ and $m_e(t)$ are periodic; if in addition $G(p)$ and $H(p)$ have low pass characteristics then $m_x(t) \simeq X \cos wt$ and $m_e(t) \simeq \xi \cos (wt + \theta)$.

From Theorem 2, represent the functionals $\eta(x, \eta_0)_t = \eta(\tilde{x} + m_x, \eta_0)_t$, $\phi(e)_t = \phi(\tilde{e} + m_e)_t$ as a series of orthogonal functionals

$$\eta(\tilde{x} + m_x, \eta_0) = \sum_{k_1,k_2=0}^{\infty} A_{k_1,k_2}\psi_{k_1}{}^{(1)}(m_x)\psi_{k_2}{}^{(2)}(\tilde{x}) \tag{19}$$

$$\phi(\tilde{e} + m_e) = \sum_{k_3,k_4=0}^{\infty} C_{k_3,k_4}\psi_{k_3}{}^{(3)}(m_e)\psi_{k_4}{}^{(4)}(\tilde{e}) \tag{20}$$

where the orthogonal polynomials $\psi_{k_i}{}^{(i)}(f)$ are of the form

$$\psi_{k_i}{}^{(i)}(f) = \sum_{j=0}^{k_i} \alpha_{k_i,j} \int \ldots \overset{t}{\underset{0}{(j)}} \ldots \int b_j{}^{(i)}(u_j) f(t - u_j)\, du_j \tag{21}$$

and the power series coefficients are given by

$$A_{k_1,k_2} = \int_{c_1}\int_{c_2} \eta(\tilde{x} + m_x, \eta_0)\psi_{k_1}{}^{(1)}(m_x)\psi_{k_2}{}^{(2)}(\tilde{x}) \frac{w(m_x)}{h_{k_1}} \frac{w(\tilde{x})}{h_{k_2}}\, d\tilde{x}\, dm_x \tag{22}$$

and

$$C_{k_3,k_4} = \int_{c_3}\int_{c_4} \phi(\tilde{e} + m_e)\psi_{k_3}{}^{(3)}(m_e)\psi_{k_4}{}^{(4)}(\tilde{e}) \frac{w(m_e)}{h_{k_3}} \frac{w(\tilde{e})}{h_{k_4}}\, d\tilde{e}\, dm_e. \tag{23}$$

Then from Eqns (15), (16), (19) and (20),

$$m_e(t) = -m_s(t) = -E\Bigg\{ \sum_{k_1,k_2=0}^{\infty} \int_0^{\infty} h(t - \tau) A_{k_1,k_2}\psi_{k_1}{}^{(1)}(m_x(\tau)) \times \psi_{k_2}{}^{(2)}(\tilde{x}(\tau))\, d\tau \Bigg\} = -E\{f(x(t))\} \tag{24}$$

and

$$x(t) = \sum_{k_3,k_4=0}^{\infty} \int_0^{\infty} g(t - \tau) C_{k_3,k_4}\psi_{k_3}{}^{(3)}\big(-Ef(x(\tau))\big) \times \psi_{k_4}{}^{(4)}\big(r(\tau) - (1 - E)f(x(\tau))\big)\, d\tau \tag{25}$$

where E is the expectation operator.

It can be seen from Eqn (25) that an explicit input/output relationship exists between the loop variable $x(t)$ and the random input $r(t)$, if the kernels that define the orthogonal polynomial functionals $\psi_{k_i}{}^{(i)}$ can be found. Seeking a solution to $x(t)$ in the form of a product of functional series of orthogonal polynomials of the same form as $\psi_{k_1}{}^{(1)}$, $\psi_{k_2}{}^{(2)}$, but with respective kernels $v_j(u_j)$, $z_i(u_i)$, which must be selected such that

$$x(t) = \sum_{i,j=0}^{\infty} V_j(m_x(t))Z_i(r(t)) \tag{26}$$

satisfies loop Eqn (25). Then substituting the above in Eqn (25), expanding terms within the series and equating terms of like order in $\tilde{x}$ and r gives the relationship between the unknown kernels $v_j(u_j)$, $z_i(u_i)$ and the open loop polynomial coefficients A_{k_1,k_2}, C_{k_3,k_4} and the linear elements of the feedback system; these relationships through Eqn (26) enable a complete description of the stochastic perturbed non-linear feedback system of Fig. 1.

In a large number of physical systems the disturbing perturbations are Gaussian distributed; then an obvious physically intuitive choice for the orthogonal polynomials $\psi_{k_2}{}^{(2)}$, $\psi_{k_4}{}^{(4)}$ of the centralised random functions of the system would be modified Hermite polynomials whose weighting function is a unit variance Gaussian density function. The filtering properties of $G(p)$, $H(p)$ will also ensure that the random functions throughout the loop are 'almost' Gaussian distributed. Similar reasoning for the periodic functions m_x, m_e lead to a choice of Chebyschev polynomials for $\psi_{k_1}{}^{(1)}$, $\psi_{k_3}{}^{(3)}$. Then Eqns (24), (25) become

$$f(x(t)) = \sum_{k_1,k_2=0}^{\infty} \int_0^{\infty} h(t-\tau)H_{ek_1}\left(\frac{\tilde{x}(\tau)}{\sigma_x}\right) \times \left[a_{k_1,k_2} \operatorname{Re} J_{k_2}\left(\frac{m_x}{X}\right) + jb_{k_1,k_2} \operatorname{Im} J_{k_2}\left(\frac{m_x}{X}\right)\right] d\tau \tag{27}$$

$$x(t) = \sum_{k_3,k_4=0}^{\infty} \int_0^{\infty} g(t-\tau)C_{k_3,k_4}H_{ek_3}\left(\frac{r-(1-E)f}{\sigma_e}\right) \times \operatorname{Re} J_{k_4}\left(-\frac{Ef(x(\tau))}{\xi}\right) d\tau\dagger \tag{28}$$

† Only the real term in the Chebyschev polynominal expansion for $\phi(e)$ is necessary, since ϕ is single-valued.

where $\sigma_x{}^2$ and $\sigma_e{}^2$ are the variance of $x(t)$ and $e(t)$ respectively and the polynomial coefficients are given by

$$a_{k_1,k_2} = \frac{2^{-(k_1-2)/2}}{k_1!\,\pi^{3/2}\sigma_x}\int_{-X}^{X}\int_{-\infty}^{\infty}\eta(\tilde{x}+m_x,\eta_0)H_{k_1}\left(\frac{\tilde{x}}{\sqrt{2}\sigma_x}\right)\operatorname{Re}J_{k_2}\left(\frac{m_x}{X}\right)$$

$$\times\exp\left\{-\left(\frac{\tilde{x}}{\sqrt{2}\sigma_x}\right)^2\right\}(X^2-m_x{}^2)^{-\frac{1}{2}}\,d\tilde{x}\,dm_x \tag{29}$$

$$b_{k_1,k_2} = \frac{2^{-(k_1-2)/2}}{k_1!\,\pi^{3/2}\sigma_x}\int_{-X}^{X}\int_{-\infty}^{\infty}\eta(\tilde{x}+m_x,\eta_0)H_{k_1}\left(\frac{\tilde{x}}{\sqrt{2}\sigma_x}\right)\operatorname{Im}J_{k_2}\left(\frac{m_x}{X}\right)$$

$$\times\exp\left\{-\left(\frac{\tilde{x}}{\sqrt{2}\sigma_x}\right)^2\right\}(X^2-m_x{}^2)^{-\frac{1}{2}}\,d\tilde{x}\,dm_x \tag{30}$$

and

$$C_{k_3,k_4} = \frac{2^{-(k_3-2)/2}}{k_3!\,\pi^{3/2}\sigma_e}\int_{-\xi}^{\xi}\int_{-\infty}^{\infty}\phi(\tilde{e}+m_e)H_{k_3}\left(\frac{\tilde{e}}{\sqrt{2}\sigma_e}\right)\operatorname{Re}J_{k_4}\left(\frac{m_e}{\xi}\right)$$

$$\times\exp\left\{-\left(\frac{\tilde{e}}{\sqrt{2}\sigma_e}\right)^2\right\}(\xi^2-m_e{}^2)^{-\frac{1}{2}}\,d\tilde{e}\,dm_e \tag{31}$$

Then from Eqn (26) the closed-loop solution for $x(t)$ becomes

$$x(t) = \sum_{i,j=0}^{\infty}\int_0^{\infty}\int_0^{\infty}\left[v_{j,1}(\tau_2)\operatorname{Re}J_j\left(\frac{m_x(t-\tau_2)}{X}\right)+jv_{j,2}(\tau_2)\right.$$

$$\left.\times\operatorname{Im}J_j\left(\frac{m_x(t-\tau_2)}{X}\right)\right]z_i(\tau_3)H_{e_i}[r(t-\tau_3)]\,d\tau_2\,d\tau_3 \tag{32}$$

Equating Eqns (28) and (32), expanding both polynomial functional series as far as $(i, j, k_1, k_2) = 1$ (i.e. linear terms only) and repeatedly substituting for $x(t)$ in the expansion for Eqn (28) from Eqn (32) gives, on equating terms in m_x/X and taking Laplace transforms,

$$V_{1,1}(p)Z_0(0) = \frac{G(p)H(p)[a_{10}c_{10}X\xi - c_{01}a_{01}]}{1+G(p)H(p)c_{10}a_{10}} \tag{33}$$

where

$$V_{1,1}(p) = \mathscr{L}v_{1,1}(t), \qquad Z_0(0) = \{\mathscr{L}z_0(t)\}_{p=0}$$

Similarly terms in $r(t)$ and $\sin(\cos^{-1}(m_x/X))$ gives respectively

$$V_{0,1}(0)Z_1(p) = \frac{c_{10}G(p)}{1 + G(p)H(p)c_{10}a_{10}} \tag{34}$$

$$V_{1,2}(p)Z_0(0) = -\frac{G(p)H(p)c_{01}b_{01}}{1 + G(p)H(p)c_{10}a_{10}} \tag{35}$$

The constant terms in the truncated linear series expansion give

$$V_{0,1}(0)Z_0(p) = V_{0,2}(0)Z_0(p) = 0 \tag{36}$$

Taking the inverse Laplace transforms of the above transfer functions enables the unknown kernels of the closed-loop solution to Eqn (32) to be found. It is not difficult to see that the process can be extended to find the higher order kernels by extending the expansions of the polynomial series for $x(t)$ in Eqns (25) and (32).

4. Least Mean Squares Statistical Model

To completely characterise the states at various points within the closed-loop system of Fig. 1 requires an infinite number of kernels $v_j(u_j)$, $z_i(u_i)$ to be found to satisfy Eqn (26). However an optimum (in the sense of minimum squared error) linear stochastic model can be found which corresponds to linear combinations of transfer functions of the unknown kernels found in Section 3 of the exact solution.

Consider the linear approximation

$$x^*(t) = \int_0^\infty K_0(\tau)m_x(t-\tau)\,d\tau + \int_0^\infty K_1(\tau)r(t-\tau)\,d\tau \tag{37}$$

to equation (32), where the kernels or coefficients of statistical linearisation $K_0(\tau)$, $K_1(\tau)$ are chosen to minimise

$$E\{[x(t) - x^*(t)]^2\} = E\left\{\left[\sum_{i,j=0}^{\infty} V_j(m_x(t))Z_i(r(t)) - x^*(t)\right]^2\right\}. \tag{38}$$

Minimising the above with respect to $K_0(\tau)$ and $K_1(\tau)$ and noting that $E[r(t)] = 0$, $E[K(t)] = K(t)$ gives

$$\int_0^\infty K_0(\tau_1)E\{m_x(t-\tau_1)m_x(t-\tau)\}d\tau$$

$$= E\left\{\sum_{i,j=0}^{\infty} V_j(m_x)Z_i(r)m_x(t-\tau_1)\right\} \tag{39}$$

and

$$\int_0^\infty K_1(\tau_1)E\{r(t-\tau_1)r(t-\tau)\}d\tau = E\left\{\sum_{i,j=0}^{\infty} V_j(m_x)Z_i(r)r(t-\tau_1)\right\} \quad (40)$$

Now $V_j(m_x)$ and $Z_i(r)$ are Chebyschev and Hermite polynomials (see Eqn (32)); then expanding the right-hand sides of Eqns (39), (40) in the linear terms only gives

$$\int_0^\infty X\,\xi\,K_0(\tau_1)\,E\{m_x(t-\tau_1)m_x(t-\tau)\}\,d\tau_1$$

$$\simeq \int_0^\infty \int_0^\infty [v_{1,1}(\tau_1)+jv_{1,2}(\tau_1)]\,z_0(\tau)E\{m_x(t-\tau)\,m_x(t-\tau_1)\}\,d\tau\,d\tau_1 \quad (41)$$

and

$$\int_0^\infty K_1(\tau_1)E\{r(t-\tau_1)r(t-\tau)\}\,d\tau_1$$

$$\simeq \int_0^\infty \int_0^\infty [v_{0,1}(\tau)+jv_{0,2}(\tau)]z_1(\tau_1)E\{r(t-\tau)r(t-\tau_1)\}\,d\tau\,d\tau_1 \quad (42)$$

but since $E\{m_x(t-\tau)m_x(t-\tau_1)\}$ and $E\{r(t-\tau)r(t-\tau_1)\}$ are non-zero, then taking the Laplace transforms of Eqns (41), (42), and utilising the transformed expressions, Eqns (33–36), for the kernels of the closed-loop solution, gives the transformed solutions for the unknown optimum kernels as

$$K_0(p) = \frac{(V_{1,1}(p)+jV_{1,2}(p))Z_0(0)}{X\xi}$$

$$= \frac{G(p)H(p)[a_{10}c_{10}X\xi - c_{01}a_{01} - jc_{01}b_{01}]}{X\xi[1+G(p)H(p)c_{10}a_{10}]} \quad (43)$$

and

$$K_1(p) = [V_{0,1}(p)+jV_{0,2}(p)]Z_1(p) = \frac{c_{10}G(p)}{1+G(p)H(p)c_{10}a_{10}} \quad (44)$$

By taking the Laplace transform of Eqn (37) and noting that $X(p) = \tilde{X}(p) + M_x(p)$ it can easily be seen that the conditions for self-oscillatory modes are $K_0(p) = 1$, i.e.

$$\frac{G(p)H(p)c_{01}[a_{01}+jb_{01}]}{X\xi} = G(p)H(p)N(\sigma_x, \sigma_e, X, \xi) = -1 \quad (45)$$

Also the transfer function between the random input $r(t)$ and the centralised random process $x(t)$ within the closed-loop is

$$\frac{\tilde{X}(p)}{R(p)} = \frac{c_{10}G(p)}{1 + G(p)H(p)c_{10}a_{10}} \tag{46}$$

The function $N(\sigma_x, \sigma_e, X, \xi)$ can be interpreted as the stochastic describing function for the closed-loop system of Fig. 1, which is identical to the conventional describing function of deterministic systems for input stochastic processes of zero variance σ_r^2. The polynomial cofficients a_{01}, b_{01}, c_{01} are functions of the variances $\sigma_x^2 = f_1(\sigma_r^2)$, $\sigma_e^2 = f_2(\sigma_r^2)$ (see Eqns (29–31)), then the non-zero variances σ_r^2 leads to a distortion of the self-oscillatory modes of the closed-loop system through the functional dependence of the conditions for oscillation (Eqn (45)) on the coefficients a_{01}, b_{01}, c_{01}. To evaluate $\sigma_x^2 = f_1(\sigma_r^2) = f_3(\sigma_e^2)$ requires the solution of Eqn (46), which is itself a function of the coefficients $a_{10}(\sigma_x^2)$, and $c_{10}(\sigma_e^2)$; by using the method of successive approximations a solution may be obtained for differing input variances σ_r^2. However it is simpler to specify various magnitudes for σ_x^2, σ_e^2 and consider the effect of each upon the Nyquist plot of $G(p)H(p)$ and $-N^{-1}$.

5. Evaluation of Characterising Coefficients a_{ij}, b_{ij}, c_{ij}

The evaluation of integrals of the form of Eqns (29)–(31) is not simple, since a closed solution cannot be found for most non-linearities of practical interest, and it is normal to express integrals of this form in a series, usually the confluent hypergeometric function (see [4], p. 310 for special cases).

5.1 *Single-valued non-linearities*

From Eqn (31),

$$c_{01}(\sigma_e) = \frac{2}{\sigma_e \pi^{3/2}} \int_0^{\pi} \cos\theta \, d\theta \int_{-\infty}^{\infty} \exp\left(-\left(\frac{\tilde{e}}{\sqrt{2}\sigma_e}\right)^2\right) \phi(\xi \cos\theta + \tilde{e}) \, d\tilde{e} \tag{47}$$

and

$$c_{10}(\sigma_e) = \frac{\sqrt{2}}{\sigma_e \pi^{3/2}} \int_0^{\pi} d\theta \int_{-\infty}^{\infty} \frac{\tilde{e}}{\sigma_e} \exp\left(-\left(\frac{\tilde{e}}{\sqrt{2}\sigma_e}\right)^2\right) \phi(\xi \cos\theta + \tilde{e}) \, d\tilde{e} \tag{48}$$

Solution of Eqns (47), (48) are well known for particular single-valued non-linearities such as saturated proportional devices, relays with dead space, etc. [4].

5.2 *Multi-valued non-linearities*

Generalised results similar to Eqns (47), (48) follow from Eqns (29), (30); however, exact solution is only possible in few examples and in general approximate methods must be used. For example, consider the relay with hysteresis shown in Fig. 2.

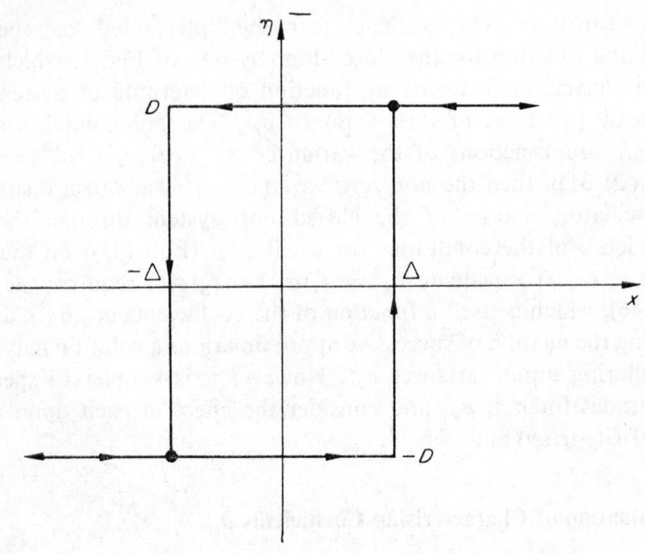

FIG. 2.

Solution of Eqns (29), (30) by Cauchy's residue theorem yields

$$a_{10} \simeq \frac{16}{\pi^2\sqrt{2\pi x}}\left\{\frac{\sigma_x\sqrt{\pi}}{2D}P\left(\frac{\sqrt{2}(X-\Delta)}{\sigma_x}\right)-\frac{(X-\Delta)}{D}\right.$$

$$\left.\times \exp\left(-\frac{(X-\Delta)^2}{{\sigma_x}^2}\right)\right\}+\sqrt{\frac{2}{\pi}}\exp\left(-\frac{(X-\Delta)^2}{{\sigma_x}^2}\right) \tag{49}$$

$$a_{01} \simeq \frac{8}{3\pi^2}P\left(\sqrt{2}\,\frac{(X-\Delta)}{\sigma_x}\right)\left(5-2\frac{{\sigma_x}^2}{X^2}\right)$$

$$+\frac{32}{3\pi^2\sqrt{\pi}}\,\frac{\sigma_x(X-\Delta)}{X^2}\,\exp\left(-\frac{(X-\Delta)^2}{{\sigma_x}^2}\right) \tag{50}$$

$$b_{01} = \frac{4\Delta}{\pi X}\,P\left(\sqrt{2}\,\frac{(X-\Delta)}{\sigma_x}\right) \tag{51}$$

where $P(f)$ is the probability integral.

The conditions for self-oscillatory modes for a closed-loop system with the above non-linearity for which $\Delta = 0{\cdot}25$, $D = 1$, in the forward loop and $G(p)\,H(p) = 2(1 - \exp(-p))/p^2(p+1)$ is shown in Fig. 3 as a function of the variance σ_x^2.

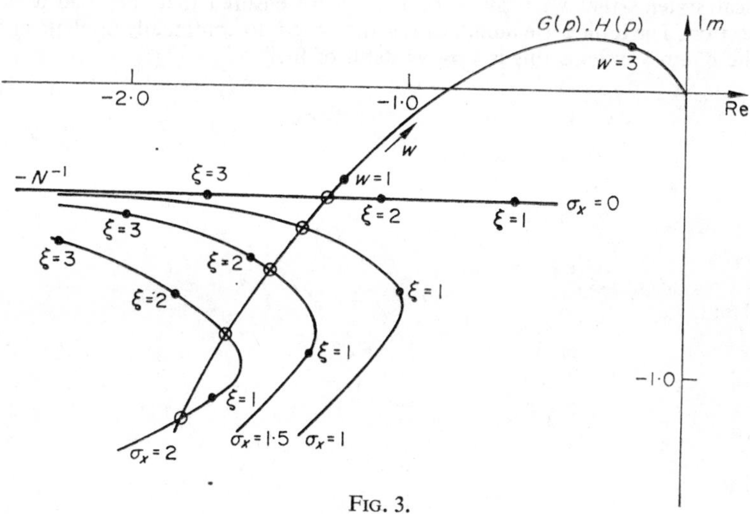

FIG. 3.

References

1. Barrett, J. F., *Int. J. Electron. Control* **16** (1) (1964), 107–113.
2. Cameron, R. H. and Martin, W. T., *Ann. Math.* **48** (2) (1947), 385–392.
3. Cheney, E. W., "Introduction to Approximation Theory". McGraw-Hill, New York (1966).
4. Pervozvanskii, A. A., "Random Processes in Nonlinear Control Systems". Academic Press, New York (1965).
5. Shen, D. W. C. and Rosenberg, A. *Trans. 3rd Prague Conf.*, 629–636 (1967).
6. Van Trees, H. L. *Proc. I.E.E.E.*, **52** (9) (1964), 894–911.
7. Wiener, N. "Nonlinear Problems in Random Theory". M.I.T. Press, Cambridge U.S.A., (1958).

Discussion

MEES (University of Cambridge). You talked about making the limit cycle less stable. Does this mean that you could actually look at the stability of a limit cycle by noting what happens when you inject noise into the system?

HARRIS. These are expected values. They always show that the effect of noise is to make the system less stable.

I mentioned that the points of intersection shown in Fig. 3 tend to turn back as an S shape. One obtains three or four points of intersection. Usually only two of these points are stable. The third one is an unstable point; it so happens that the ones we have considered always occur in-between stable limit cycles.

With just two intersections, trajectories enter an invariant limit cycle annuli, which excludes the origin of state space. For three intersections they tend to congregate along the edges of the two extreme limit cycle annuli. I think it worthwhile to do more work to see the effect on practical control systems, although in the practical systems that we were working with we ensured that they had a single intersection. The limit cycle annuli causes the system to continually oscillate and in satellite control systems this is very wasteful of fuel.

4. Some Control System Stability and Optimality Results obtained via Functional Analysis

E. A. FREEMAN

Sunderland Polytechnic, Sunderland, England.†

1. Introduction

The functional analytic approach to control system problems appears to have begun filtering into the literature some fifteen years ago [1], so, even when attention is restricted to the fields of stability and optimization, contributions are too numerous to review in a meaningful way here. Instead of a review therefore, this paper will focus attention on the essential ideas involved in what, in the writer's view, have been fundamental contributions to the stability theory (via functional analysis) of nonlinear multivariable systems. It will then be shown that the application of these ideas leads to a rigorous justification of the engineering describing function method and to a recent generalization of Rosenbrock's diagonal dominance stability criterion.

The application of contraction mappings, so useful in stability analysis, also yields results when examining the problem of optimizing linear multivariable systems with quadratic cost. Here, the results take the form of iterative computing algorithms for which one can bound convergence rates. Indeed, the concepts of conicity and posivity, also introduced for stability work, can be used to derive conditions for the optimum optimizing algorithm.

Finally, it is shown that the optimization of nonlinear systems can also be studied using the contraction mapping theorem and the resulting algorithm, which indirectly establishes the Maximum Principle, can be given a priori convergence rate bounds.

2. Stability

The main contributions in this area are attributed to I.W. Sandberg [2, 3] and G. Zames [4, 5].‡ Both authors have considered essentially the same

† Present address: Department of Engineering Science, University of Oxford, Oxford, England.

‡ Editor's note: See also NARENDRA, K. S. and GOLDWYN, R. H. *I.E.E.E. Trans.* CT–11 (1964) 406–408.

problems and obtained very similar results. The problem considered concerns systems which are defined by functional equations of the form

$$f = g - KQf \tag{1a}$$

or

$$Kf = g - Qf \tag{1b}$$

Here K and Q are causal [6] (non anticipative) operators which may be time varying. Q is also allowed to be nonlinear. f and g are column vector valued functions of time t. The actual feedback structure represented by equation (1a) is shown in Fig. 1.

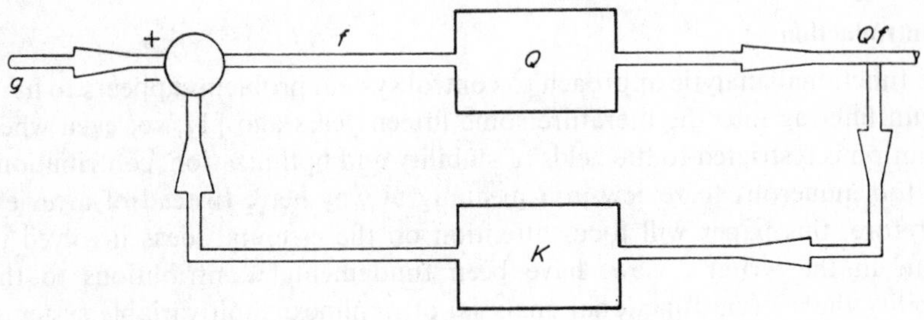

FIG. 1. System representation of $f = g - KQf$.

2.1 *Bounded input–Bounded output stability*

The problem posed is as follows:

> Suppose the input g is square integrable over an *infinite* interval, i.e. $g \in L_{2n}[0, \infty)$, and that f is square integrable over any *finite* interval, i.e. $f \in L_{2n}[0, y]$ (for any finite y); under what conditions will f be integrable over the infinite interval $[0, \infty)$? In effect what we are saying is, suppose we have a system whose response does not explode in finite time (in the L_{2n} sense); what additional conditions will guarantee that the response does not explode ever?

It has been shown [2] that if a scalar x exists such that

(i) $(I + xK)^{-1}$ exists and is causal

and

(ii) $$r = \|(I + xK)^{-1} K\| \sup_y \frac{\|(Qf_y)_y - xf_y\|}{\|f_y\|} < 1$$

where f_y is the truncation of f at time y defined by

$$\begin{aligned} f_y &= f \qquad 0 \leqslant t \leqslant y \\ &= 0 \qquad t > y \end{aligned}$$

then

$$\|f\| \leqslant \frac{1}{(1-r)} \|(I + xK)^{-1} g\| \tag{2}$$

Here $\|.\|$ designates the $L_{2n}[0, \infty)$ norm. Clearly, because $r < 1$, f is L_{2n} bounded when g is L_{2n} bounded.

The approach to establishing the above conditions is first to transform equation (1a) by adding xKf to both sides. This gives

$$(I + xK)f = g - (KQ - xK)f \tag{3}$$

and with condition (1a) we may write

$$f = (I + xK)^{-1} g - (I + xK)^{-1} (KQ - xK)f \tag{4}$$

Because at this stage we only know that a truncated f is bounded we equate the truncated components of the above equation (causality allows us to do this) using a truncation operator P defined by $Pf = f_y$. This gives

$$f_y = P(I + xK)^{-1} g_y - P(I + xK)^{-1} PKP(Q - xI)f_y. \tag{5}$$

Because the new "loop operator" $P(I + xK)^{-1} PKP(Q - xI)$ has a norm $r < 1$ it follows that,

$$\|f_y\| \leqslant \frac{1}{1-r} \|(I + xK)^{-1} g\| \tag{6}$$

for arbitrary y. Hence, the result that f is bounded if g is bounded.

Now sufficient conditions under which (i) and (ii) obtain are

(a) $\operatorname{Re}\langle Qf_y, f_y\rangle \geqslant k_1\|f_y\|^2, \qquad k_1 \geqslant 0.$

(b) $\|Qf_y\|^2 \leqslant k_2\|f_y\|^2, \qquad k_2 > 0.$

and

(c) $\operatorname{Re} < Kh, h > \geqslant c\|h\|^2, \qquad c \geqslant 0.$

When this is applied to a nonlinear *multivariable* feedback system defined by

$$g(t) = f(t) + \int_{t_0}^{t} K(t - \tau)\psi\{f(\tau), \tau\}\, d\tau \tag{7}$$

for which the nonlinearity ψ is diagonal and

$$\alpha f \leqslant \psi_{ii}\{f(\tau), \tau\} \leqslant \beta f \tag{8}$$

then conditions (i) and (ii) become

1. $\det\left[I + \frac{\beta+\alpha}{2}K(s)\right] \neq 0 \quad \text{for} \quad \operatorname{Re} s \geqslant 0$

2. $\frac{(\beta-\alpha)}{2}\sup_{\omega}\Lambda\left\{\left[I + \frac{\beta+\alpha}{2}K(i\omega)\right]^{-1}K(i\omega)\right\} < 1$

Further specialising this to the scalar case one obtains the well known circle criterion illustrated in Fig. 2. In this case if the $K(i\omega)$ locus does not pass through or encircle the shaded circle then conditions 1 and 2 are satisfied and square integrable inputs will produce square integrable outputs. Actually

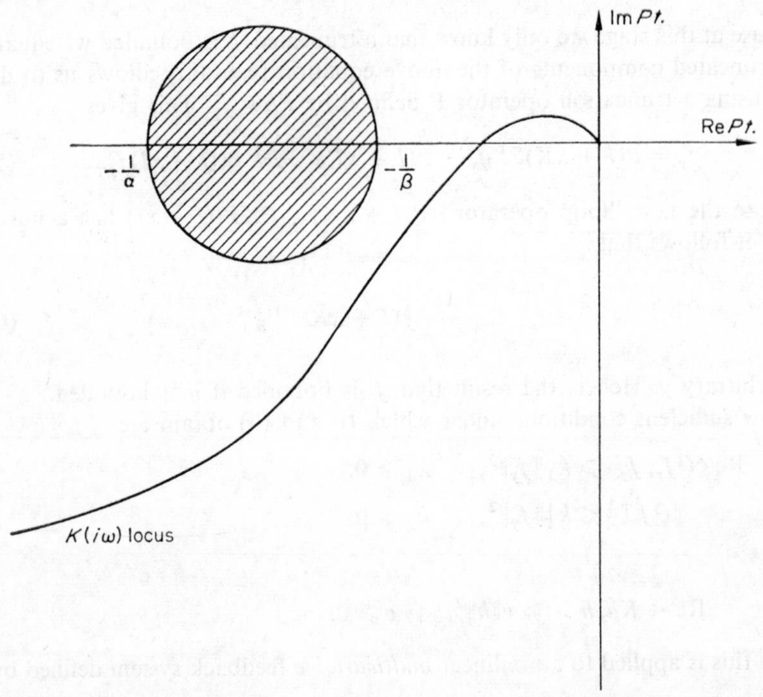

Fig. 2. Circle stability criterion for single input–single output systems for the case $\beta > 0$, $\alpha > 0$.

one can say a little more [7] for, if $g(t) \to 0$ as $t \to \infty$ then so does $f(t)$ provided $K(i\omega)$ is square integrable, with respect to ω, over $(-\infty, +\infty)$ i.e. provided $K(i\omega) \in L_2(-\infty, \infty)$. Thus we have a Liapunov type of stability with this additional proviso.

In cases where the nonlinearity of the system satisfies an incremental gain condition (a Lipschitz condition) of the form

$$\alpha_i \leqslant \frac{\psi(x, t) - \psi(y, t)}{(x - y)} \leqslant \beta_i, \qquad (x - y) \neq 0, \tag{9}$$

arguments almost identical to those presented above lead to a condition which guarantees that the output depends continuously on the input. The condition is that the $K(i\omega)$ locus does not pass through or encircle the continuity disc shown in Fig. 3. This continuity condition is relevant when one wishes to avoid features like jump resonance in system performance but it is applicable to a wider class of signals than sinusoids.

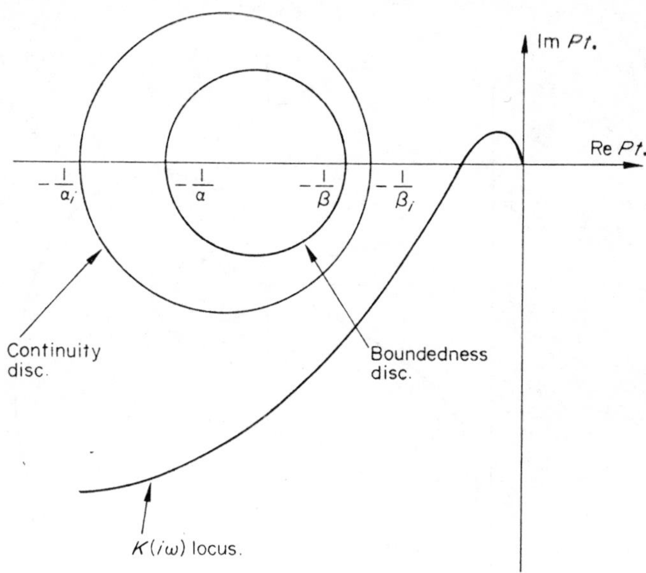

FIG. 3. Continuity circle representation for single input–single output systems for the case $\alpha > 0, \alpha_i > 0, \beta > 0, \beta_i > 0$.

One of the important drawbacks of the circle criteria is that whilst the conditions they give are sufficient for boundedness or continuity they are sometimes overly pessimistic. This can be overcome, to a certain extent by

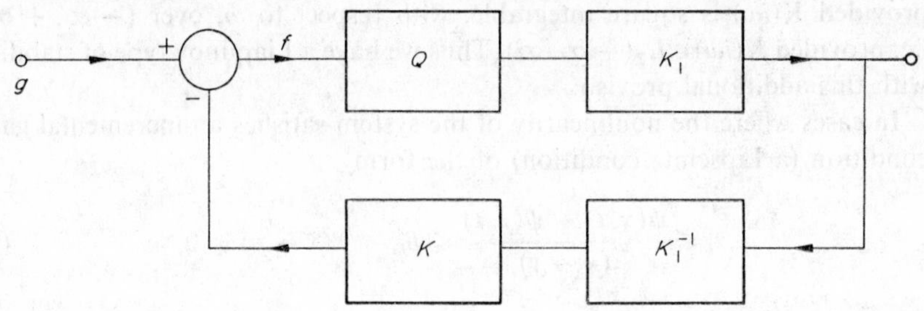

FIG. 4. System factorisation to improve stability bounds.

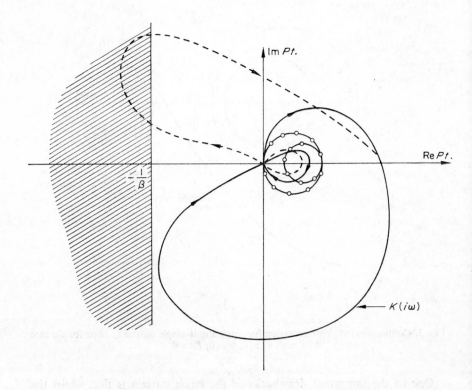

FIG. 5. Improving stability bounds by factorisation for the case $\beta > 0, \alpha = 0$; —— $K(i\omega)$; - - - - Popov's multiplier; -○-○-○- Zames' K_1 Multiplier

what is essentially a generalization of Popov's criterion [5, 8]. The improvement is achieved by factoring the system as shown in Fig. 4. With the factoring shown Q is replaced by QK_1 and K by KK_1^{-1}, and the idea is to choose K_1 so that the circle corresponding to the new nonlinear operator QK_1 is little or no different from the old circle whilst the frequency locus of KK_1^{-1}, is shaped by K_1^{-1} to avoid the circle. A real advantage accrues from having a reasonably free choice of K_1 as illustrated in Fig. 5. Here we see that when restricted to factoring with Popov's multiplier the danger of instability appears to be simply shifted from the low frequency range to a higher one. Zames' multiplier on the other hand demonstrates clearly that the system is stable.

2.2 *Describing function justification*

The describing function method of analysis may be used to calculate the closed loop frequency response of a nonlinear system and also to examine such systems for limit cycles. It is based on the assumption that the nonlinearity is adequately characterized by its fundamental component response to a harmonic free sinusoidal input. In 1963 Sandberg [9] gave some justification for the method when used to calculate the closed loop frequency response of a system. The system considered is shown in Fig. 6. It is assumed that y is periodic with period T and that the nonlinearity satisfies a Lipschitz condition of the form

$$\alpha(x - y) \leqslant (Nx - Ny) \leqslant \beta(x - y) \text{ for all } x \text{ and } y. \tag{10}$$

The question posed is:

> Under what conditions is x periodic (and of period T) and a continuous function of y and how accurate is the describing function method of calculating such periodic x?

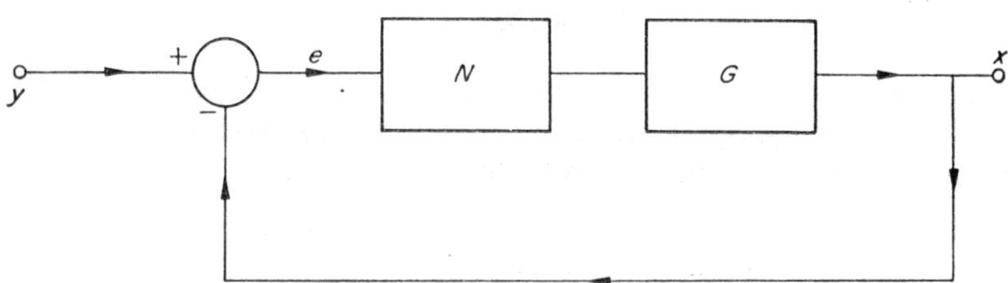

FIG. 6. Single-loop nonlinear system, N is a time invariant nonlinear operator; G is a bounded linear operator on the space of periodic functions.

The approach once again is to transform the system equation so as to obtain a contraction and then invoke the fixed point theorem of functional analysis to prove existence and uniqueness. The system equation is

$$x = GN(y - x) \tag{11}$$

and by letting $N(y - x) = N_0(y - x) + \bar{N}(y - x)$, this becomes

$$x = (I + GN_0)^{-1} GN_0 y + (I + GN_0)^{-1} G\bar{N}(y - x). \tag{12}$$

Now if for all x_1 and x_2 in the space K of periodic signals,

$$\|(I + GN_0)^{-1} G\{\bar{N}(y - x_1) - \bar{N}(y - x_2)\}\| \leqslant r\|x_1 - x_2\| \tag{13}$$

and $r < 1$, then the operator $(I + GN_0)^{-1} G\bar{N}$ is a contraction on K and the fixed point theorem assures us that there is a unique periodic x for each $y \in K$. Moreover this x may be calculated from

$$x_{n+1} = (I + GN_0)^{-1} GN_0 y + (I + GN_0)^{-1} G\bar{N}(y - x_n). \tag{14}$$

The minimum value of r namely r^* is obtained by making [10] $N_0 = (\beta + \alpha)/2$ and in this case

$$r^* = \sup_n \left| \frac{G_n}{1 + \dfrac{\alpha + \beta}{2} G_n} \right| \frac{\beta - \alpha}{2} \tag{15}$$

where $G_n e^{in2\pi t/T}$ is the output of G with input $e^{in2\pi t/T}$. Thus for a unique periodic x we require $r^* < 1$. It is not difficult to deduce that $r^* < 1$ if and only if the $G(i\omega)$ locus does not pass through or encircle the circle in Fig. 2. This result is closely related to the continuity condition depicted in Fig. 3. The only difference being the space of signals under consideration.

In the describing function method the exact solution of the system equation is approximated by the solution of the approximate equation

$$\hat{x} = PGN(y - \hat{x}) \tag{16}$$

where P is a projection onto the subspace of K spanned by $\exp(\pm i2\pi t/T)$ (i.e. the fundamental component). By choosing $\hat{x}$ as x_0 in the iterative scheme for the solution of the exact equation it is easy to show that

$$\|x^* - \hat{x}\| \leqslant \frac{1}{1 - r} \left\{ \sum_{n \neq \pm 1} \left| \frac{G_n p_n}{1 + G_n} \right|^2 \right\}^{\frac{1}{2}} \tag{17}$$

where p_n is the nth Fourier coefficient of $N(\hat{x} - y)$.

This result shows that the error between $\hat{x}$ and the exact solution, x^*, will be small when the nonlinearity produces little harmonic distortion (i.e. p_n small) or the system acts as a good low pass filter or both. These are exactly the qualitative assumptions implicitly made by engineers using the describing function method.

Finally it should be noted that a necessary condition for jump resonance is $r > 1$.

A drawback of Sandberg's approach is that it cannot be used to justify the describing function method when it is applied to the examination of a system for limit cycles. In such cases $y = 0$ and if we assume $x_0 = 0$, which is admissible, the iterative scheme, (14), gives $x^* = 0$ when $N(o) = 0$. Bergen and Franks [11] have recently overcome this difficulty in quite a new way which was paralleled by Kudrewicz [12] at about the same time†. In this approach the ideas of contraction mappings [13] and the homotopic invariance of the degree [14] of a nonlinear mapping play central roles. Consider the system of Fig. 6, but with $y = 0$ and recall we are seeking a periodic solution of the system equation

$$e = -GNe. \tag{18}$$

In the describing function method we solve

$$\hat{e} = -PGN\hat{e} \tag{19}$$

which is equivalent to

$$\hat{e} = -PGP\,N\hat{e}. \tag{20}$$

Because of the isomorphism [15] between the space of periodic signals K and the space l_2 of infinite dimensional vectors representing periodic signals, we may solve the above equation in l_2. The equivalent equation is

$$r_{\hat{e}} + gn\,r_{\hat{e}} = A = 0 \tag{21}$$

in which n is equivalent to $N(n \sim N)$ and $g \sim G$ and the vector $r_{\hat{e}}$ is the vector of Fourier coefficients of $\hat{e}$; formally we write $r_{\hat{e}} = T\hat{e}$ and call T the isomorphism.

Now the exact solution, as distinct from the describing function solution, must satisfy Eqn (18) which is equivalent to the two coupled equations

$$Pe = -PGNe = e_p \tag{22}$$

$$(I - P)e = -(I - P)GNe = e_p{}^{\perp}. \tag{23}$$

† A. Mees has also effected a solution to this problem by a slightly different route which is reported in these proceedings.

In Eqns (22) and (23) we are simply equating the fundamental component and the harmonics respectively. Equation (22) is similar to Eqn (19). The difference is that the e in the right-hand side of Eqn (22) contains components e_p and $e_p^{\perp}$ and that e must also satisfy Eqn (23) simultaneously. To guarantee the unique solution of (23) for any e_p we assume that

$$\|(I-P)\,G\{N(e_p+e_{p1}^{\perp})-N(e_p+e_{p2}^{\perp})\}\| \leqslant r\|e_{p1}^{\perp}-e_{p2}^{\perp}\| \tag{24}$$

with $r < 1$ and invoke the fixed point contraction mapping theorem. Then for each e_p there is a unique $e_{p*}^{\perp}$. Substituting this into Eqn (22) gives

$$e_p + PGN(e_p + e_{p*}^{\perp}) = 0. \tag{25}$$

The equivalent equation in l_2 is

$$r_{e_p} + gTPN(T^{-1}r_{e_p} + e_{p*}^{\perp}) = E = 0. \tag{26}$$

Returning now to the describing function justification, we require to know under what conditions a periodic solution found by the describing function method implies the existence of a periodic solution of the exact equation. This is equivalent to the question: under what conditions does a solution of Eqn (21) imply a solution of Eqn (26)? To answer this question we use two further results. The first is that of the degree of a nonlinear mapping A {designated $d(A, 0, \Omega)$} measures the number of solutions of $A = 0$ in Ω, and the second is that the degree of a homotopy $\Phi = tA + (1-t)(E-A)$ is invariant for $t \in [0, 1]$, provided Φ is not zero on the boundary of Ω. Since the degree of Φ is invariant we have

$$d(\Phi, 0, \Omega) = d(A, 0, \Omega) = d(E, 0, \Omega). \tag{27}$$

This means that if $A = 0$ in Ω then $E = 0$ in Ω exactly the same number of times as $A = 0$. To obtain $A = 0$ we simply seek intersections of the describing function locus and the $-1/G(i\omega)$ locus, whilst to ensure $\Phi \neq 0$ on $\partial\Omega$ it may be shown that it is sufficient that $\|A - E\| < \|A\|$; this latter condition therefore establishes the size of Ω. It is a straightforward matter to show that the condition is satisfied for

$$|1/N(r) + G(i\omega)| > B(\omega)\sqrt{\|N(r\xi_1)\|^2/|r\,N(r)|^2 - 1} \tag{28}$$

where $\xi_1 = \sin \omega t$ and r is its amplitude,

$$B(\omega) = \rho_\omega |G(i\omega)|/(1-\rho_\omega) \tag{29}$$

and we require $\rho_\omega = \|(I-P)\,GN\| < 1$.

The formula for ρ_ω depends on the nonlinearity in the system. If this has a Lipschitz constant M then

$$\rho_\omega = \sup_k M|G(i\omega k)|, \qquad k > 1$$

where M is defined by

$$\|Nx - Ny\| \leqslant M\|x - y\|.$$

In the case of a nonlinearity with a hysteresis type of characteristic no finite M will do so that one defines an M' by $\|Nx - Ny\| \leqslant M'\|x - y\|_\infty$, in which $\|f\|_\infty$ is the maximum value of the function f, and then

$$\rho_\omega = \frac{M'\pi}{\sqrt{8}} \max k|G(i\omega k)|, \qquad k > 1.$$

Thus to verify describing function predictions one simply calculates the range of r and ω for which (28) is satisfied, say $\langle r \rangle$ and $\langle \omega \rangle$, and concludes that an exact solution of the system equation will have an amplitude in $\langle r \rangle$ and a fundamental frequency in $\langle \omega \rangle$ if there is a describing function solution in these ranges. Conversely, if there is no describing function solution in these ranges there is no exact solution in the ranges.

2.3 *Linear multivariable system stability via contraction mappings*

It is interesting to see how the notions of contraction mappings and operator transformations may be used to examine the stability of linear systems. Recently Freeman [16] has examined the system structure shown in Fig. 7 in which K, G, and H are linear operators whose transfer function matrices have elements which are holomorphic in the right half plane.

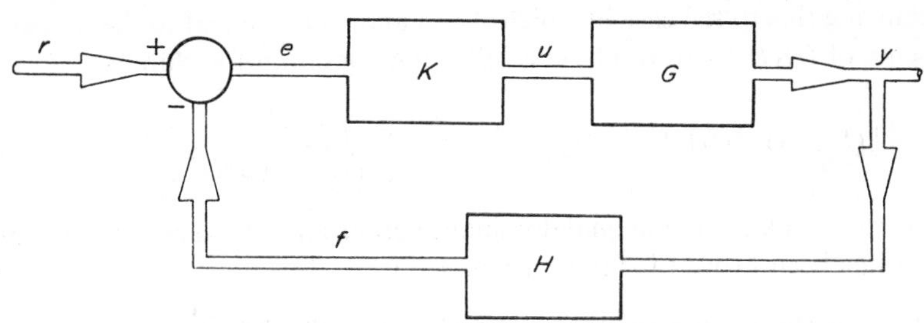

FIG. 7. Linear constant multivariable system.

In this study we assume there exists a class of input functions bounded by

$$|r_i(t)| \leqslant M_i e^{-\sigma_i t} \qquad \sigma_i > 0 \qquad \text{and} \qquad \forall\, i.$$

Then a system is defined to be stable if the out puts $y_i(t)$ are bounded by

$$|y_i(t)| \leqslant N_i\, e^{-\alpha_i t} \qquad \text{for some} \qquad \alpha_i \geqslant \sigma > 0$$

and for all i; here $\sigma = \min \sigma_i$. With inputs and outputs bounded in this way it follows that their Laplace transforms are holomorphic functions in the right half plane. The stability question therefore reduces to: under what conditions does the system map inputs with holomorphic transforms into outputs with holomorphic transforms?† It is not difficult to show that

$$y(s) = GKr(s) + z_0(s) - GKH\, y(s), \tag{30}$$

in which $z_0(s)$ is the transform of an initial condition term, and if we define a norm function on the complete space of holomorphic functions on the right half plane, Ω, by

$$\|y(s)\| = \max_i \sup_{s\in\Omega} |y_i(s)|$$

the norm of the operator GKH can be shown to be

$$\|GKH\| = \max_i \sup_{s\in\partial\Omega} \sum_j |GKH|_{ij}$$

Therefore, if $\|GKH\| < 1$ the fixed point theorem guarantees that the $y(s)$ calculated iteratively from Eqn (30) will converge to a unique holomorphic function on Ω.

This stability criterion is not particularly useful but by transforming Eqn (30) we obtain

$$\begin{aligned} y(s) = {} & (I + A)^{-1} GK\, r(s) + (I + A)^{-1}\, z_0(s) \\ & - (I + A)^{-1}\, (GKH - A)\, y(s) \end{aligned} \tag{31}$$

and a contraction now obtains when $\|(I + A)^{-1}(GKH - A)\| < 1$. Specializing this to the case in which A is diagonal and equal to the diagonal elements of GKH there results the following condition for stability

$$M = \|(I + A)^{-1}(GKH - A)\| = \max_i \sup_{s\in\partial\Omega} \sum_{j\neq 1} \left|\frac{(GKH)_{ij}}{\{1 + (GKH)_{ii}\}}\right| < 1. \tag{32}$$

This is the well known diagonal dominance criterion of Rosenbrock [17] but now it has been proved for more general transfer functions and inputs and,

† Editor's note: In order to prove that analyticity implies stability it has been found necessary to restrict the class of system considered. See Desoer, C. A. and Wu, M. Y. *J. Math. Anal. Appl.* **23** (1968) 121–129.

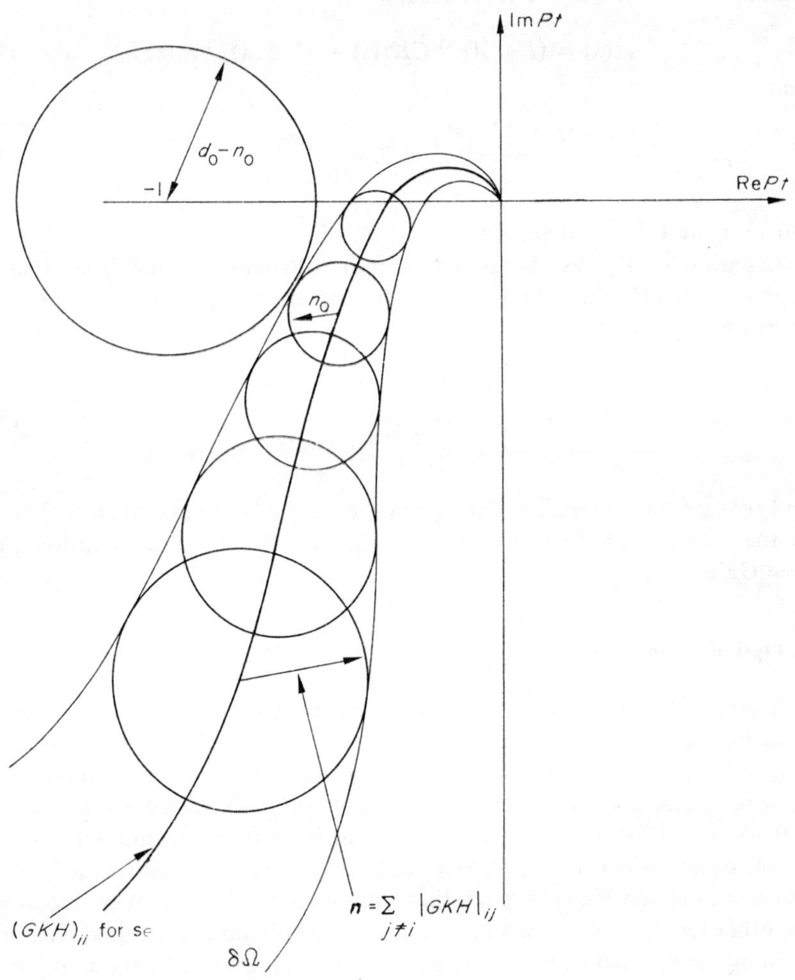

FIG. 8. Stability criterion and error measure for linear constant multivariable system.

moreover, because A is not necessarily diagonal, one has the possibility of effecting more accurate designs. By invoking a corollary of the contraction mapping theorem the error arising from such designs may also be assessed.

We have

$$\|y^* - y^n\| \leqslant \frac{M^n}{1 - M} \|y^1 - y^0\|. \tag{33}$$

Putting $y^0 = 0$ and $n = 1$ there results

$$y^1(s) = (I + A)^{-1} GKr(s) + (I + A)^{-1} z_0(s) \tag{34}$$

and

$$\frac{\|y^* - y^1\|}{\|y^1\|} \leqslant R = \frac{n_0}{(d_0 - n_0)} \tag{35}$$

where n_0 and d_0 are defined in Fig. 8.

Equation (34) gives the system output calculated on the basis that A represents GKH whilst (35) gives a measure of the error arising from this representation. Notice that

$$R = \frac{\max\limits_{i} \sup\limits_{s \in \partial\Omega} |y_i^*(s) - y_i^1(s)|}{\max\limits_{i} \sup\limits_{s \in \partial\Omega} |y_i^1(s)|} \tag{36}$$

and that we have therefore obtained a measure of the peak fractional error in the closed loop frequency response arising from the assumption that $A = GKH$.

3. Optimisation

A great deal of effort has been spent on studying optimization problems utilizing the concepts of functional analysis. Broadly, the problems studied may be divided into those in which the cost is non-quadratic (i.e. studies in Banach space) and quadratic cost problems or problems in Hilbert space. The former class of problems includes minimum time and minimum effort problems and often results are obtained using some form (either analytic or geometric) of the Hahn–Banach theorem. With quadratic cost problems on the other hand, as we shall see, it is possible to obtain a functional equation defining the optimal control. The existence and uniqueness of the solution of this equation can then be deduced using the fixed point theorem, demonstrating once again the great utility of the contraction mapping concept. We begin with the consideration of the linear quadratic cost problem [18].

3.1 *Linear quadratic cost problem*

We shall examine a system defined by

$$\dot{x} = A(t)x + B(t)u \tag{37}$$

$$c = C(t)x + D(t)u \tag{38}$$

for which it is required to find a control that is square integrable which minimizes a cost J given by

$$J = \int_{t_0}^{T} (e'\lambda e + u'Ru)\, dt$$
$$= \langle e, \lambda e\rangle + \langle u, Ru\rangle. \tag{39}$$

Here λ and R are matrices of appropriate dimensions, $\langle\cdot,\cdot\rangle$ the inner product operation and

$$e(t) = c_d(t) - c(t) \tag{40}$$

where $c_d(t)$ is the desired output.

By noting that

$$x(t) = \Phi(t, t_0)\, x(t_0) + \int_{t_0}^{t} \Phi(t, \tau)\, B(\tau)\, u(\tau)\, d\tau \tag{41}$$

in which $\Phi(t, t_0)$ is the system transition matrix, the system equations may be written in the form

$$e = z - Lu \tag{42}$$

where

$$z = c_d(t) - C(t)\, \Phi(t, t_0)\, x(t_0)$$

and

$$Lu = D(t)u + \int_{t_0}^{t} C(t)\, \Phi(t, \tau)\, B(\tau)\, u(\tau)\, d\tau. \tag{43}$$

The cost J then becomes

$$J = \langle z, \lambda z\rangle - 2\langle u, L^*\lambda z\rangle + \langle (L^*\lambda L + R)u, u\rangle \tag{44}$$

where L^* is the adjoint of L.

By perturbing the control from an assumed optimum it is easy to deduce [18] that the control minimizing equation (44) must necessarily satisfy

$$u^* = R^{-1}L^*\lambda z - R^{-1}L^*\lambda L u^* \tag{45}$$

Now for $\|R^{-1}L^*\lambda L\| < 1$ we have a contraction and are guaranteed the existence and uniqueness of a u^* which may be calculated iteratively by

$$u_{n+1} = R^{-1}L^*\lambda z - R^{-1}L^*\lambda L\, u_n \tag{46}$$

This algorithm has been implemented by Freeman and Brown [19] and used to calculate optimal controls for low order systems. The principal drawback of the method, as presented, is that $\|R^{-1}L^*\lambda L\|$ is not always less than unity

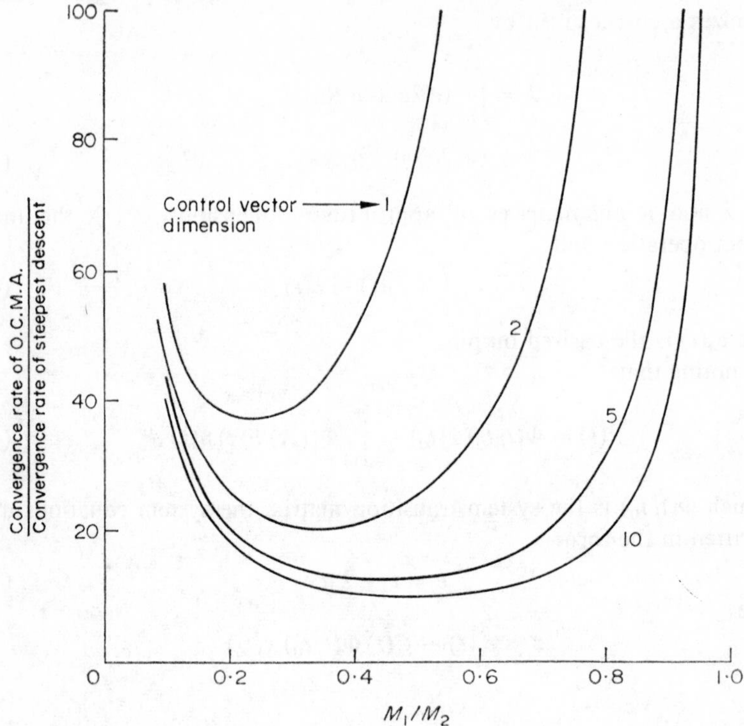

FIG. 9. Comparison of optimum contraction mapping algorithm with the steepest descent method.

and as illustrated when $\|R^{-1}L^*\lambda L\| \to 1$ the algorithm converges slowly. Operator transformations of the type proposed by Zames or by Freeman and Todd [20] may be used to speed convergence as follows.

Multiply Eqn (45) by R, add ku^* to both sides to obtain

$$u^* = (R + kI)^{-1} L^*\lambda z - (R + kI)^{-1} (L^*\lambda L - kI)u^* \qquad (47)$$

In this case a contraction results if

$$\|(R + kI)^{-1} (L^*\lambda L - kI)\| < 1$$

and Allwright [21] has shown that by choosing k large enough one can always obtain a contraction. By using the notions of conicity Freeman [22] has shown that there is an optimum choice of k given by

$$k = (M_L + m_L)/2.$$

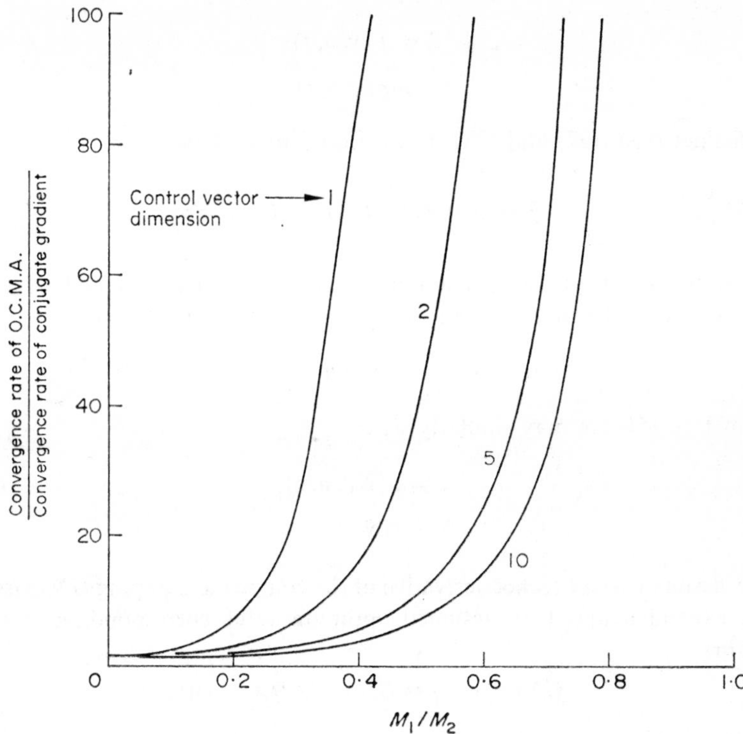

FIG. 10. Comparison of optimum contraction mapping algorithm with the conjugate gradient method.

where M_L and m_L are bounds of the operator $L^*\lambda L$ defined by

$$m_L\langle u, u\rangle \leqslant \langle L^*\lambda L u, u\rangle \leqslant M_L\langle u, u\rangle. \tag{49}$$

Using the optimal value of k in Eqn (47) results in an optimal contraction mapping algorithm which has been shown to converge significantly faster than the steepest descent and conjugate gradient methods for a given computer storage, see Figs 9 and 10. For high order systems accurate controls can also be calculated more rapidly than by the Matrix Riccati approach.

3.2 *Non-linear multivariable system optimisation via contraction mappings*

We now turn our attention to a nonlinear system whose control is to be chosen so as to minimize the same quadratic measure of cost as given by

Eqns (39) and (40). The system equations are now

$$\dot{x} = f(x, u, t) \tag{50}$$

$$c = g(x, u, t) \tag{51}$$

as distinct from (37) and (38). From Eqn (50) we have

$$x = x(0) + \int_0^t f\{x(\tau), u(\tau), \tau\}\, d\tau \tag{52}$$

Continuity of f guarantees a unique $x(t)$ for given $x(0)$ and $u(\tau)$. We therefore let N be the mapping $\{x(0), u\} \to x$, so we may write

$$x = Nu \tag{53}$$

From Eqn (51) we may similarly write

$$\begin{aligned} c &= g\{Nu, u, t\} \\ &= GNu. \end{aligned} \tag{54}$$

Now assuming the Fréchet derivative of the composite mapping GN exists we may expand it about an assumed optimum value corresponding to u^* to obtain

$$GN(u^* + v) = GN\, u^* + D_f v + O(v) \tag{55}$$

in which D_f is the Fréchet derivative of GN. This may be used in the expansion of the cost function J of Eqn (39) from which it may be deduced [23] that for a minimum cost we require

$$u^* = R^{-1} D_f^* \lambda\, c_d - R^{-1} D_f^* \lambda\, GNu^* \tag{56}$$

Equation (56) defines the optimal control and detailed interpretation of it reveals that we have established the Maximum Principle. To use the contraction mapping theorem we require

$$\| R^{-1} D_f^* \lambda\, GN \| < 1.$$

Transformations can of course be used as before to give a new equation

$$u^* = (R + kI)^{-1} D_f^* \lambda\, c_d - (R + kI)^{-1} (D_f^* \lambda\, GN - kI) u^* \tag{57}$$

defining the optimal control and for convergence of the corresponding contraction algorithm we require

$$\| (R + kI)^{-1} (D_f^* \lambda\, GN - kI) \| < 1.$$

It is interesting to note that the above approach, apart from producing algorithms for calculating optimum controls, provides a means of establishing a priori estimates of convergence rates for the usual algorithms based on the Maximum Principle. It also points to the need to investigate transformations which will in some sense optimize these algorithms.

The problem of including hard constraints on the control u can be studied in the above framework by replacing each element of u, say u_c, which is so constrained by a function $\gamma_c(u_c)$ in the system equations, where

$$\left.\begin{aligned} \gamma_c(u_c) &= u_c, & a \leqslant u_c \leqslant b \\ &= b & u_c > b \\ &= a & u_c < a \end{aligned}\right\} \tag{58}$$

We then repeat the above analysis with a modified function $f\{x, \gamma(u), t\}$ to obtain the same algorithm. Clearly the optimal control calculated will satisfy the hard constraints on u.

An apparent mathematical difficulty arises because the Fréchet derivative of G does not exist at the corners a and b. This difficulty is not real however, for the "corners" may be rounded-off, arbitrarily sharply to give a control set which produces a trajectory set dense in the trajectory set attainable with the actual controls subject to hard constraints. Because of this, any real world calculation of optimal controls will produce a control indistinguishable from optimal controls based on other theoretically "exact" methods.

4. Conclusions

By focussing attention on some of the methods of functional analysis as well as on the results obtained by them, an attempt has been made to show the relevance of this branch of applied mathematics to control engineers. Indeed, the great utility of contraction mappings in stability and optimization studies and of projection theory in Hilbert space in approximation methods, optimization, modelling and identification and sub-optimal control implies a relevance of this mathematics to many engineers. As a consequence, one question seems to urge itself persistently forward. It is: is it not timely for mathematicians and engineers to come together with a view to rethinking the mathematics courses for inclusion in the curricular of engineering undergraduates and perhaps to provide some means of updating the mathematics possessed by the more mature engineer?

References

1. Porter, W. A. "Modern Foundations of System Engineering". Collier–MacMillan, London (1966).

2. Sandberg, I. W. On the L_2-boundedness of solutions of nonlinear functional equations. *Bell Syst. Tech. J.* **43** (1964), 1581–1599.
3. Sandberg, I. W. A frequency-domain condition for the stability of feedback systems containing a single time-varying nonlinear element. *Bell Syst. Tech. J.* **43** (1964), 1601–1608.
4. Zames, G. On the input-output stability of time-varying nonlinear feedback systems—Part I: Conditions derived using concepts of loop gain, conicity, and positivity, *I.E.E.E. Trans. on Automatic Control* AC. **11** (1966), 228–239.
5. Zames, G. On the input-output stability of time-varying nonlinear feedback systems—Part II: Conditions involving circles in the frequency plane and sector nonlinearities, *I.E.E.E. Trans, Automatic Control* AC **11** (3) (1966), 465–476.
6. Willems, J. C. "The analysis of Feedback Systems" (Massachusetts Institute of Technology, Research Monograph No. 62).
7. Zadeh, L. A. and Polak, E., (Ed.), "System Theory". McGraw-Hill, New York. (1969).
8. Zames, G. and Falb, P. L. Stability conditions for systems with monotone and slope-restricted nonlinearities. *S.I.A.M. J. Control* **6** (1) (1968), 89–108.
9. Sandberg, I. W. On the response of nonlinear control systems to periodic input signals. *Bell Syst. Tech. J.* **43** (1964), 911–926.
10. Sandberg, I. W. A note on the application of the contraction-mapping fixed-point theorem to a class of nonlinear functional equations. *S.I.A.M. Review* **7** (2) (1965), 199–204.
11. Bergen, A. R. and Franks, R. L. Justification of the describing function method. *S.I.A.M. J. Control* **9** (4) (1971), 568–589.
12. Kudrewicz, J. Theorems on the existence of periodic vibrations based upon the describing function method. Proc. International Federation on Automatic Control, Warsaw, Poland (1969).
13. Holtzman, J. M. "Nonlinear system theory—A Functional Analysis Approach". Prentice-Hall (1970).
14. Ortega, J. and Rheinboldt, W., "Iterative Solution of Nonlinear Equations in Several Variables". Academic Press (1970).
15. Kolmogorov, A. N. and Fomin, S. V. "Elements of the Theory of Functions and Functional Analysis". Graylock Press, New York (1961).
16. Freeman, E. A. "Stability of Linear Constant Multivariable Systems—Contraction Mapping Approach". Internal Report. See also *Proc. I.E.E.* **120** (3) (1973), 379–384.
17. Rosenbrock, H. H. "Multivariable circle theorems". This Volume, pp. 345–365.
18. Freeman E. A. "On the optimization of linear time variant, multivariable control systems using the contraction mapping principle", *J. Optimization Theory and Applic.* **3** (1969), 416–443.
19. Freeman, E. A. and Brown, G. L. "Digital calculations of optimum controls for time varying linear multivariable systems", *Proc. I.E.E.* **117** (2) (1970), 449–458.
20. Freeman, E. A. and Todd, A. "Sub-optimal control for multivariable systems based on an orthogonal decomposition technique", *Int. J. Control* **13** (1971), 737–761.
21. Allwright, J. C. "Contraction mapping algorithm with guaranteed convergence", presented at 5th IFAC Congress, Paris, 1972, session 34.
22. Freeman, E. A. "Optimization of contraction-mapping algorithm for calculating optimal control", *Proc, I.E.E.* **199** (9), (1972).

23. Freeman, E. A. "Optimization of Nonlinear Multivariable Systems, Using Contraction Mapping". Internal Report.

Discussions

BRODSKY (U.S. Office of Naval Research). Would you say if the comparisons you referred to regarding convergence rates of the algorithms for calculating optimal controls were for particular practical examples?

FREEMAN. The rates are theoretical in the sense that I have compared the best available estimates for the upper bounds on the convergence rates of the three methods; namely the contraction mapping algorithm, the steepest descent algorithm and the conjugate gradient algorithm. However these comparisons shown in Figs. 9 and 10 cover all practical systems for, irrespective of the system parameters, the value of M_1/M_2 lies between 0 and 1.

BRODSKY. Have you actually tried the algorithms on a practical system and if so, how do the theoretical convergence rates compare with the actual ones. Was there not some difference?

FREEMAN. Yes, what you say is true. I've tried the optimal contraction mapping algorithm out on a couple of low order problems and because the theory presented gives upper bounds for convergence rates actual convergence is much faster except when the norm of the contraction operator approaches unity. In this case theoretical and actual rates agree closely.

MEES (University of Cambridge). A point of clarification. When you were talking about describing function justification you had a factor M' which was the norm of the nonlinear operator in some sense. How do you find the value of M'?

FREEMAN. There are two norms depending on the conditions the nonlinearity satisfies. For example, if N satisfies

$$||Nx - Ny|| \leqslant M\,||x - y||$$

for all odd periodic x and y we use this value of M. In effect, M is the maximum slope of the nonlinearity. The above relationship does not define a finite M for nonlinearities with hysteresis such as friction controlled backlash or a relay with hysteresis. In this case one seeks an M' which satisfies

$$||Nx - Ny|| \leqslant M'\,||x - y||_\infty$$

for all odd periodic x and y. Here $||z||_\infty$ is the peak value of the odd periodic z. Bergen and Franks show how M' can be chosen for the case of friction controlled backlash.

MEES. I interpret this last condition to mean that an infinitesimal difference between the amplitudes of x and y should give change in the mean square value of the output of N of the same order. Therefore I cannot see how the last condition helps with a relay having hysteresis and zero switching times.

FREEMAN. I agree there is difficulty but the condition can help; however, we do need to extend a little the work of Bergen and Franks. Consider the nonlinearity shown in Fig. a. We let the nonlinearity be represented by

$$y = Nx = N'x + N_0 x$$

where N' is the nonlinearity shown in Fig. b. Equation 18 then becomes

$$e = -(1 + GN_0)^{-1} GN'e$$
$$e = -\ G'N'e$$

which is exactly the same form but with N' replacing N and $(1 + GN_0)^{-1}G$ replacing G.

However, we now have a nonlinearity whose output can be bounded by the Bergen and Franks formulae. In fact, one can show that the condition

$$\rho_\omega = \| (I - P)\, G'N' \| < 1$$

implies that the $G'(i\omega)$ locus must not lie in a circle whose centre is on the real axis and which intersects the real axis at $1/a$ and $1/\beta$.

MEES. That's odd because the describing function locus lies in this circle and we require G' to enter it to obtain an intersection.

FREEMAN. That's right but the above condition has to be satisfied only for those harmonics outside the space PK so that there's no contradiction.

HARRIS (*University of Hull*). As I recall, there was a paper covering similar work and Franks but which allowed one to deal with to Bergen infinite slope nonlinearities.

[*Editor's note*: Dr. Harris has since stated that an alternative approach to Professor Lighthill's and Dr. Freeman's work can be made by considering the sector inequality

$$0 \leqslant f(x) \leqslant Kx^2$$

for the nonlinearity $f(x)$, where K is a positive constant, which is part of a closed loop system with linear plant $G(p)$. Garber† has shown for such systems that a necessary and sufficient condition for oscillation with frequency

$$\omega = \frac{2\pi}{T}$$

is

$$\sum_{i=0}^{\infty} \{\mathrm{Re}\{(1 + qij\omega)\, G(ij\omega)\} + (1/N)\}\, B_i^{\,2} \leqslant 0$$

where B_i = Fourier coefficients of $f(x)$.

Similarly, Harris‡ has shown that the criterion for pulsed systems is

$$|B_0|^2\, [G(jo) + 1/N] + 2 \sum_{i=1}^{\infty} |B_i|^2 \{\mathrm{Re}\{G(ij\omega)(1 + q(1 - e^{-ij\omega}))\} + 1/N\} \leqslant 0$$

where q, N = positive constants.]

† E. D. GARBER: *Automation and Remote Control* No. **11**, 1776–1780(1967).
‡ C. J. HARRIS: Ph. D. Thesis, Southampton University, 1971.

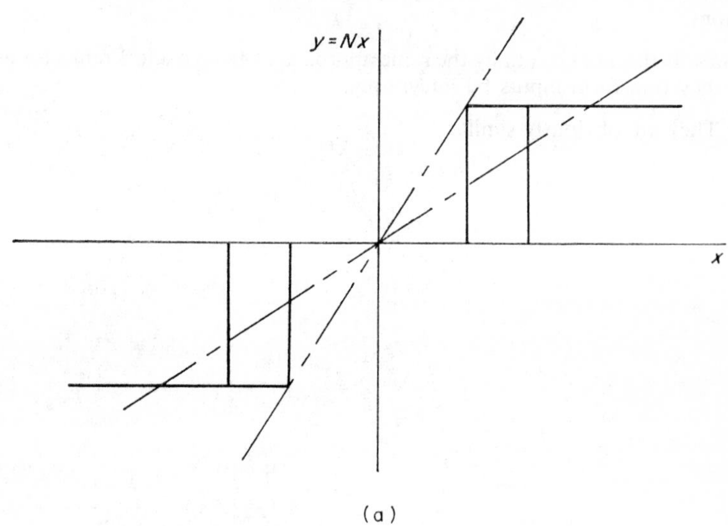

(a)

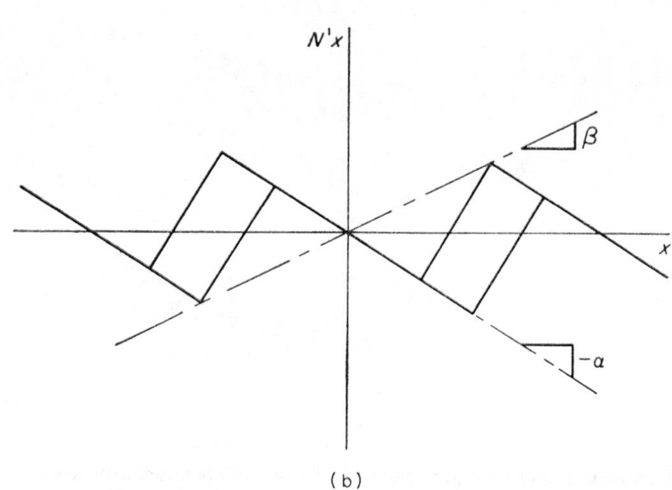

(b)

FREEMAN. I am aware of the work of Kudrewicz which is similar to what you suggest but your reference is unknown to me.

MEES. There was yet another contribution on this topic at the Paris IFAC.† In it the authors overcome the difficulty by transforming the equations to integral equations.

FREEMAN. Is this not essentially the same approach as Bergen and Franks for as you know they transform inputs by integrating.

MEES. They are obviously similar.

† FREY, T., SOMLÓ, J. and VAN QUY, NGUYEN. The use of operators with degenerated kernel for nonlinear system investigation. *Proc. IFAC* (Paris) 1972, Paper 32·5.

5. Optimal Control

J. H. WESTCOTT

Department of Computing and Control
Imperial College of Science and Technology,
London, England

Introduction

The topic of optimal control is today a broad one. It is proposed to restrict these remarks on recent developments to that part of the field most familiar to the author, namely deterministic optimal control. The march of progress here has been conditioned more by the pressing need by engineers for tangible answers to practical control engineering problems than by the desire for elegant mathematical formulations. By way of background on the subject there are now excellent text books available; Lee and Markus, Athans and Falb, Bryson and Ho, to mention but a few. The days of the grandoise sweep forward as exemplified for example in books by Bellman that rarely needed to descend to the difficult matter of tangible answers, is over. No longer are forces sweeping unopposed across open plains; today it is a war of attrition; sturdier opposition is met in the foothills requiring the bringing into action of the heavier armoury in computational weapons. A cross-fire of direct questions is on all sides and is being answered sturdily, but progress is that much more deliberate and advances that much more costly. There are a few more commanding heights due for capture before the war can be said to be won, but in another sense the ground on which the contest is being fought is itself shifting.

Critics have claimed that optimal control has an intrinsic defect in its dependence on use of state variable formulations. Recent work on the use of a weighting function approach to optimal control has shown among other advantages that dependence on state variable formulations is not essential. A further criticism has been that optimal control over designs to an extent that frequently makes it uneconomic in practice. To the extent that optimal control gives the best scientific answer and cannot give less this is true. There is work that can be reported on ways of using the same techniques of minimisation that leads to a result that is less than optimal.

Progress there certainly has been with the result that we now have practical procedures for use of the optimal control concept in solving a number of engineering problems. There remains the question is the game then worth the candle; in an increasing number of applications we are able to demonstrate that it is.

A difficult area where there has been little progress is optimal control distributed parameter systems and this is no doubt due to the outstanding difficulty of problems, bad enough when control is restricted to a boundary region but far worse where the control is itself diffused throughout the system. A number of potentially interesting systems are of this type but progress in controlling them can be expected to be slow. Interesting mathematical problems are involved.†

Statement of the Problem

The problem of optimal control is to minimise a functional, whose minimisation is to represent the performance desired, subject to constraints. If the constraints were no more than satisfaction of boundary values then the calculus of variations would provide an adequate answer to the problem; which would then be solved in two parts of equal importance, namely the finding of a curve by implicit minimisation and the matching of the given boundary condition by the 'correct' curve.

The constraints in optimal control, however, arise from the dynamics of the system, and in the calculus of variations this fact would appear as a side constraint on the problem of minimisation which on the face of it seems an odd way of looking at a control problem. Then again the Lagrange multipliers associated with this constraint turn out to have a dynamic of their own, whose physical significance is difficult to explain.

So progress in solving control problems has been concentrated on seeking more natural formulations of the minimisation problem subject to dynamic constraints for a performance index V, say

$$V(x_0, u, t_0) = \int_{t_0}^{t_1} L(x, u, t)\, dt$$

The constraints statements come in two forms:

(a) those in which the constraint condition is expressed as the satisfaction of a differential equation $\dot{x} = f(x, u, t)$. This leads to necessary conditions

† See for example Proceedings of a conference on "P.D.E. & Distributed Parameter Control Systems". Univ. of Warwick, July 1971.

of optimality in the form of canonical equations, an expression of the Pontriagin minimum principle

$$\dot{x} = \frac{\partial H}{\partial \lambda}; \qquad \dot{\lambda} = -\frac{\partial H}{\partial x}; \qquad \text{where} \qquad H = L + \lambda^T f.$$

In this formulation a physical significance can be attributed to λ; it represents the gradient of the cost function V with respect to the state x, thus

$$\lambda(t) = \frac{\partial V}{\partial x}(t).$$

A decent rigorous proof of the Pontriagin minimum principle (see Halkin) has previously necessitated appeal to a fixed point theorem and other mathematical artefacts that could hardly be considered a part of the engineers' standard kit of techniques. A simple constructive algebraic proof of the minimum principle for the linear continuous case needing only Riemann integration has been found.†

(b) those in which the constraint condition is expressed by a causal convolution operator rather than a differential equation; that is to say, when the dynamics of the system are given by a weighting function (Impulse response). This leads to a less familiar set of relationships, but nevertheless a well-structured set of expressions in the form of operator conditions for optimality. In a certain sense they are less mysterious than the previous set in that the counterpart of the adjoint variables have a much more direct relationship to the physical dynamics now. There can be more than conceptual simplicity in this direct connection with time responses; it is a very practical matter for example to measure them on a system, which is more than could be said of differential relationships.

Theoretical Approaches

The range of theoretical formulations of the optimal control problem continues to grow; even among the now classical ones improvements continue to flourish.

The linear dynamics, quadratic cost case (LQP)

The solution to this special case devolves around Riccati theory and has the great practical advantage that the optimal control can be obtained in feedback form through the solution of a matix Riccati differential equation.

† by G. F. Bryant, to be published.

A successive substitutional method of achieving this due to Kleinman [1] has quadratically convergent iterations.

However, the canonical equations of optimality for this case take the simple form

$$\begin{bmatrix} \dot{x} \\ \dot{\lambda} \end{bmatrix} = \begin{bmatrix} A & -BR^{-1}B^T \\ -Q & -A \end{bmatrix} \begin{bmatrix} x \\ \lambda \end{bmatrix}$$

where Q and R come from

$$L = \tfrac{1}{2}\langle x, Qx\rangle + \tfrac{1}{2}\langle u, Ru\rangle$$

whose solution is given directly by the state transition matrix $\Phi(t, \tau)$ in closed form:

$$\begin{bmatrix} x(t) \\ \lambda(t) \end{bmatrix} = \Phi(t, \tau) \begin{bmatrix} x(\tau) \\ \lambda(\tau) \end{bmatrix}$$

Due to the implied difficulty of eigenvector/eigenvalue analyses involved, one has fought shy of this direct path to a solution. However it can be reported [21] that through the use of a good eigen-analysis computer programme it has proved feasible to take this route to a solution for a 30 state system having a range of eigenvalues of 10^5.

Another surprise is to hear of a method used in certain problems in astrophysics drawn to our attention by Kailath [2] which solves the Riccati equation as the simultaneous solution of a pair of non-linear equations in a smaller number of variables than those involved in the Riccati solution.

Dynamic programming

Dynamic programming achieves optimal control by a process of "imbedding" sets of implicit optimal solutions obtained as functions of initial conditions over a small tract of solution time and then increasing the solution time by increments until the given boundary conditions are matched. The counterpart of the optimality conditions is the rather obvious statement that continuation of an optimal solution is optimal (known as the principle of optimality). A partial differential equation expresses this continuity condition which, subject to certain smoothness requirements, is as follows:

let

$$V(x(t), t) = \min_{u(t)} \int_t^{t_1} L(x, u, t)\, dt$$

then

$$-\frac{\partial V}{\partial t} = \min_{u(t)} \left[L(x, u, t) + \frac{\partial V}{\partial x} \dot{x} \right]$$

$$-\frac{\partial V}{\partial t} = \min_{u(t)} \left[L(x, u, t) + \frac{\partial V}{\partial x} f(x, u, t) \right]$$

which expresses the required condition on u for optimality.

Computational methods

From a computational point of view Dynamic Programming as such is fairly useless since it involves solving problems in segments as functionals of initial conditions and only in the assembly stages can the many superfluous segments of trajectory be discarded. But the viewpoint of dynamic programming has proved quite valuable as a pointer for derivation of methods leading to useable algorithms. A number of special cases have led directly to computable solutions. The minimum time solution between given initial and final states has been given by Neustadt [3]. The minimum control energy to unit radius of a target has been solved by Polak [4] and the minimum quadratic departure from a target by Ho [5].

A variant on these methods are general boundary iteration schemes [6] which also deal in a currency of optimal trajectories only and seek to force a match of boundary conditions. State and adjoint variable equations are integrated together in the forward direction involving the integration of exponentially increasing modes so that computational difficulties can easily arise if the time epoch is much longer than the time constant of dynamics. The rate of improvement towards a matching of the boundary is assisted by the use of sensitivity matrices, which are themselves given by the solution of further differential equations known as accessory equations. This then permits the use of Newton's algorithm in n dimensions to force a match of boundary values. The method which is very sensitive to initial estimates of boundary values can give very precise answers in suitable cases. The method cannot converge on incorrect solutions as can happen with some trajectory improvement methods.

Differential dynamic programming [7]

This is the most successful general method that has come out of the dynamic programming viewpoint [8], that is to say embodying the partial differential relationship given in the previous section. The method is started by assuming a nominal control vector $\bar{u}(t)$ which allows a nominal state vector $\bar{x}(t)$ to be obtained by forward integration of the dynamic equations of state. This state vector $\bar{x}$ will not be consistent with the requirements of the p.d.e. so we now find a δu to be added to $\bar{u}$ that does satisfy it:

$$-\frac{\partial V}{\partial t} = \min_{\delta u} \left[H(\bar{x}, \bar{u} + \delta u, V_x; t) \right];$$

δu is not necessarily small. The resulting u, u^* say, is such that $H_u(\bar{x}, u^*, V_x; t) = 0$.

So far so good, but now it is necessary if we want approximations better than first variations to expand about $\bar{x}$ and u^* when terms involving δu will be:

$$\langle \delta u, (H_{ux} + f_u{}^T V_{xx})\delta x \rangle + \tfrac{1}{2}\langle \delta u, H_{uu}\delta u \rangle \qquad \text{where} \qquad H = L + V_x f.$$

The condition for minimality obtained by differentiating w.r.t. δu now becomes

$$(H_{ux} + f_u{}^T V_{xx})\delta x + H_{uu}\delta u = 0;$$

that is

$$\delta u = -\beta \delta x \qquad \text{where} \qquad \beta = -H_{uu}{}^{-1}(H_{ux} + f_u{}^T V_{xx})$$

Thus we have obtained a feedback controller for maintaining the necessary condition of optimality, but note that it only holds locally for δx sufficiently small. By contrast the previous minimisation holds globally since δu there is not necessarily small.

Finally then the new control to be applied to the system is

$$u = u^* + \beta \delta x$$

where u^* hold globally and $\beta \delta x$ hold locally.

This is a very rough heuristic way of explaining things. What is interesting about the method is that an exact statement of change in Performance Index can be made depending on changes in control u, thus:

$$V(x, u, t) - V(x, \bar{u}, t) = \int_{t_0}^{t_1} \left[H(\bar{x}, u, V_x(\bar{x}); t) - H(\bar{x}, \bar{u}, V_x(\bar{x}); t) \right] dt$$

which is negative if u minimises $H(\bar{x}, u, V_x(\bar{x}); t)$ but $V_x(\bar{x})$ is unfortunately unknown, although λ (the solution of the usual adjoint equation) can be regarded as an estimate of $V_x(\bar{x})$ with an error norm proportional to

$$\int_{t_0}^{t_1} \| u - \bar{u} \| \, dt.$$

Note the dependence on the time integral of u rather than u itself. Thus sharp changes in u over small time intervals need not effect the cost very much. If λ replaces $V_x(\bar{x})$ then the resultant change in performance cost is proportional to the square of this integral which is much smaller still.

The method has all sorts of pleasing features, second order convergence is a good start so that *LQP* problems are solved in one step and non-*LQP* problems converge rapidly in the neighbourhood of the optimum due to second order expansions of functions. In some problems solution of Riccati equations become unbounded along nominal trajectories but will not using this method. It also has less differential equations to integrate than most other second order methods. It can handle bang-bang solutions, terminal manifold and implicit time cases. Mayne [9] has shown that the approach leads to a unified theory for the optimisation of dynamic systems.

Use of weighting functions

In considering the *LQP* problem we have seen that a feasible method makes use of the state transition matrix $\phi(t, \tau)$ and deals in a currency of eigenvalues and eigenvectors. It is not a long jump from this to ask the question of whether it was ever necessary to put up with the manifest difficulties of dealing with a differential equation as such at all. Why can we not state the entire problem in convolution terms using time weighting functions?

Conceptually, relationships are much more direct as shown in Fig. 1; instead of going by the indirect route by way of the state space x we go straight from input u to output y.

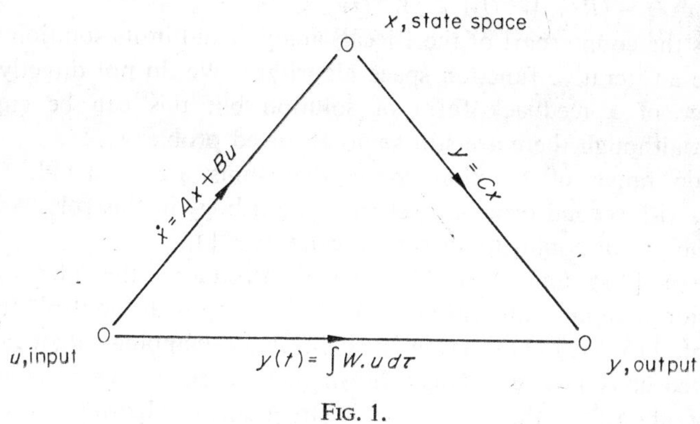

FIG. 1.

Particular numerical advantage should accrue by use of the direct input-output map provided by the weighting function where u and y are of low dimension relative to x. For example in plant simulations for testing out various control schemes or for calculating open-loop optimised controls using function-space algorithms.

On the face of it this could have two interesting advantages:

(a) it gets away from the finite dimensional restrictions of the state vector itself and allows the possiblity of handling pure time delay and distributed parameter systems;
(b) it gets away from the non-trivial task of finding state-space realisations of the system. The adjoint system then becomes simply the direct transpose of the weighting function.

The statement of the problem in these terms works out very happily in a Hilbert (function) space setting:

Performance $\quad V = \frac{1}{2}\langle y, Qy\rangle + \frac{1}{2}\langle u, Ru\rangle$

Dynamics $\quad y = y_s + Wu$

where y_s represents initial condition disturbances.

$$\begin{aligned} V(u) &= \tfrac{1}{2}\langle y_s + Wu, Q(y_s + Wu)\rangle + \tfrac{1}{2}\langle u, Ru\rangle \\ &= \tfrac{1}{2}\langle y_s, Qy_s\rangle + \langle u, W^*Qy_s\rangle + \tfrac{1}{2}\langle u, W^*QWu\rangle + \tfrac{1}{2}\langle u, Ru\rangle \\ &= \tfrac{1}{2}\langle y_s, Qy_s\rangle + \langle u, W^*Qy_s\rangle + \tfrac{1}{2}\langle u, (R + W^*QW)u\rangle \end{aligned}$$

$$\frac{\partial V}{\partial u} = 0 = W^*Qy_s + (R + W^*QW)u$$

whence $u^0 = -(R + W^*QW)^{-1}W^*Qy_s$

This is the counterpart of the Riccati analysis and in its solution we have to utilise an iterative function space algorithm. We do not directly get the advantage of a feedback form of solution but this can be engineered [10, 11], although there are still some unsolved problems.

A wide range of function space algorithms are to hand. Gradient methods with second order convergence are a bogy in this role. Advantage lies in the use of conjugate direction techniques [12].

Freeman [13] and Mee [14] have introduced the idea of using contraction-mapping algorithms for finding optimal control functions. Allwright [15, 16] has given a contraction-mapping algorithm with guaranteed convergence but has shown that in fact for every contraction-mapping algorithm there is a conjugate-gradient algorithm with better convergence properties.

Iterative algorithms for control function optimisation usually require (theoretically) the exact solution of the system differential equations for a sequence of non-optimal control functions. This seems to be an undesirable waste of effort. Balakrishnan [17] has discussed the use of penalty-function techniques whereby the optimisation subject to dynamic constraints is converted to a sequence of unconstrained optimisations. By using an integral description of the system's dynamics, duality can be utilised to give an algorithm which avoids direct solution of the system equations for non-optimal control functions but does not suffer from the disadvantages of the penalty-function method. Frick [18] has employed the integral description of the system dynamics to obtain a new penalty-function method which can have advantages over the earlier one.

Duality concepts can be used also to give lower bounds for minimal performance index which are useful since they can be used as stopping conditions for iterative control function optimisations. By finding open-loop optimised control for a selection of the important initial conditions a feedforward optimal controller can be constructed. The method as with Riccati analysis gives constant gains for the infinite-time case.

The weighting function method has many practical virtues and we shall, no doubt, see further useful results coming from this direction.

Applicability of Optimal Control

Optimal control is most suitable where complexity of dynamics is modest so that controllers are not too complicated, scheduling of steel cold rolling is an example [19], or alternatively the importance of the application is such as to justify a complicated controller with feedback from every component of the state vector. A number of cases in the aerospace field have justified this. In the case of the moon landing, optimal control has been utilised for three of the operations—docking, mid-course guidance and landing. These are all cases where transient behaviour is all important. In a number of less esoteric applications it has been used to obtain feedforward designs, heat distribution in a reactor [20] for example, which is at the other extreme of the frequency spectrum as a steady-state design. Successful application has been made in the middle ground; design of controls for a nuclear reactor [21] is an example where both steady-state and transient phenomena are important. This case is interesting in that freely available controls are few and the matching of key steady-state gain using these proves to be an important feature of the design.

An interesting experience was gained from the scheduling problem in steel cold rolling. Each stand of the mill has two available controls; namely the force applied to reduce the thickness of the strip by means of screws situated in the top of the main frame and the speed of motors driving the rolls dragging or pushing the strip through the gap between them. The scheduling problem consists of choosing the screw positions and motor speeds on the sequence of stands which for a given strip width and material achieve the desired output thickness and flatness of sheet. It is not difficult to see that this problem can be formulated as a non-linear sample data boundary value problem where states are sampled at intervals of space, corresponding to the stands of the mill, rather than time. However once an optimal control problem has been properly formulated the solution emerges without any explanations as to why it is as it is. Considerable incredulity can be expressed by those associated with the day-to-day running of the plant at these results. At first sight it seems unrewarding to spend time in trying to rationalise what the mathematics is trying to tell, but let it be said that the practical rewards for making this intellectual effort can pay off most handsomely in the insight given into the underlying mechanism of good control and what to do about it. In this case it was a most rewarding exercise and it may well be that this struggle for physical insight will become the general pattern of experience of industrial application of optimal control.

Reference has been made to the criticism that frequently optimal control represents an expensive best in controller design and there is a practical need for something designed down from this. Mee [11] in work using weighting function methods describes an algorithm for obtaining an optimal matrix of feedback gains for the reduced dimensionality of description that this form of analysis allows. Cumming [22] and, independently, Levine and Athans [23], describe a parameter optimisation method using state vector descriptions resulting in simplified controllers. Both methods utilise a Kleinman type algorithm with second order convergence. Further work along these general lines can be expected to give results of interest in engineering applications where cost of implementing solutions plays a significant role. Some of these developments, however, are not without their theoretical interest.

References

1. Kleinman, D. L. An iterative technique for Riccati equation computations, *I.E.E.E. Trans.* **AC-13** (1968) 114-115.
2. Kailath, T. "Some Chandrasekhar-type algorithms for Quadratic Regulators", Department of Computing and Control publication, Imperial College, 72/25.
3. Neustadt, L. W. Synthesising time optimal control systems, *J. Math. Anal. Appl.* **1,** (1960) 454.
4. Polak, E. On the removal of Ill-conditioning effect in the computation of optimal controls. IFAC Symposium on Dynamic Systems, Sydney 1969.
5. Ho, Y. C. A successive approximation technique for optimal control systems subject to input saturation. *Trans. A.S.M.E. J. Basic Eng.* **84,** (1962) 33.
6. Plant, J. B. "Some Iterative Solutions in Optimal Control". MIT Press, (1968).
7. Jacobson, D. H. and Mayne, D. Q. "Differential Dynamic Programming". Elsevier (1970).

8. Mayne, D. Q. "A second order gradient method for determining trajectories of nonlinear discrete-time systems". *Int. J. Control,* **3,** (1966) 85.
9. Mayne, D. Q. "Differential Dynamic Programming—A Unified Approach to Optimization of Dynamic Systems". Harvard Division of Engineering and Applied Physics Technical Report No. 623, (Sept. 1971).
10. Allwright, J. C. Optimal control synthesis using function decomposition techniques. Paper 7.3 Preprints of 4th IFAC World Congress, (1969).
11. Mee, D. "Optimal feedback gains for the linear system-quadratic cost problem", *Int. J. Control,* **13** (1), (1971). 179–187
12. Hestenes, M. R. and Stiefel, E. Methods of conjugate gradients for solving linear systems, *J. of Research of the National Bureau of Standards,* **149** (6) (1952).
13. Freeman, E. A. On the optimization of linear time-variant and multivariable control systems using the contraction-mapping principle. *J. Opt. Theory and Applications* **3,** (1969).
14. Mee, D. "A Weighting Function Approach to Linear Control System Design", Ph.D. Thesis, University of London (1969).
15. Allwright, J. C. Contraction-mapping algorithms with guaranteed convergence Paper 34.4, Preprints of 5th IFAC World Congress, (1972).
16. Allwright, J. C. "Contraction-mapping and Conjugate-gradient Algorithms for Optimal Control". Department of Computing and Control publication, Imperial College, 72/12, to be published.
17. Balakrishnan, A. V. On a new computing technique in optimal control. *SIAM J. Control,* **6** (1968).
18. Frick, P. A. Iterative optimum control function determination without directly solving the system dynamical equations. Paper 34.2, Preprints of 5th IFAC World Congress, 1972.
19. Bryant, G. F., Halliday, J. M. and Spooner, P. D. Optimal scheduling of a tandem cold rolling mill. Paper 2.4, Preprints of 5th IFAC World Congress, 1972.
20. Rosenbrock, H. H., Storey, C. Computational Techniques for Chemical Engineers, Pergamon Press (1966).
21. Tepper, L. "Application of Modern Control Theory to the Control of a Nuclear Power Reactor", Ph.D. Thesis, University of London (1972).
22. Cumming, S. D. S. "Observer Theory and Control System Design", Ph.D. Thesis, University of London (1969).
23. Levine, W. S. and Athans, M. On the design of optimal linear systems using only output variable feedback, Proc. of 6th Allerton Conference on Circuits and Systems Theory, 1969, 661–670.

Discussion

REVINGTON (C.E.G.B.). How can such mathematical methods be modified to incorporate any physical insight that is available, particularly the human operator's ability to pick out and concentrate on the few relevant variables that dominate a current aspect of the overall problem of controlling multivariable plant?

WESTCOTT. In one way it is out of order to mention the matter here because the mathematical techniques themselves are not relevant. They are, after all, only the framework of the computations from which come the answers. The effort involved is to try to puzzle out why they come out that way. This is not a mathematical procedure at all. The reason why I mention it is that we could not do this for ourselves. You are not the chap who is likely to understand the process best. So you are forced to collaborate with someone who does understand it very well but can't understand the mathematics. There is no doubt whatever that when trying to apply optimal control to industrial problems this aspect did prove to be very important.

I think our feeling now is there never will be a purely handle-turning way of solving these problems. You have to reckon with more than that to be successful. Not for any reason that the solutions were different at the end of it. But in the understanding of it there was a vast difference.

REVINGTON. A further comment on this is that you advocate the use of weighted quadratic functions which in fact often require a lot of trial and error to produce reasonable weighting terms. It would be useful if methods could be developed which get down directly to satisfying the actual plant constraints rather than limiting ourselves to trying to satisfy them by using a quadratic cost function.

WESTCOTT. You always try to utilize a quadratic form of the cost as far as you possibly can and you would be prepared to go to a lot of trouble to find something of that form, since the analysis is so congenial then. Much effort is now devoted to weighting the terms of the quadratic cost function in order to satisfy all parties. I think this is one avenue where the answer could come; from using what we've got rather than inventing new forms. On the whole, other methods of measuring performances are very difficult to use in practice.

LEIGH (British Steel Corporation). It is well known that most industrial control systems are very far from optimal in their performance. However, this does not necessarily represent a deplorable state of affairs, since in many of these cases, very careful studies have ensured that investment in the control system has been sufficient to guarantee adequate though not optimal performance.

At present then, there is something of a dichotomy. Control theorists are coming forward with a single optimal solution that can rarely if ever be implemented. Meanwhile the industrial control system is designed by ad-hoc methods and the industrial designer cannot benefit from the optimal control studies. The problem could be overcome if the control theorist were to produce a graph showing attainable system performance (cost function) against investment in the control system. This would allow engineers to assess the relative merits of different levels of investment in control. Putting this another way, optimal control theory identifies the summit of the cost function hill but tells little about the shape of the slopes below the summit where most applications necessarily lie.

WESTCOTT. This is a real problem and you have put your finger on the exact issue. If you are a purist and stick rigidly to your optimal control then you must have your full state vector and do the whole thing properly. You could argue as a purist that anything outside this is not optimal control.

Increasingly, however, one is applying the mathematical techniques of optimal control to cases which, strictly speaking, are not optimal because perhaps they only control the y-output and not the x-output. Now it is a question of deciding the balance of trade-off. As one does this so the performance gets worse and worse. But from an engineering point of view many of these other solutions are very interesting and the trade-off is very useful as an alternative to the full-blown thing. This is exactly what I had in mind and I think we shall see a lot more of it—it is very valuable. You might say that optimal control today it not what it used to be!

LEIGH. I am pleased that you agree that more attention should be given to the design of sub-optimal systems. You may be interested in a method that I developed to allow the shape of the cost function hill to be assessed. The problem was concerned with the design of a control system for a hot rolling mill and I first simplified the problem until at each of the six stands there were four discrete design choices (giving the possibility of 4^6 distinct designs). I then used a method analogous to dynamic programming to produce a whole spectrum of design solutions and corresponding system performances ranging from "optimal" to "no control at all". Each design solution was of course the minimum cost solution for the corresponding value of performance index, all other solutions having been eliminated during the computation as soon as they were identified.

6. A Naïve Approach to the Optimal Control of Nonlinear Stochastic Systems

B. L. JOFFE† AND R. W. H. SARGENT

Department of Chemical Engineering and Chemical Technology, Imperial College of Science and Technology, London, England

1. Introduction

This paper presents an approximate solution to the problem of determining the optimal control policy for a system subjected to random perturbations. More specifically, we shall consider systems whose behaviour can be described in terms of the solution of a set of differential equations of type

$$\dot{x}(t) = f(t, x(t), u(t), \xi(t)) \tag{1.1}$$

where $x(t) \in E^n$ is a vector describing the "state" of the system at time t, $u(t) \in E^m$ is a control vector, $\xi(t) \in E^r$ is a vector-valued stochastic process, and $f(t, x, u, \xi)$ is an n-dimensional vector-valued function of its arguments.

For the analogous deterministic system, $\dot{x}(t) = f(t, x(t), u(t))$, it is well known that under suitable conditions on $u(t)$ and $f(t, x, u)$ there exists a unique solution $x(t)$ for $t \geqslant t_0$ corresponding to a given control $u(t), t \geqslant t_0$, and given initial conditions, $x(t_0)$. In fact, a large part of control theory rests on this fundamental property of deterministic systems, that the current state provides sufficient information about the past to make possible the prediction of the future.

In the presence of random perturbations the analogous property is that $x(t)$ should be a continuous Markov process, and for processes described by (1.1) this in turn requires that $\xi(t)$ itself should be a continuous Markov process. Now it is well known—see for example [2]—that a large class of continuous Markov processes can be generated as solutions to stochastic differential equations of the type

$$d\xi(t) = \mu(t, \xi(t))\, dt + \sigma(t, \xi(t))\, dw(t) \tag{1.2}$$

where $w(t)$ is an r-dimensional continuous Brownian motion.

†Present address: Computer Sciences S.A. Limited, 51 Juta Street, Braamfontein, Johannesburg, South Africa.

If $\xi(t)$ is adjoined to $x(t)$ to form an extended state vector, it is clear that equations (1.1) and (1.2) together form a special case of the system

$$\mathrm{d}x(t) = f(t, x(t), u(t))\,\mathrm{d}t + \sigma(t, x(t))\,\mathrm{d}w(t) \tag{1.3}$$

which can thus be taken as a fairly general model for a controlled continuous Markov process. Almost all the published work on continuous stochastic systems is restricted to systems of this type, and Fleming [3] has given an excellent review of the progress made.

However, such systems present a number of theoretical and practical problems. First, in view of the properties of $w(t)$, Eqns.(1.2) or (1.3) fall outside the scope of Riemann or Lebesgue integration and require the definition of a special stochastic integral; several definitions have been proposed but the best known are those due to Stratonovitch and Ito. With such a definition, the solution of (1.2) has a transition density under suitable conditions, and this satisfies forward and backward equations of Fokker–Planck–Kolmogorov type. Secondly, a continuous Markov process is not strictly speaking physically realizable and the question arises as to how good a model (1.3) is for an actual physical process. Wong and Zakai [12] studied piece-wise linear approximations to the Brownian motion $w(t)$ in (1.2) when $\xi(t)$ and $w(t)$ are scalar, and Clark [1] extended their work to the vector case, studying the limit as the approximation tends to Brownian motion for both the Ito and Stratonovich integrals. It turns out that, except under certain conditions for the Stratonovich integral, additional terms appear in the limiting Markov equations, so care is needed in formulating the correct Markov model for a given physical process.

In the context of optimal control, application of the dynamic programming formalism yields the familiar Bellman-type partial differential equation, which provides a theoretical solution to the problem, but which cannot be solved in practical terms except for simple cases. Kushner [8] has derived a stochastic maximum principle for fixed end-time and the case where $\sigma(t, x(t))$ in (1.3) does not depend explicitly on $x(t)$, but again this is useless for practical computation without further simplifications.

In the face of these difficulties several authors have investigated approximate solutions based on the assumption that the random perturbations are small. Stratonovich [11] proposed a method of successive approximation which involves solving a sequence of partial differential equations, convergent for small noise, while Kushner [9] and Fleming [4] have both considered expansions about the corresponding optimal deterministic trajectory in the absence of noise. Kushner's approach is designed for random impulses at a few isolated instants of time, and the system of differential equations for the coefficients in the expansion rapidly grows with the number of impulses, becoming computationally intractable for a continuous stochastic process. Fleming considers singular perturbations of the Hamilton–Jacobi equation and the coefficients in his expansion are obtained by solving partial differential equations. Both approaches are also restricted in their scope by mathematical assumptions which need to be made in their development; for example, both authors take σ as independent of $x(t)$, Kushner can deal only with unbounded controls, while Fleming only considers system equations linear in the controls and expansions in regions of strong regularity.

In the present paper we start from Eqn (1.1) as a basis, taking $\xi(t)$ to be a physically realizable stochastic process, and develop an approximate solution for small noise by the expansion of the right-hand side of (1.1). The basic assumptions which ensure the existence of the requisite solutions, together with the precise statistical assumptions concerning the noise, are given in Section 2, followed by development of the approximate solution in Section 3. The analogous results for approximate correction of the deterministic trajectory are given in Section 4, and finally the application to optimal control problems is discussed in Section 5.

2. Basic Assumptions

We begin by stating conditions which ensure the existence of solutions to the system of differential equations (1.1) over the time interval $[t_0, t_f]$ for a given control function $u(t)$ and a given realization of the stochastic process $\xi(t)$ over this interval.

We suppose that all such functions $u(t)$, $\xi(t)$, of interest are Lebesgue-measurable functions of t, and for each $t \in [t_0, t_f]$ we have $x(t) \in E^n$, $u(t) \in U \subseteq E^m$, and $\xi(t) \in V \subseteq E^r$, where U is a given set and V is a suitable convex set. The function $f(t, x, u, \xi) \in E^n$ is defined for all $t \in [t_0, t_f]$, $x \in E^n$, $u \in U$, $\xi \in V$, and has the properties:

(i) For each fixed set of values of $x, u, \xi, f(t, x, u, \xi)$ is a measurable function of t.

(ii) For each fixed $t, f(t, x, u, \xi)$ is continuous in x, u, ξ.

(iii) There exists a function $S(t)$ summable on $[t_0, t_f]$, and a function $\phi(z)$, positive and continuous for $z \geqslant 0$ but not summable on $[0, \infty)$, such that $\|f(t, x, u, \xi)\| \leqslant S(t)/\phi(\|x\|)$ for each t, x, u, and ξ.

(iv) There exists a function $M(t)$ summable on $[t_0, t_f]$ such that $|(x - x^*)^T (f(t, x, u, \xi) - f(t, x^*, u, \xi))| \leqslant M(t) \|x - x^*\|^2$.

Note: In the above ,and throughout the paper, $\|.\|$ denotes the Euclidean norm: for a vector $\|x\| = [\sum_{i=1}^n (x^i)^2]^{\frac{1}{2}}$ and for a matrix

$$\|A\| = \sup_{\|x\| \leqslant 1} \|Ax\|.$$

Vector or matrix elements will be denoted by superscripts, and subscripts will be used to denote differentiation; thus f_x is the Jacobian matrix whose (i,j) element is $\partial f^i/\partial x^j$, while $f_{x\xi}{}^k$ is the matrix with (i,j) element $\partial^2 f^k/\partial x^i \partial \xi^j$.

Under the above conditions, as shown for example by McShane [10], for given functions $u(t), \xi(t)$, and given initial condition $x(t_0) = x_0$, there exists a unique function $x(t) \in E^n$ which satisfies the system of differential equations (1.1) for almost all $t \in [t_0, t_f]$. Moreover the family of solutions $x(t)$ corresponding to all admissible functions $u(t), \xi(t)$, is absolutely continuous in t and uniformly bounded on the interval $[t_0, t_f]$.

We define the set $W(t_f, x_0)$ as the set of all points $x(t)$ reachable from $x(t_0) = x_0$ in the interval $[t_0, t_f]$ with admissable functions $u(t)$, $\xi(t)$, $t \in [t_0, t_f]$. Then if $X(t_0)$ is a given set we define the set $X(t_f)$ as the convex hull of the union of all sets $W(t_f, x_0)$ corresponding to $x_0 \in X(t_0)$; it follows from the above that if $X(t_0)$ is bounded then $X(t_f)$ is also bounded.

We are now in a position to specify further conditions which will enable us to establish error bounds on the approximate solution to be developed in Section 3. It is assumed that these conditions hold for all $t \in [t_0, t_f]$; $x, x^* \in X(t_f)$; $u \in U$; $\xi, \xi^* \in V$:

(a) For each t and u the second partial derivatives of the elements of $f(t, x, u, \xi)$ with respect to x and ξ exist, and there is a constant $L < \infty$ such that all these derivatives satisfy Lipschitz conditions of type

$$\left.\begin{aligned} \|f_{x\xi}{}^k(t, x, u, \xi) - f_{x\xi}{}^k(t, x^*, u, \xi)\| &\leqslant L\|x - x^*\| \\ \|f_{x\xi}{}^k(t, x, u, \xi) - f_{x\xi}{}^k(t, x, u, \xi^*)\| &\leqslant L\|\xi - \xi^*\| \end{aligned}\right\} \tag{2.1}$$

(b) There exist finite constants C_1, C_2, C_3 such that:

$$\left.\begin{aligned} &\|f_x(t, x, u, \xi)\| \leqslant C_1, \qquad \|f_\xi(t, x, u, \xi)\| \leqslant C_2 \\ &n^{\frac{1}{2}}\|f_{xx}{}^k(t, x, u, \xi)\| \leqslant C_3, \qquad n^{\frac{1}{2}}\|f_{x\xi}{}^k(t, x, u, \xi)\| \leqslant C_3 \\ &\qquad\qquad n^{\frac{1}{2}}\|f_{\xi\xi}{}^k(t, x, u, \xi)\| \leqslant C_3 \\ &\text{From which:} \\ &\|f(t, x, u, \xi) - f(t, x^*, u, \xi)\| \leqslant C_1\|x - x^*\| \\ &\|f(t, x, u, \xi) - f(t, x, u, \xi^*)\| \leqslant C_2\|\xi - \xi^*\| \\ &\|f_x(t, x, u, \xi) - f_x(t, x^*, u, \xi)\| \leqslant C_3\|x - x^*\| \\ &\|f_x(t, x, u, \xi) - f_x(t, x, u, \xi^*)\| \leqslant C_3\|\xi - \xi^*\| \end{aligned}\right\} \tag{2.2}$$

Finally, we list the assumptions concerning the stochastic process $\xi(t)$, and the initial state $x(t_0)$. Each realization $\xi(t)$, $t \in [t_0, t_f]$ is assumed to be a measurable function of t, with $\xi(t) \in V$ for each t, with probability 1. Similarly the initial state $x(t_0) \in X(t_0)$ with probability 1.

We define

$$\left.\begin{aligned} \xi^*(t) &= E\{\xi(t)\} \qquad & \delta\xi(t) &= \xi(t) - \xi^*(t) \\ x_0^* &= E\{x(t_0)\} \qquad & \delta x_0 &= x(t_0) - x_0^* \end{aligned}\right\} \tag{2.3}$$

and assume the existence of all expectations and integrals in the following inequalities, which are assumed to hold for some $\varepsilon < \infty$ and $A < \infty$:

$$\left.\begin{aligned} &\int_{t_0}^{t_f} E\{\|\delta\xi(t)\|^3\}\,dt \leqslant T\varepsilon^3 & &E\{\|\delta x_0\|^3\} \leqslant A^3\varepsilon^3 \\ &\int_{t_0}^{t_f} \|E\{\delta\xi(t).\delta\xi^T(t')\}\|\,dt \leqslant T\varepsilon^3 & &\|E\{\delta x_0.\delta\xi^T(t')\}\| \leqslant A\varepsilon^3 \\ &\int_{t_0}^{t_f} E\{\|\delta\xi(t)\|.\|\delta\xi(t')\|^2\}\,dt \leqslant T\varepsilon^3 & &E\{\|\delta x_0\|.\|\delta\xi(t')\|^2\} \leqslant A\varepsilon^3 \\ &\int_{t_0}^{t_f} E\{\|\delta\xi(t)\|^2.\|\delta\xi(t')\|\}\,dt \leqslant T\varepsilon^3 & &E\{\|\delta x_0\|^2.\|\delta\xi(t')\|\} \leqslant A^2\varepsilon^3 \end{aligned}\right\} \tag{2.4}$$

where $T = t_f - t_0$ and $t' \in [t_0, t_f]$.

It is worth noting that we can take $X(t_0) = E^n$ and $V = E^r$, but the conditions in (2.1) and (2.2) then unduly restrict the permissible forms of $f(t, x, u, \xi)$ so it is preferable to take $X(t_0)$ and V as bounded subsets of their respective spaces. It then follows that $X(t_f)$ is bounded and conditions (2.2) are satisfied as a consequence of (2.1).

A simple example of a stochastic process $\xi(t)$ satisfying the above conditions, which can be used as a general model, is the piece-wise constant process

$$\xi(t) = \xi_i, \qquad t \in (t_{i-1}, t_i] \tag{2.5}$$

where the t_i form a finite sequence such that

$$t_0 < t_1 < t_2 < \ldots < t_{i-1} < t_i < \ldots < t_f,$$

with $t_i - t_{i-1} \leqslant \varepsilon$, and the ξ_i are a sequence of independent random vectors such that $E\{\|\delta\xi_i\|^3\} \leqslant \varepsilon^3$.

For $x(t_0)$ and the ξ_i there are, of course, a number of well known distributions which restrict the realizations to a bounded set, and for most distributions negligible error in expectations will result by truncating the distribution so that $Pr(\|\delta\xi\| > c) = 0$ for some suitably large c.

Thus, apart from the small noise assumptions of (2.4) themselves, the basic assumptions do not seem to be unduly restrictive.

3. The Approximate Solution

In this section we shall show that a good approximation to $E\{x(t)\}$ is given by $x^*(t)$, the solution of the system:

$$\left.\begin{aligned}
&\dot{x}^*(t) = f(t, x^*(t), u(t), \xi^*(t)) + z(t), \qquad && x^*(t_0) = x_0^* \\
&\dot{P}(t) = f_x P(t) + P(t) f_x^T, && P(t_0) = E\{\delta x_0 . \delta x_0^T\} \\
&\text{where} \\
&z^k(t) = \tfrac{1}{2}\,\text{trace}[f_{xx}^k P(t) + f_{\xi\xi}^k Q(t)] \\
&Q(t) = E\{\delta\xi(t)\,\delta\xi^T(t)\}
\end{aligned}\right\} \tag{3.1}$$

In (3.1), and throughout the rest of this section it is assumed that where arguments are omitted the function in question is evaluated at the appropriate point of the trajectory $(t, x^*(t), u(t), \xi^*(t))$.

It is clear from the assumptions of the previous section that the system (3.1) uniquely defines $x^*(t)$ and $P(t)$ over the interval $[t_0, t_f]$, and that $x^*(t)$ is absolutely continuous in t and uniformly bounded over this interval for all admissible $u(t)$.

We start by establishing bounds for certain quantities appearing in the sequel, and will make repeated use of the following "generalized Gronwall lemma" due to Halkin [5]:

LEMMA. Given a real-valued function $y(t)$, defined and absolutely continuous on $[t_0, t_f]$, and a real-valued function $g(t)$, summable on $[t_0, t_f]$, such that for almost all $t \in [t_0, t_f]$, $\dot{y}(t)$ exists and satisfies

$$\dot{y}(t) \leqslant Cy(t) + g(t)$$

where C is a constant, then the following holds for all $t \in [t_0, t_f]$:

$$y(t) \leqslant e^{C(t-t_0)}\left\{y(t_0) + \int_{t_0}^{t} e^{-C(\tau - t_0)} g(\tau)\, d\tau\right\}.$$

Further, if $C \geqslant 0, y(t_0) \geqslant 0$, and $g(t) \geqslant 0, t \in [t_0, t_f]$, this relation simplifies to

$$y(t) \leqslant e^{C(t_f - 0)}\left\{y(t_0) + \int_{t_0}^{t_f} g(\tau)\, d\tau\right\}. \tag{3.2}$$

Defining $\delta x(t) = x(t) - x^*(t)$, we have, with probability one and for almost all $t \in [t_0, t_f]$:

$$\frac{d}{dt}\|\delta x(t)\| \leqslant \left\|\frac{d}{dt}\delta x(t)\right\| = \|\dot{x}(t) - \dot{x}^*(t)\|$$

and using (3.1) and (2.2) this yields

$$\frac{d}{dt}\|\delta x(t)\| \leqslant C_1\|\delta x(t)\| + C_2\|\delta\xi(t)\| + \frac{n}{2}C_3(\|P(t)\| + \|Q(t)\|). \tag{3.3}$$

Now from (3.1) and (2.4), using the Hölder inequality for expectations:

$$\left.\begin{aligned} &\|P(t_0)\| \leqslant E\{\|\delta x_0\,\delta x_0^T\|\} = E\{\|\delta x_0\|^2\} \leqslant A^2\varepsilon^2 \\ &\int_{t_0}^{t_f}\|Q(t)\|\,dt \leqslant \int_{t_0}^{t_f} E\{\|\delta\xi(t)\|^2\}\,dt \leqslant T\varepsilon^2 \end{aligned}\right\}. \tag{3.4}$$

Thus, from (3.1), (3.2) and (3.4), for almost all $t \in [t_0, t_f]$:

$$\|P(t)\| \leqslant \exp(2C_1T)\,\|P(t_0)\| \leqslant \bar{A}^2\varepsilon^2 \tag{3.5}$$

where $\bar{A} = A\exp(C_1T)$.

Similarly, applying (3.2) to (3.3) and using (3.4) and (3.5):

$$\|\delta x(t)\| \leqslant \exp(C_1T)\,\{\|\delta x_0\| + C_2\int_{t_0}^{t_f}\|\delta\xi(\tau)\|\,d\tau$$

$$+ \frac{n}{2}C_3T\,\varepsilon^2(1 + \bar{A}^2)\}. \tag{3.6}$$

Taking expectations of (3.6) and using (2.4) we have

$$E\{\|\delta x(t)\|\} \leqslant K\varepsilon \qquad \text{where} \qquad K = \bar{A} + \bar{C}_2 + \frac{n}{2}\bar{C}_3\varepsilon(1 + \bar{A}^2) \tag{3.7}$$

and we have written: $\bar{C}_1 = T\exp(C_1T)$, $\bar{C}_2 = \bar{C}_1C_2$, $\bar{C}_3 = \bar{C}_1C_3$. Also, from (3.6) and (2.4), using the Hölder and Minkowski inequalities for both expectations and integrals, it is straightforward to establish the following inequalities:

$$\left.\begin{aligned}
&E\{\|\delta x(t)\|^2\,\|\delta\xi(t')\|\} \leqslant \varepsilon^3\left[\bar{A} + \bar{C}_2 + \frac{n}{2}\bar{C}_3\right.\\
&\qquad\left.\times\,(1 + \bar{A}^2)(\varepsilon E\{\|\delta\xi(t')\|\})^{\frac{1}{2}}\right]^2\\
&E\{\|\delta x(t)\|^3\} \leqslant K^3\varepsilon^3;\\
&\int_{t_0}^{t_f} E\{\|\delta x(t)\|^2\,\|\delta\xi(t)\|\}\,dt \leqslant TK^2\varepsilon^3;\\
&\int_{t_0}^{t_f} E\{\|\delta x(t)\|\,\|\delta\xi(t)\|^2\}\,dt \leqslant TK\varepsilon^3.
\end{aligned}\right\} \tag{3.8}$$

Now in view of (2.1) and the convexity of the sets V and $X(t_f)$ we can expand $f(t, x, u, \xi)$ about the trajectory $(t, x^*(t), u(t), \xi^*(t))$ and hence obtain from (1.1) and (3.1), with probability one:

$$\begin{aligned}
\frac{d}{dt}\delta x^k(t) &= f_x{}^k\delta x(t) + f_\xi{}^k\delta\xi(t) + \tfrac{1}{2}\delta x^T(t)f_{xx}{}^k\delta x(t) + \delta x^T(t)f_{x\xi}{}^k\delta\xi(t)\\
&\quad + \tfrac{1}{2}\delta\xi^T(t)f_{\xi\xi}{}^k\delta\xi(t) + \eta^k(t) - z^k(t)
\end{aligned} \tag{3.9}$$

where $|\eta^k(t)| \leqslant \frac{1}{6}L(\|\delta x(t)\| + \|\delta\xi(t)\|)^3$. Similarly from (2.2) we obtain the lower order expansion:

$$\frac{d}{dt}\delta x(t) = f_x\delta x(t) + f_v\delta\xi(t) + v(t) - z(t) \tag{3.10}$$

where $\|v(t)\| \leqslant \frac{1}{2}C_3(\|\delta x(t)\| + \|\delta\xi(t)\|)^2$.

Taking expectations of (3.9) and using (2.3) and (3.1)

$$\begin{aligned}
\frac{d}{dt}E\{\delta x^k(t)\} &= f_x{}^kE\{\delta x(t)\} + \tfrac{1}{2}\,\text{trace}\,\delta P(t)\\
&\quad + \text{trace}\,[f_{x\xi}{}^kE\{\delta\xi(t)\delta x^T(t)\}] + E\{\eta^k(t)\}
\end{aligned} \tag{3.11}$$

where $\delta P(t) = E\{\delta x(t)\delta x^T(t)\} - P(t)$. Then taking norms of (3.11) and using (2.2) we have for almost all $t \in [t_0, t_f]$:

$$\frac{d}{dt}\|E\{\delta x(t)\}\| \leqslant C_1\|E\{\delta x(t)\}\| + \frac{n}{2}C_3\|\delta P(t)\| + nC_3\|E\{\delta\xi(t)\delta x^T(t)\}\|$$

$$+ \frac{n^{\frac{1}{2}}}{6}LE\{(\|\delta x(t)\| + \|\delta\xi(t)\|)^3\}. \tag{3.12}$$

Similarly from (3.10), again using (2.3) and (3.1) we obtain

$$\tfrac{1}{2}\frac{d}{dt}\|\delta P(t)\| \leqslant C_1\|\delta P(t)\| + C_2\|E\{\delta\xi(t)\delta x^T(t)\}\| + \frac{n}{2}C_3(\|P(t)$$

$$+ Q(t)\|)\,E\{\|\delta x(t)\|\} + \tfrac{1}{2}C_3E\{\|\delta x(t)\|\,(\|\delta x(t) + \|\delta\xi(t)\|)^2\} \tag{3.13}$$

and for a fixed $t' \in [t_0, t_f]$:

$$\frac{d}{dt}\|E\{\delta x(t)\,\delta\xi^T(t')\}\| \leqslant C_1\|E\{\delta x(t)\,\delta\xi^T(t')\}\| + C_2\|E\{\delta\xi(t)\,\delta\xi^T(t')\}\|$$

$$+ \frac{n}{2}C_3(\|P(t) + Q(t)\|)\,E\{\|\delta\xi(t')\|\}$$

$$+ \tfrac{1}{2}C_3E\{\|\delta\xi(t')\|\,(\|\delta x(t)\| + \|\delta\xi(t)\|)^2\}. \tag{3.14}$$

The Minkowski inequality can be used to reduce the last term in each of Eqns (3.12)–(3.14), then applying (3.2) to each equation in turn and using the previously established inequalities (2.4), (3.5), (3.7) and (3.8) we obtain successively:
From (3.14):

$$\|E\{\delta x(t)\,\delta\xi^T(t')\}\| \leqslant \varepsilon^3(\bar{A} + \bar{C}_2) + \frac{n}{2}\bar{C}_3(1 + \bar{A}^2)\,\varepsilon^2E\{\|\delta\xi(t')\|\}$$

$$+ \tfrac{1}{2}\bar{C}_3\varepsilon^3\left[1 + \bar{A} + \bar{C}_2 + \frac{n}{2}\bar{C}_3(1 + \bar{A}^2)(\varepsilon E\{\|\delta\xi(t')\|\})^{\frac{1}{2}}\right]^2. \tag{3.15}$$

Whence:

$$\int_{t_0}^{t_f}\|E\{\delta x(t)\,\delta\xi^T(t)\}\|\,dt \leqslant TK'\varepsilon^3$$

where (3.16)

$$K' = \bar{A} + \bar{C}_2 + \frac{n}{2}\bar{C}_3(1 + \bar{A}^2) + \tfrac{1}{2}\bar{C}_3(1 + K)^2.$$

From (3.13), using (3.16):

$$\|\delta P(t)\| \leqslant \varepsilon^3\{2\bar{C}_2K' + n\bar{C}_3K(1 + \bar{A}^2) + \bar{C}_3K(1 + K)^2\}. \tag{3.17}$$

From (3.12), using (3.16) and (3.17):

$$\|E\{\delta x(t)\}\| \leqslant \varepsilon^3 \left\{\frac{n}{2}\bar{C}_3[2K'(1 + \bar{C}_2) + n\bar{C}_3K(1 + \bar{A}^2) + \bar{C}_3K(1 + K)^2] + \frac{n}{6}L\bar{C}_1(1 + K)^3\right\}. \tag{3.18}$$

Equation (3.18) establishes the required bound on the error in taking $x^*(t)$ as $E\{x(t)\}$ and shows it to be $O[\varepsilon^3]$. It then follows from the definition of $\delta P(t)$ and equation (3.16) that $P(t)$ is an approximation to the covariance matrix of $x(t)$, again with error of $O[\varepsilon^3]$. However, $P(t)$ simply represents the propagation of the uncertainty in the initial state, since the effects of $\xi(t)$ are only $O[\varepsilon^3]$.

The result could have been generalized by allowing different coefficients of ε^3 in the various inequalities in (2.4), but this would merely have complicated the expressions without altering the order of the approximations. The bounds in (3.15)–(3.17) are in fact quite crude, due to the use of the generalized Gronwall lemma, and are not intended for use in estimating the error.

The method could also be extended to produce higher order approximations if desired, provided that the assumptions in Section 2 are appropriately extended; in particular it would be necessary to assume the corresponding higher order for the bounds on the autocorrelation integrals of $\xi(t)$ and the cross-correlations of this with the initial state in (2.4). However the equations become increasingly cumbersome and the second-order approximation of (3.1) should be adequate for most situations where the noise is small.

It is of interest to note that if Eqn. (1.1) is linear in $x(t)$ and $\xi(t)$ we have $C_3 = 0$, $L = 0$ and $x^*(t) = E\{x(t)\}$ exactly; it also follows from (3.1) that $x^*(t)$ is the equivalent deterministic trajectory since $z(t) = 0$. However, $P(t)$ still contains error terms of $O[\varepsilon^3]$, due to the autocorrelation effects of $\xi(t)$ and its crosscorrelation with $x(t_0)$.

4. Approximate Correction to the Deterministic Trajectory

Evidently an analysis similar to that in Section 3 can be carried out based on expansions about the equivalent deterministic trajectory, which is the solution of

$$\dot{\bar{x}}(t) = f(t, \bar{x}(t), u(t), \xi^*(t)), \quad \bar{x}(t_0) = x_0^* \tag{4.1}$$

We again denote this approximate trajectory by $x^*(t) = \bar{x}(t) + \delta x^*(t)$ where $\delta x^*(t)$ is the solution of the system

$$\left.\begin{aligned} &\frac{d}{dt}\,\delta x^*(t) = f_x \delta x^*(t) + z(t), \qquad \delta x^*(t_0) = 0 \\ &\dot{P}(t) = f_x P(t) + P(t) f_x{}^T, \qquad P(t_0) = E\{\delta x_0 \delta x_0{}^T\} \\ &\text{with} \\ &z^k(t) = \tfrac{1}{2}\,\text{trace}[f_{xx}{}^k P(t) + f_{\xi\xi}{}^k Q(t)] \end{aligned}\right\} \tag{4.2}$$

The similarity to (3.1) is obvious, except that in (4.2) the functions are evaluated along the trajectory (t, $\bar{x}(t), u(t), \xi^*(t)$) obtained by solving (4.1).

It can again be established that $x^*(t)$ and $P(t)$ are approximations with errors of $O[\varepsilon^3]$, although the bounds analogous to (3.15)–(3.18) contain a few extra terms which arise from omission of terms of $O[\varepsilon^3]$ containing f_{xx} in the first equation of (4.2).

These correction equations are of use if the deterministic trajectory has been established for a certain control $u(t)$ and it is desired to study the effect of various noise-levels on the trajectory, using the same control. Equations (4.2) are linear in $\delta x^*(t)$ and $P(t)$ and hence general solutions may be obtained by superposition.

Conversely, if the noise parameters are fixed and various controls are to be compared it is more efficient, and also slightly more accurate, to solve the system (3.1) directly for each control.

5. Application to Optimal Control

We suppose that the system behaviour is described by Eqn (1.1) with a suitably specified stochastic process $\xi(t)$ satisfying the assumptions of Section 2.

The class of admissible control functions $u(t)$ must similarly satisfy the conditions of Section 2, where the control region U is specified for each particular problem, and t_f is not necessarily fixed.

An optimal control function $\hat{u}(t)$ is defined as a member of the class of admissible controls which minimizes the value of the objective function

$$J(x(t_f)) = c^T x(t_f) \tag{5.1}$$

where c is a constant n-vector, subject to any further conditions imposed in particular cases. This choice of objective function is made for convenience, and it is well known that a wide class of problems can be cast in this form, if necessary by adjoining suitably defined variables to the state vector $x(t)$ and corresponding differential equations to (1.1).

Within this framework several types of control problem arise, depending on the further information made available. Because of the stochastic effects the best we can hope to do is to minimize the expected value of $J(x(t_f))$, taking account of this information, and in view of the linear form of $J(x(t_f))$ in (5.1) we have immediately

$$E_c J\{(x(t_f))\} = J(E_c\{x(t_f)\}) \tag{5.2}$$

where $E_c\{.\}$ denotes the conditional expectation $E\{.|\text{ available information}\}$.

Clearly solution of the control problem implicitly defines the optimal control function $\hat{u}(t)$ as a functional on the available information, and we may be interested in the control policy $\hat{u}(t)$ for specific cases or in the general form of the functional for particular classes of information.

Now examination of equations (3.1) shows that this additional information affects the solution of $x^*(t)$ only through the initial conditions $x^*(t_0), P(t_0)$ and through the control policy $u(t)$, $t \in [t_0, t_f]$. Thus, if the appropriate values for these are substituted, the solution of (3.1) generates an approximation to the appropriate conditional expectation $E_c\{x(t_f)\}$ required in (5.2), so that

$$E_c\{J(x(t_f))\} = J(x^*(t_f)) + O[\varepsilon^3] \tag{5.3}$$

This provides us with a general technique for converting an optimal stochastic control problem into an equivalent deterministic problem.

For example, in the simplest case only the statistical parameters for the initial state, x_0^* and $P(t_0)$ will be available. The optimal control can then only be a function of time and $\hat{u}(t)$ is obtained by solving the optimal control problem with (3.1) as it stands, using (5.3) for the objective function.

If the initial state is known exactly, the corresponding optimal "feed-forward" policy is again a function of time only, obtained similarly, with $x^*(t_0)$ set to the known initial state and $P(t_0) = 0$. Clearly we then have $P(t) = 0$, $t \in [t_0, t_f]$ so that the solution of (3.1) is considerably simplified.

It is worth noting that if in any problem the uncertainty in the initial state is such that $\|E\{\delta x_0 \delta x_0{}^T\}\| = O[\varepsilon^3]$, we also have $P(t) = O[\varepsilon^3]$, $t \in [t_0, t_f]$ and again the terms and equations in $P(t)$ can be dropped from (3.1).

If information is made available during the implementation of the control the optimal solution is a feed-back control, and there is no escape from finding the general form of the functional dependence of the control on the information.

In the deterministic case, and also if $x(t)$ is a Markov process, it is known that the current state $x(t)$ provides sufficient information about the past, so if the state is continuously measured without error it suffices to consider a feed-back control law of the form

$$u(t) = \phi(t, x(t)) \tag{5.4}$$

The process we are considering is non-Markovian, but it seems reasonable to suppose that, since the autocorrelation effects are $O[\varepsilon^3]$, any past information in addition to the current state can only make possible an improvement to $E_c\{J(x(t_f))\}$ of $O[\varepsilon^3]$. We can then assume the same form of control law as for perfectly observed states.

Unfortunately the problem of finding an explicit representation of the form (5.4) for the control law for a general non-linear deterministic system is still unsolved, so for the present we are no nearer solving the corresponding stochastic problem. However, if the form of the control law can be postulated on heuristic grounds it is possible to use the above approach to determine optimal values of any unknown parameters in the law. Thus if the law is given by

$$u(t) = \phi(t, x(t), \alpha(t)) \tag{5.5}$$

where $\alpha(t), t \in [t_0, t_f]$, is a vector of parameters, one simply substitutes (5.5) into (3.1), treating $\alpha(t)$ as the control in the resulting problem.

Such an approach has been used with complete success by the authors on a chemical reactor control problem [6, 7] and a special case of the approach is the determination of the optimal linear feed-back controller for a non-linear system.

With the general problem for perfectly observed states still unsolved, it is evident that no solution is in sight for partially observed systems. Again it may be possible to postulate an explicit form of control law for the dependence on the observed data, or it may be reasonable to assume approximate validity of a "separation principle" for the state estimation and optimal control problems.

It is in checking the effects of noise for such heuristically derived controls that the equations of Section 4 are of use. Such checks were made for the chemical reactor example mentioned above, and the results were compared with those from Monte Carlo simulations. For this particular example the noise levels were such that excellent agreement was obtained between the two sets of results, and it would seem that this situation is likely to be typical of chemical process control applications.

Acknowledgements

The authors would like to thank Dr. J. M. C. Clark and Dr. H. Halkin for very helpful discussions on the content of this paper. Dr. Halkin in particular suggested the introduction of the subsets V and $X(t_f)$ to provide greater generality of the results.

References

1. Clark, J. M. C., "The Representation of Non-linear Stochastic Systems with Applications to Filtering", Ph.D. Thesis, University of London, (1966).
2. Doob, J. L., "Stochastic Processes". Wiley, New York, (1953).
3. Fleming, W. H., Optimal continuous-parameter stochastic control. *SIAM Review* **11**, (4) (1969), 470.
4. Fleming, W. H., Stochastic control for small noise intensities. *SIAM J. Control* **9**, (3) (1971), 473.
5. Halkin, H., Mathematical foundations of system optimization. *In* G. Leitmann (ed.) "Topics in Optimization". Academic Press, London and New York (1967).
6. Joffe, B. L., "A design strategy for the optimal control of a non-linear stochastic system". Ph.D. Thesis, University of London (1971).
7. Joffe, B. L. and Sargent, R. W. H., The design of an on-line control scheme for a tubular catalytic reactor. *Trans. Instn. Chem. Engrs.* **50** (1972), 270.
8. Kushner, H. J., On the stochastic maximum principle: Fixed time of control. *J. Math. Anal. Appl.* **11** (1965), 78.
9. Kushner, H. J., Near optimal control in the presence of small stochastic perturbations. *Trans. ASME (D), J. Basic. Engg.* **87** (1965), 103.
10. McShane, E. J., "Integration". Princeton University Press, (1944).
11. Stratonovich, R. L., On the theory of optimal control. An asymptotic method for solving the diffusive alternative equation. *Automation and Remote Control* (translation) **23** (1962), 1352.
12. Wong, E. and Zakai, M., On the convergence of ordinary integrals to stochastic integrals. *Ann. Math. Statist.* **36**, (1965), 1560.

Discussion

STOREY (*University of Loughborough*). If you postulate that the controls are a general linear function of the current state, your approach would enable you to calculate the optimum linear feed-back controller for the system.

SARGENT. That is perfectly correct, and of course this is an important special case.

FULLER (*University of Cambridge*). What is the order of your example?

SARGENT. The answer isn't straightforward. There are two actual time-varying state variables, the current catalyst activity and the rate constant for the catalyst decay. There are also two controls, the feed temperature and the feed-flow rate. However, in the describing equations we had to integrate along the reactor at each instant of time in order to get the instantaneous yield and the measurements were samples from the temperature profile along the reactor. So, depending on what you mean by state, the order is either infinite or two!

7. The Numerical Solution of a Problem in the Calculus of Variations

P. E. GILL AND W. MURRAY

Division of Numerical Analysis and Computing
National Physical Laboratory
Teddington, Middlesex, England

1. Introduction

Numerical methods for the computation of optimal controls have long lagged behind the theoretical advances of mathematical control theory. Although a vast amount is known about the conditions for the existence of optimal controls of quite complex systems, good algorithms for their computation have occurred relatively rarely. Existing algorithms have tended to be tailored to suit particular small scale problems rather than designed to solve general problems. The ultimate aim of numerical methods for optimal control must be, like other branches of numerical analysis, the provision of subroutines which will approximate the solution of a general large-scale problem to within some specified accuracy.

If the algorithm is to be sufficiently useful as a general purpose tool, it should satisfy a few basic properties.

(a) If an answer to a problem is computed it must be known how close the computed solution $\bar{u}(t)$ (say) is to the optimal solution $\overset{*}{u}(t)$. This would imply knowledge, either *a-priori* or *a-posteriori* of a scalar M such that

$$\|\overset{*}{u}(t) - \bar{u}(t)\| \leqslant M, \text{ for some suitable norm.}$$

In control theory, the problem of finding such an M figures prominently since even if the solution $\overset{*}{u}(t)$ is known analytically, the very nature of digital computation implies that the function $\overset{*}{u}(t)$ must be represented on the computer by a finite data set. In general the representation of $\overset{*}{u}(t)$ itself will give rise to a non-zero value of the scalar M. The representation of functions is a common problem in many other branches of numerical analysis and considerable work has been done on "best" approximation. One efficient method of approximating a wide class of functions by a data set is by piecewise polynomials. We shall see later that this particular choice has a number of important advantages when applied to control.

Many algorithms are made up of several numerical techniques used independently as "black boxes". For example, a control may be computed using a differential equation solver, a quadrature routine and a finite-dimensional optimization method. Often it is assumed that each separate algorithm evaluates the required solution exactly. Estimating the value of M in these circumstances becomes exceedingly difficult.

(b) The algorithm should solve a wide class of problems yet still remain relatively efficient on the more specialized problems.

(c) The algorithm should rely on as little *a-priori* theoretical analysis of the problem as possible. This would not be true for example if a finite-dimensional optimization algorithm required global convexity of the objective function since even if this were true it would usually be impossible to verify. Ideally, the only conditions necessary to compute the solution should be the necessary and sufficient conditions on the optimal solution itself.

In this paper we will consider the numerical solution of one of the simplest problems in the Calculus of Variations. Although we are mainly concerned with optimization of the discretized problem, we will demonstrate that the methods considered here are easily generalized to solve more difficult problems. The particular way in which the properties (a) (b) and (c) hold will be discussed at a later date.

2. The Calculus of Variations Problem

First we consider the problem:

P1. Find a strong local minimum $\overset{*}{x}(t)$ of the functional

$$J(x(t)) = \int_0^1 f(t, x(t), \dot{x}(t))\, dt$$

over the set of piecewise differentiable curves with boundary conditions $x(0) = x(1) = 0$.

We have specifically taken $x(t)$ as a scalar function in order to clarify the description, but once the basic algorithm has been understood it is easy to see how it can be generalized to the case where $x(t)$ is a vector of functions.

One method of discretizing P1 is to approximate $\overset{*}{x}(t)$ by functions $x(t)$ which are finite linear combinations of basis functions. This concept can be formalized [14] by defining $W_1{}^0$ to be the Banach space of continuous functions $x(t)$ defined on $[0, 1]$ and vanishing at 0 and 1, with derivatives $\dot{x}(t)$

almost everywhere in $L_2[0, 1]$ and norm

$$\|x(t)\|_1 = \left(\int_0^1 [(\dot{x}(t))^2 + (x(t))^2]\, dt\right)^{\frac{1}{2}}.$$

We approximate $\overset{*}{x}(t)$ by functions which are members of a finite-dimensional subspace W_n of $W_1{}^0$, spanned by n basis functions $\{w_i(t)\}_{i=1}^n$. The discretized problem then becomes

$$\underset{x(t)\in W_n}{\text{minimize}} \int_0^1 f(t, x(t), \dot{x}(t))\, dt.$$

Since the $\{w_j(t)\}_{j=1}^n$ are basis functions for W_n each $x(t)$ can be expressed by a vector of coefficients c_j, $j = 1, 2, \ldots, n$. This finally gives an *unconstrained* minimization problem:

DP1. $\quad \underset{x\in E^n}{\text{minimize}} \quad F(x)$

where we have defined

$$F(c) = \int_0^1 f\left(t, \sum_{j=1}^n c_j\, w_j(t), \sum_{j=1}^n c_j\, \dot{w}_j(t)\right) dt.$$

We have substituted x for c since this is the vector normally used in the literature when minimizing in Euclidean space.

2.1 *Gradient methods for unconstrained optimization*

Let $x^{(0)}$ be some approximation to a local minimum $\overset{*}{x}$ of $F(x)$. If the function $F(x)$ is twice continuously differentiable over a sufficiently large region containing $x^{(0)}$ we can use a gradient method to locate $\overset{*}{x}$. These methods, which are the most reliable and efficient available, involve an iteration of the following form:

kth *iteration*

Set

$$x^{(k+1)} = x^{(k)} + \alpha^{(k)} p^{(k)}$$

where $\alpha^{(k)}$ is a scalar chosen to satisfy certain requirements such as

$$F(x^{(k+1)}) < F(x^{(k)}) \tag{1}$$

and the *direction of search* $p^{(k)}$ is an n-vector which is a function of the first and second derivatives of $F(x)$ at $x^{(k)}$. The choice of $p^{(k)}$ characterizes a particular method.

The derivative information available is the gradient vector $g(x)$ defined by

$$g_j(x) = \frac{\partial F(x)}{\partial x_j}, \quad j = 1, 2, \ldots, n$$

and the $n \times n$ symmetric Hessian matrix of second derivatives $G(x)$, with elements

$$g_{ij}(x) = \frac{\partial^2 F}{\partial x_i \, \partial x_j}, \qquad i, j = 1, 2, \ldots, n.$$

Particular choices of $p^{(k)}$ that have been used in the past are given as follows [12]. For convenience we will define $G^{(k)} = G(x^{(k)})$ and $g^{(k)} = g(x^{(k)})$.

(i) *Steepest descent*

$$p^{(k)} = -g^{(k)}$$

(ii) *Newton's method*

The vector $p^{(k)}$ satisfies the set of equations

$$G^{(k)} p^{(k)} = -g^{(k)} \tag{2}$$

(iii) *Quasi-Newton method*

The vector $p^{(k)}$ satisfies

$$B^{(k)} p^{(k)} = -g^{(k)}$$

where $B^{(k)}$ is now some approximation to the matrix $G^{(k)}$. We are particularly concerned with "steepest descent" algorithms in a wider sense in which each individual $p^{(k)}$ is the solution to the problem:

$$\text{minimize } \{p^T g\}$$

subject to

$$\|p\| = \beta, \qquad \text{a constant.}$$

These algorithms have favourable theoretical properties and are such that an $\alpha^{(k)} > 0$ can be found satisfying (1). Algorithms (ii) and (iii) are steepest descent under norms

$$\|y\| = (y^T G^{(k)} y)^{\frac{1}{2}}, \qquad \|y\| = (y^T B^{(k)} y)^{\frac{1}{2}}$$

respectively, provided $G^{(k)}$ and $B^{(k)}$ are positive-definite matrices.

The gradient vector and Hessian matrix of the particular problem DP1 are as follows:

$$g_j = \int_0^1 \{f_x(t, x(t), \dot{x}(t))\, w_j(t) + f_{\dot{x}}(t, x(t), \dot{x}(t))\, \dot{w}_j(t)\}\, dt$$

$$j = 1, 2, \ldots, n$$

and

$$g_{ij}^{(k)} = \int_0^1 \{f_{xx}(t, x(t), \dot{x}(t))\, w_i(t)\, w_j(t) + f_{x\dot{x}}(t, x(t), \dot{x}(t))\, (\dot{w}_i(t)\, w_j(t) + w_i(t)\, \dot{w}_j(t)) + f_{\dot{x}\dot{x}}(t, x(t), \dot{x}(t))\, \dot{w}_i(t)\, \dot{w}_j(t)\}\, dt$$

where f_{xx}, $f_{x\dot{x}}$ and $f_{\dot{x}\dot{x}}$ denote the usual partial derivatives with respect to x and $\dot{x}$ evaluated along $x(t)$ and $\dot{x}(t)$ and

$$x(t) = \sum_{j=1}^{n} x_j\, w_j(t), \qquad c_j = x_j, \qquad j = 1, 2, \ldots, n.$$

Clearly these quantities are critically dependent upon the choice of basis functions. An ideal choice would be those functions such that $G^{(k)} = I$, the identity matrix, for all k. Unfortunately this is not feasible computationally. However, if the elements of the set $\{w_j(t)\}_{j=1}^{n}$ are defined as spanning some space of piecewise polynomials, the matrix $G^{(k)}$ becomes *sparse* for all k.

There are many different types of piecewise polynomials. Since we are mainly concerned with the numerical solution of the resulting sparse optimization problem we will choose the piecewise polynomial subspace $H_0^{(\mu)}(\pi)$, [2], where π is the partition $0 = t_0 < t_1 < t_2 < \ldots < t_N < t_{N+1} = 1$, and

$$H_0^{(\mu)}(\pi) = \{x(t) \mid x(t) \in C^{\mu-1}[0, 1],\ x(0) = x(1) = 0,\ x(t) \text{ a polynomial of degree at most } 2\mu - 1 \text{ on } [t_j, t_{j+1}] \text{ for } 0 \leqslant j \leqslant N\}.$$

This space $H_0^{(\mu)}(\pi)$ is spanned by the $\mu(N+2) - 2$ functions $\phi_{i,r}(t)$ for $1 \leqslant i \leqslant N$, $0 \leqslant r \leqslant \mu - 1$ and, for $i = 0, N+1$, $1 \leqslant r \leqslant \mu - 1$ where

$$D^s \phi_{i,r}(t_j) = \delta_{ij}\,\delta_{rs} \quad \text{for} \quad 0 \leqslant s \leqslant \mu - 1.$$

For any $x(t) \in H_0^{(\mu)}(\pi)$ there exist constants $d_i^{(r)}$ for $1 \leqslant i \leqslant N$, $0 \leqslant r \leqslant \mu - 1$ and $i = 0, N+1$, $1 \leqslant r \leqslant \mu - 1$ such that

$$x(t) = \sum_{r=1}^{\mu-1} d_0^{(r)} \phi_{0,r}(t) + \sum_{r=1}^{\mu-1} d_{N+1}^{(r)} \phi_{N+1,r}(t) + \sum_{i=1}^{N} \sum_{r=0}^{\mu-1} d_i^{(r)} \phi_{i,r}(t).$$

If we order these basis functions in the form

$$w_1(t) = \phi_{0,1}(t),\ w_2(t) = \phi_{0,2}(t),\ \ldots,\ w_{\mu-1}(t) = \phi_{0,\mu-1}(t)$$

$$w_\mu(t) = \phi_{1,0}(t),\ \ldots \qquad \ldots,\ w_{\mu(N+2)-2}(t) = \phi_{N+1,\mu-1}(t)$$

and the coefficients

$$c_1 = d_0^{(1)}, c_2 = d_0^{(2)}, \ldots, c_{\mu(N+2)-2} = d_{N+1}^{(\mu-1)},$$

we have

$$x_n(t) = \sum_{j=1}^{n} c_j w_j(t)$$

where

$$n = \mu(N+2) - 2.$$

Since the functions $\phi_{i,r}(t)$ are non-zero only in $[t_{i-1}, t_{i+1}]$, the Hessian matrix has the form

$$G = \begin{bmatrix} G_{11} & G_{21}^T & & & & \\ G_{21} & G_{22} & G_{32}^T & & & \\ \cdot & G_{32} & \cdot & & & \\ & \cdot & \cdot & \cdot & & \\ & & \cdot & \cdot & \cdot & \\ & & & \cdot & \cdot & \cdot \\ & & & & \cdot & G_{N+2,N+1}^T \\ & & & & G_{N+2,N+1} & G_{N+2,N+2} \end{bmatrix}$$

where the order of each G_{ij} is the number of parameters associated with the jth mesh point in the interval $0 \leqslant t \leqslant 1$, i.e. $G_{22}, G_{32}, \ldots G_{N+1,N+1}$ are of order μ and G_{11}, $G_{N+2,N+2}$ are of order $\mu - 1$. This essentially gives a matrix with bandwidth $4\mu - 1$.

We now describe a gradient algorithm that can take full advantage of this structure in the Hessian matrix.

2.2 *A modified Newton method*

Assume for the moment that $G^{(k)}$ is positive definite for all $x^{(k)}$ (this condition will be relaxed later). We can factorize the matrix $G^{(k)}$ by Cholesky's method into the product of a unit lower-triangular matrix $L^{(k)}$ and diagonal matrix $D^{(k)}$ such that

$$G^{(k)} = L^{(k)} D^{(k)} L^{(k)T}.$$

The matrix $L^{(k)}$ has the structure

$$L^{(k)} = \begin{bmatrix} L_{11}^{(k)} & & & & \\ L_{21}^{(k)} & L_{22}^{(k)} & & & \\ & L_{32}^{(k)} & L_{33}^{(k)} & & \\ & & \ddots & \ddots & \\ & & & \ddots & \ddots \\ & & & L_{N+2,N+1}^{(k)} & L_{N+2,N+2}^{(k)} \end{bmatrix},$$

with $L_{jj}^{(k)}$, $2 \leqslant j \leqslant N+1$ unit lower-triangular matrices of order μ, $L_{11}^{(k)}$, $L_{N+2,N+2}^{(k)}$ unit lower-triangular matrices of order $\mu - 1$ and $L_{ij}^{(k)}$ the same order as $G_{ij}^{(k)}$, $2 \leqslant i \leqslant N+2$, $1 \leqslant j \leqslant N+1$. If we assume that $L^{(k)}$ is strictly a band matrix with bandwidth $4\mu - 1$, the jth step of Cholesky's method becomes [10]

$$d_j^{(k)} = g_{jj}^{(k)} - \sum_{r=j-m}^{j-1} d_r^{(k)} l_{jr}^{(k)2}$$

$$l_{ij}^{(k)} = \left(g_{ij}^{(k)} - \sum_{r=i-m}^{j-1} d_r^{(k)} l_{ir}^{(k)} l_{jr}^{(k)}\right) \Big/ d_j^{(k)}, \qquad i = j+1, \ldots, j+m$$

where $m = 2\mu - 1$. The vector $p^{(k)}$ can be determined by successively solving

$$L^{(k)} y = -g^{(k)}$$
$$L^{(k)T} p^{(k)} = D^{(k)-1} y.$$

Explicitly we have

$$y_i = -g_i^{(k)} - \sum_{r=i-m}^{i-1} l_{ir}^{(k)} y_r, \qquad i = 1, 2, \ldots, n$$

and

$$p_i^{(k)} = y_i / d_i^{(k)} - \sum_{r=i+1}^{i+m} l_{ri}^{(k)} p_r^{(k)}, \qquad i = n, n-1, \ldots, 1.$$

The factorization involves approximately $\mu n(2\mu + 1)$ multiplications and the back substitution requires $4\mu n$ multiplications. In practice only low values of μ are used; for example, with piecewise cubic polynomials $\mu = 2$ and the amount of work for one iteration is essentially of the order of n multiplications. This is especially important when increased accuracy is sought by increasing the value of n. With this algorithm the computational effort increases *linearly* with n.

When $G^{(k)}$ is singular, or more commonly indefinite, the algorithm defined by (2) is no longer a "steepest descent algorithm", since $y^T G^{(k)} y$ no longer satisfies the properties of a norm. In addition the method of Cholesky is no longer numerically stable, even if the factorization exists.

In order to allow for these difficulties we now propose a modified Newton method in which $G^{(k)}$ is replaced by a related positive definite matrix $\bar{G}^{(k)}$ when $G^{(k)}$ is not sufficiently positive definite. With this method the amount of work necessary to compute $p^{(k)}$ is essentially the same and the vector $p^{(k)}$ is determined in a numerically stable way.

The modified Newton method is based upon the modified Cholesky factorization of the matrix $G^{(k)}$. This factorization gives the lower triangular factors of a matrix

$$\bar{G}^{(k)} = G^{(k)} + E^{(k)}$$

where $E^{(k)}$ is a diagonal matrix determined during the factorization process such that $\bar{G}^{(k)}$ is positive definite and equal to $G^{(k)}$ when $G^{(k)}$ is sufficiently positive definite. Since $E^{(k)}$ is a diagonal matrix, $\bar{G}^{(k)}$ has the same zero/non-zero structure as $G^{(k)}$. The reader is referred to [7] for further details. In the algorithm we now propose, the scalar $\alpha^{(k)}$ is determined using a safeguarded cubic linear search algorithm (SCLS), [6].

2.3 *A basic iteration for the solution of the discretized Calculus of Variations problem*

The kth iteration for performing unconstrained optimization is as follows.

Step 1. Calculate $G^{(k)}$, the Hessian matrix of $F(x)$ at $x^{(k)}$.

Step 2. Form the modified Cholesky factorization of $G^{(k)}$ which gives

$$L^{(k)} D^{(k)} L^{(k)T} = G^{(k)} + E^{(k)}.$$

Step 3. If $\|g^{(k)}\|_2 \leqslant \varepsilon$, where ε is a small positive scalar and $\|E^{(k)}\| = 0$ then $x^{(k)}$ is an adequate approximation to $\overset{*}{x}$ and the algorithm is terminated.

If $\|g^{(k)}\|_2 > \varepsilon$ then $p^{(k)}$ is determined by solving the equations

$$L^{(k)} D^{(k)} L^{(k)T} p^{(k)} = -g^{(k)}.$$

Otherwise (i.e. $\|g^{(k)}\|_2 \leqslant \varepsilon$ but $\|E^{(k)}\| \neq 0$) $p^{(k)}$ is determined by the alternative search procedure described in [6]; that is, solve

$$L^{(k)T} y = e_j$$

and set

$$p^{(k)} = \begin{cases} -\operatorname{sign}(y^T g^{(k)})y, & \|g^{(k)}\|_2 \neq 0 \\ y \quad , & \|g^{(k)}\|_2 = 0. \end{cases}$$

The index j is such that

$$d_j^{(k)} - E_j^{(k)} \leqslant d_i^{(k)} - E_i^{(k)}, \qquad i = 1, 2, \ldots, n.$$

Step 4. Determine the step $\alpha^{(k)}$ generated by the SCLS algorithm [5, 6]. This algorithm first determines a scalar $\bar{\alpha}$ such that

$$|g(x^{(k)} + \bar{\alpha}\, p^{(k)})^T p^{(k)}| < -\eta g^{(k)T} p^{(k)}$$

and

$$F(x^{(k)} + \bar{\alpha}\, p^{(k)}) < F^{(k)}$$

where η $(0 < \eta < 1)$ is a preassigned scalar. If $\bar{\alpha}$ is such that

$$F^{(k)} - F(x^{(k)} + \bar{\alpha}\, p^{(k)}) > -\sigma\bar{\alpha} g^{(k)T} p^{(k)}$$

where σ $(0 < \sigma \ll \eta)$ is another preassigned scalar, then $\alpha^{(k)} = \bar{\alpha}$. Otherwise if $w^{(k)}$ is the first number of the sequence $\{(\frac{1}{2})^j\}_{j=1}^{\infty}$ such that the step $\bar{\alpha} w^{(k)}$ satisfies

$$F^{(k)} - F(x^{(k)} + \bar{\alpha} w^{(k)} p^{(k)}) > -\sigma\bar{\alpha} w^{(k)} g^{(k)T} p^{(k)}$$

then

$$\alpha^{(k)} = \bar{\alpha} w^{(k)} \quad \text{and} \quad x^{(k+1)} = x^{(k)} + \alpha^{(k)} p^{(k)}.$$

Step 5. If it is not found in the process of finding $\alpha^{(k)}$, compute $g^{(k+1)}$. Set $k = k + 1$ and return to step 1.

2.4 *Approximating second derivatives by finite differences*

The algorithm described in the last section requires an auxiliary subroutine to compute the first and second variation of the objective function. It is frequently the case, especially when $x(t)$ is a vector of functions, that determining and coding the second variation is prohibitively difficult. A means of alleviating this problem is to approximate the Hessian matrix by finite differences. In [7] the matrix $\bar{G}^{(k)}$ computed by symmetrizing the matrix $V^{(k)}$ which has columns

$$v_j^{(k)} = \big(g(x^{(k)} + h_j e_j) - g^{(k)}\big)/h_j, \qquad j = 1, 2, \ldots, n \tag{3}$$

can be used instead of $G^{(k)}$ in step 1 of the modified Newton algorithm. Although this algorithm can be shown to be convergent for a large class of functions it has the disadvantage of requiring $n + 1$ gradient subroutine calls per iteration. This implies that the algorithm is competitive with quasi-Newton methods, which utilize the same gradient information, only for values of n up to about $n = 20$. However, when solving problems in which the Hessian matrix is always banded, this condition can be removed. There are two reasons for this:

(a) Since the finite difference approximation (3) preserves the band structure of the matrix G the algorithm requires only of the order of n multiplications each iteration. The class of quasi-Newton algorithms require $O(n^2)$ multiplications each iteration.

(b) The number of gradient calls per iteration can be reduced to $\bar{m}+1$, (see, for example, [3]) where $\bar{m}$ is the bandwidth of the Hessian matrix. This saving is achieved by differencing the gradient along the $\bar{m}$ vectors $e^{(i)}, i = 1, 2, \ldots, \bar{m}$ where

$$e^{(i)} = \sum_{j=0}^{r} e_{j\bar{m}+i}, \qquad r = \left[\frac{n-i}{\bar{m}}\right],$$

e_r is the rth column of the identity matrix and $[x]$ denotes the integer part of x. A single gradient evaluation with $x^{(k)}$ incremented by $e^{(i)}$ gives

$$\left[\frac{n-i}{\bar{m}}\right]$$

columns of the approximate Hessian at once.

2.5 *Illustrative numerical computations*

All computational examples were carried out in single precision on an ICL KDF9 with a 39 bit mantissa. The facility on the KDF9 for the double length accumulation of inner-products was deliberately not used. In every example uniformly spaced knots were used with piecewise cubic polynomial basis functions. Finite difference approximations to the Hessian matrix were used with h_j (see section 2.4) fixed at $(2^{-39})^{\frac{1}{2}}$. It has been demonstrated in [6] that this scheme gives almost identical results to a pure Newton method in most practical applications. A set of coefficients was accepted when the corresponding gradient satisfied $\|g\|_2 < 10^{-6}$. The values η and σ required were set at 0·5 and 10^{-4} respectively. On all examples two separate starting vectors were used. These were the vector $s_1 = (0, 0, \ldots, 0)^T$, and the vector s_2 whose elements were generated randomly over the interval $[-0{\cdot}1, +0{\cdot}1]$.

Problem 1. Compute the function $\overset{*}{x}(t)$, $0 \leqslant t \leqslant 1$ with $\overset{*}{x}(0) = \overset{*}{x}(1) = 0$ which minimizes the functional

$$\int_0^1 \{\tfrac{1}{2}(\dot{x}(t))^2 + e^{x(t)} - 1\}\, dt.$$

The unique solution of this problem is

$$\overset{*}{x}(t) = -\ln 2 + 2 \ln \{c \sec(t - \tfrac{1}{2})/2\}, \qquad c = 1.3360557.$$

Problem 2. Minimize the functional

$$\int_0^1 \{(\dot{x}(t))^2 - (x(t))^2 - 2tx(t)\}\, dt$$

with the end conditions $x(0) = x(1) = 0$. The exact solution to this problem ([1], p. 160) is $\overset{*}{x}(t) = \sin t/\sin 1 - t$.

Problem 3. Minimize the functional

$$\int_0^1 \{(\dot{x}(t))^2 + (x(t))^2 + 2x(t)e^{2t}\}\, dt$$

with the end conditions $x(0) = 1/3$, $x(1) = e^2/3$. The exact solution [1], p. 237) is $\overset{*}{x}(t) = e^{2t}/3$. Note that in this case the boundary conditions are non-homogeneous; these can be easily handled by extending the set of basis functions during the computation of a current approximation $x(t)$.

In any practical implementation of the Ritz method it is important how the integrals involved in the definition of the objective functional, first variation and second variation are approximated. In the examples considered a four-point Gaussian quadrature scheme was used over each interval $[t_i, t_{i+1}]$, $i = 0, 1, \ldots, N$. With this rule it can be shown that the quadrature errors are consistent (in the sense of [8]) with the errors of the Ritz approximation scheme.

For Problems 1, 2 and 3 it can be shown that an error bound of the form

$$\sup_{0 \leqslant t \leqslant 1} |\overset{*}{x}(t) - x_n(t)| = \|\overset{*}{x}(t) - x_n(t)\|_\infty = O(h^4)$$

holds for a given value of n. In Table 1 we give the values of

$$\|\overset{*}{x}(t) - x_n(t)\|_\infty^\pi = \max_{0 \leqslant i \leqslant N+1} |\overset{*}{x}(t_i) - x_n(t_i)|$$

for all the examples computed.

TABLE 1. Observed errors in the solution of problems 1, 2 and 3.

Problem	h	Dimension of approximating subspace	$\|\|\overset{*}{x}(t) - x(t)_n\|\|_\infty^\pi$
1	1/5	10	$1{\cdot}81 \times 10^{-6}$
	1/10	20	$3{\cdot}63 \times 10^{-7}$
	1/20	40	$8{\cdot}36 \times 10^{-9}$
2	1/5	10	$1{\cdot}49 \times 10^{-6}$
	1/10	20	$1{\cdot}09 \times 10^{-7}$
	1/20	40	$2{\cdot}60 \times 10^{-8}$
3	1/5	10	$4{\cdot}63 \times 10^{-5}$
	1/10	20	$3{\cdot}58 \times 10^{-6}$
	1/20	40	$2{\cdot}71 \times 10^{-7}$

In Tables 2—4 we give the results for the values of h and n given in Table 1. The symbol n_f denotes the total number of evaluations of the objective functional, n_g the total number of gradient subroutine calls and n_{it} the total number of iterations.

TABLE 2. Problem 1.

Starting point	n	n_f	n_g	n_{it}
s_1	10	3	24	2
	20	3	24	2
	40	3	24	2
s_2	10	3	24	2
	20	3	24	2
	40	3	24	2

TABLE 3. Problem 2.

Starting point	n	n_f	n_g	n_{it}
s_1	10	2	16	1
	20	2	16	1
	40	2	16	1
s_2	10	2	16	1
	20	3	24	2
	40	3	24	2

TABLE 4. Problem 3.

Starting point	n	n_f	n_g	n_{it}
	10	3	24	2
s_1	20	3	24	2
	40	3	24	2
	10	3	24	2
s_2	20	3	24	2
	40	3	24	2

As examples of problems with more than a mild non-linearity we have the following.

Problem 4. Minimize

$$\int_0^1 \{e^{-2x(t)^2}(\dot{x}(t)^2-1)\}\,dt$$

with the boundary conditions $x(0)=0$, $x(1)=1$.

Problem 5. Minimize

$$\int_0^1 \{x(t)^2+\dot{x}(t)\arctan\dot{x}(t)-ln(1+\dot{x}(t)^2)^{\frac{1}{2}}\}\,dt$$

with the boundary conditions $x(0)=1$, $x(1)=2$.

These problems are taken from [1] and have an Euler equation of the form

$$\ddot{x}(t)=2x(t)\,(1+\dot{x}(t)^2).$$

The test results for these problems are given in Tables 5 and 6.

TABLE 5. Problem 4.

Starting point	n	n_f	n_g	n_{it}
	10	7	49	5
s_1	20	8	50	5
	40	8	50	5
	10	6	41	4
s_2	20	6	48	5
	40	18	67	6

TABLE 6. Problem 5.

Starting point	n	n_f	n_g	n_{it}
	10	8	50	5
s_1	20	9	58	6
	40	12	61	6
	10	8	50	5
s_2	20	9	58	6
	40	16	79	8

3. Extensions to Linearly Constrained Problems

The techniques described in the previous sections can be generalized to solve the more difficult problem where there are linear dynamics:

LCP1. Find a strong local minimum of

$$\int_0^1 f(x(t), u(t), t)\, dt$$

subject to the constraints

$$\dot{x}(t) = A(t)\, x(t) + B(t)\, u(t)$$

$$x(0) = x_0. \tag{4}$$

We are assuming that $x(t)$ and $u(t)$ are scalar functions but we stress again that the techniques are easily applicable to larger dimensional problems.

The discretization of LCP1 is performed by requiring that the optimization is performed over some finite-dimensional subspaces; i.e. we require

$$\left.\begin{aligned} x(t) \in X_n \Rightarrow \exists\, c_1, c_2, \ldots, c_n \quad \text{such that} \quad x(t) &= \sum_{r=1}^{n} c_r\, w_r(t) \\ u(t) \in U_{\bar{n}} \Rightarrow \exists\, \bar{c}_1, \bar{c}_2, \ldots, \bar{c}_{\bar{n}} \quad \text{such that} \quad u(t) &= \sum_{s=1}^{\bar{n}} \bar{c}_s\, \bar{w}_s(t) \end{aligned}\right\}. \tag{5}$$

Note that in general X_n and $U_{\bar{n}}$ are not the same finite-dimensional space. In order to completely characterize feasible points for the discrete problem we must restrict the constraint set further. We will construct X_n such that for all $x(t) \in X_n$, $x(t)$ satisfies the boundary condition $x(0) = x_0$. Let $S = \{\tau_j; 0 \leqslant \tau_1 < \tau_2 < \tau_3 < \ldots < \tau_n \leqslant 1\}$ be a set of points in $[0, 1]$ on which a pair of arbitrary functions $x_n \in X_n$, $u_{\bar{n}} \in U_{\bar{n}}$ satisfy the constraint (4) exactly. We have deliberately chosen the number of points in S to be the same as the number of coefficients representing the state variable. Although not essential, this implies that we can perform a transformation which gives an optimization problem of the same dimension as the number of coefficients of the control variable. This gives n linear relations in the coefficients $\{c_r\}_{r=1}^{n}$ and $\{\bar{c}_r\}_{r=1}^{\bar{n}}$. Before we define S more specifically it will be necessary to choose X_n and $U_{\bar{n}}$. Assuming $x(t) \in C^2[0, 1]$ we define X_n as the space $H_1^{(2)}(\pi)$ of piecewise cubic polynomials satisfying the boundary condition. Similarly we will use $U_{\bar{n}} = H^{(\mu)}(\pi)$, the value of μ being fixed by the degree of continuity of the class of admissible control functions. Note that the finite-dimensional subspaces $H_1^{(2)}(\pi)$ and $H^{(\mu)}(\pi)$ are different from $H_0^{(\mu)}(\pi)$ defined earlier since elements of $H_1^{(2)}(\pi)$ satisfy only one boundary condition (at $t = 0$) and elements of $H^{(\mu)}(\pi)$ can take on arbitrary values at $t = 0$ and $t = 1$. The basis functions for these spaces are easily obtained from those defined earlier (alternatively the reader is referred to [2]). It is important that $U_{\bar{n}}$ and X_n are defined upon the same set of knots π. With this scheme the number of coefficients to be determined is $n + \bar{n}$ where

$$n = 2(N + 2) - 1 \quad \text{and} \quad \bar{n} = \mu(N + 2)$$

and the set S is chosen to contain the points $\{t_{ij}, i = 0, \dots, N+1, j = 0, 1, 2\}$ where

$$t_i = t_{i0} < t_{i1} < t_{i2} = t_{i+1}, \quad \text{for} \quad 0 \leqslant i \leqslant N + 1.$$

For convenience the t_{ij} are usually chosen to be equally spaced. The linear relationships between the coefficients then become

$$\dot{x}(t_{ij}) = A(t_{ij})\, x(t_{ij}) + B(t_{ij})\, u(t_{ij}), \qquad i = 0, 1, \dots, N + 1;\ j = 0, 1, 2;$$

$$x(t) \in X_n, \qquad u(t) \in U_{\bar{n}}.$$

Substituting for $\dot{x}(t_{ij})$, $x(t_{ij})$ and $u(t_{ij})$ from Eqn (5) gives

$$\sum_{r=1}^{n} c_r\, \dot{w}_r(t_{ij}) = \sum_{r=1}^{n} A(t_{ij})\, c_r\, w_r(t_{ij}) + \sum_{s=1}^{\bar{n}} B(t_{ij})\, \bar{c}_s\, \bar{w}_s(t_{ij}).$$

These equations are arranged so that the first is the boundary condition at $t_0 = 0$, the next are the collocation conditions for the points of S in $[t_0, t_1]$ followed by the collocation conditions for points of S in $[t_1, t_2]$ etc. If the variables $\{c_r\}_{r=1}^{n}$, $\{\bar{c}_r\}_{s=1}^{\bar{n}}$ are reordered and combined into the $n + \bar{n}$ vector x such that

$$x = (c_1, \bar{c}_1, \bar{c}_2, \ldots, \bar{c}_\mu, c_2, c_3, \bar{c}_{\mu+1}, \ldots,$$
$$\bar{c}_{\mu j}, c_{2j}, c_{2j+1}, \bar{c}_{\mu j+1}, \quad \ldots, \bar{c}_{\mu(N+2)})^T \tag{6}$$

the discretized problem becomes

DLCP1. Find a strong local minimum of the functional

$$F(x)$$

subject to the constraints

$$Ax = b$$

where

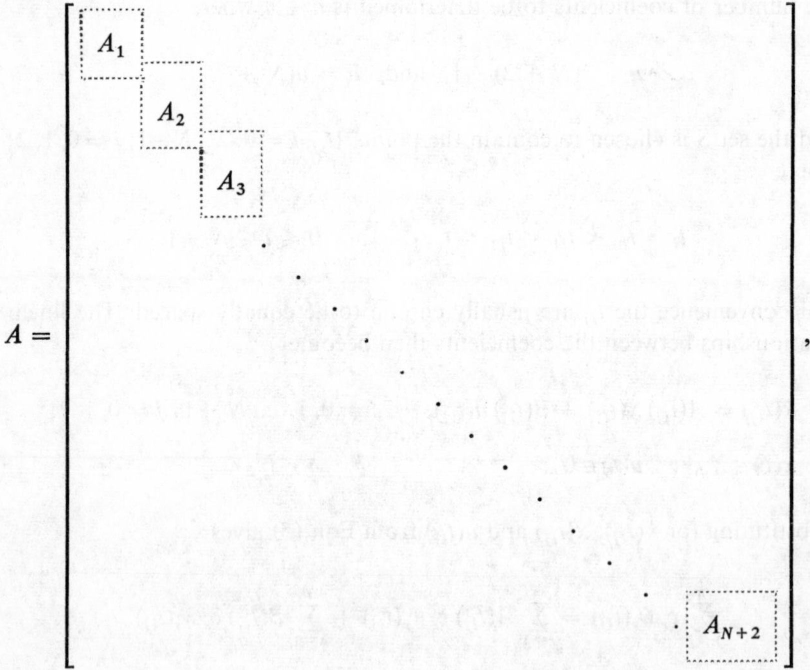

where A_1 is a $2 \times (\mu + 1)$ matrix, A_{N+2} a $2 \times (\mu + 2)$ matrix and the remaining N matrices $A_2, \ldots A_{N+1}$ are of order $3 \times (\mu + 2)$. Similarly the Hessian matrix with respect to the combined variables x has the form

$$G(x) = \begin{bmatrix} G_{11} & {G_{21}}^T & & & & \\ G_{21} & G_{22} & {G_{32}}^T & & & \\ & G_{32} & \cdot & \cdot & & \\ & & \cdot & \cdot & \cdot & \\ & & & \cdot & \cdot & \cdot \\ & & & & \cdot & \cdot & \cdot \\ & & & & & \cdot & \cdot & \cdot \\ & & & & & & \cdot & \cdot & G^T_{N+2,N+1} \\ & & & & & & & G_{N+2,N+1} & G_{N+2,N+2} \end{bmatrix}$$

where G_{11} and G_{21} are of orders $(\mu+1)\times(\mu+1)$ and $(\mu+2)\times(\mu+1)$ respectively, and the remaining matrices are of order $(\mu+2)\times(\mu+2)$. We define the linear manifold

$$M(A) = \{x \mid Ax = b\}$$

and make the following assumptions

a1. The set of rows $\{{a_i}^T\}_{i=1}^n$ of the matrix A are linearly independent.

a2. All minima of DLCP1 are strong local minima.

By a strong local minimum $\overset{*}{x} \in M(A)$ we mean

$$\exists\, \delta > 0 \quad \text{such that} \quad F(\overset{*}{x}) < F(x), \qquad \forall\, x \in M(\overset{*}{x}, \delta), \qquad x \neq \overset{*}{x}$$

where

$$M(x, \delta) = \{y \mid M \cap \{y \mid \|y - x\|_2 \leqslant \delta\}\}.$$

Problem DLCP1 can be written as

$$\underset{x \in M(A)}{\text{minimize}}\ F(x). \tag{7}$$

If we define the $(n + \bar{n}) \times \bar{n}$ matrix Z which spans the nullspace of A, then

$$AZ = 0$$

If $x^{(0)}$ is a point satisfying the constraints the problem defined at (7) can be redefined as

$$\underset{v \in E^{\bar{n}}}{\text{minimize}}\ F(x^{(0)} + Zv)$$

which is now an unconstrained problem in the variables $v_1, v_2, \ldots v_{\bar{n}}$. The matrix of first and second partial derivatives with respect to these variables are given by the vector $Z^T g$ and the matrix $Z^T GZ$ respectively, where g and G are the gradient and Hessian matrix with respect to the x variables. These quantities form the basis of an algorithm given in [7] for the minimization of a non-linear function subject to linear inequality constraints. This algorithm is a modified Newton algorithm in the space $M(A)$ and, in the form previously given, requires the explicit calculation of the matrix $Z^T GZ$. Although A and G are sparse for all x, the matrices Z and $Z^T GZ$ are generally dense and any algorithm which fails to maintain sparsity becomes untenable for large values of n and $\bar{n}$. In what follows it will be shown how the advantages of the method can be maintained without explicitly forming the matrix Z and $Z^T GZ$, all the operations requiring these matrices being replaced or performed in an alternative manner. For instance, the projected gradient $Z^T g$ can be computed very efficiently as follows. Let the orthogonal factorization of A be of the form

$$A = [L_A \,\vdots\, 0]Q \tag{8}$$

where L_A is an $n \times n$ lower-triangular matrix, and Q an $(n + \bar{n}) \times (n + \bar{n})$ orthogonal matrix such that $QQ^T = I$, the identity matrix. If Q is partitioned in the form

$$Q = \begin{bmatrix} Q_1 \\ \cdots \\ Q_2 \end{bmatrix}$$

where Q_1 is an $n \times (n + \bar{n})$ matrix and Q_2 is an $\bar{n} \times (n + \bar{n})$ matrix, then

$$AQ_2{}^T = 0 \quad \text{and} \quad Q_2 Q_2{}^T = I.$$

Clearly $Q_2{}^T$ satisfies the conditions on Z and we can write $Q_2{}^T = Z$. The factorization (8) always exists and can be found using either Householder or Givens reduction [15]. This is effected by reducing A to lower triangular form using a sequence of orthogonal matrices

$$P_1 P_2 \ldots P_\gamma$$

where γ, the number of orthogonal matrices, is dependent upon the sparsity of A and the type of transformation used. If Givens matrices are used, each

P_j is of the form

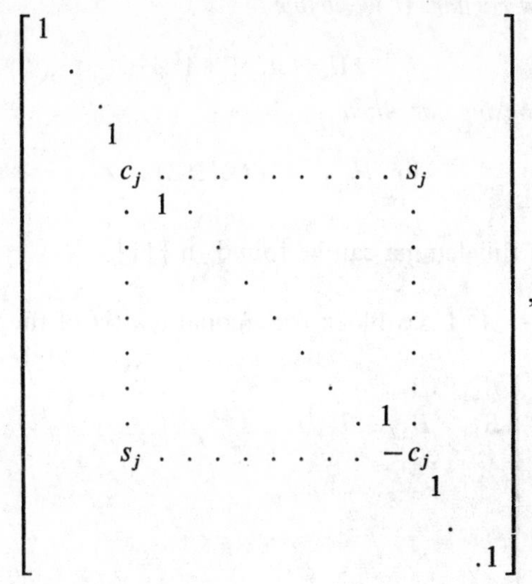

where $s_j^2 + c_j^2 = 1$.
The alternative matrices, which are Householder matrices, are of the form

$$P_j = I - \tau\, w^{(j)} w^{(j)T}$$

where $w^{(j)}$ is an $(n + n)-$vector with a very small number of non-zero elements. If, instead of storing Q explicitly, the c_j, s_j or non-zero $w_i^{(j)}$ are retained, any product $Z^T v$ or Zu can be computed very efficiently. This is because

$$Z^T v = Q_2 v$$

which can be obtained by computing the last $\bar{n}$ elements of $Q\,v$ and similarly

$$Zu = Q_2{}^T u = Q^T \begin{bmatrix} 0 \\ \cdots \\ u \end{bmatrix}.$$

We still have the problem of how to avoid explicit formation of the matrix $Z^T GZ$. The following lemmas will prove useful.

LEMMA 1. *Suppose A is an $n \times (n + \bar{n})$ matrix, G is an $(n + \bar{n}) \times (n + \bar{n})$ symmetric matrix and Z an $(n + \bar{n}) \times \bar{n}$ matrix such that $AZ = 0$. If the matrix $Z^T GZ$ is positive definite, there exists a constant v such that $G + vA^T A$ is positive definite.*

The reader is referred to [9, pp. 10–11] for a proof of this lemma.

LEMMA 2. *Let H be an $n \times n$ symmetric matrix and let A and Z be those matrices defined earlier. If we define*

$$H_\nu = H + \nu A^T A$$

and $Z^T HZ$ is nonsingular, then

$$\lim_{\nu \to \infty} H_\nu^{-1} = Z(Z^T HZ)^{-1} Z^T.$$

The proof of this lemma can be found in [11].

Note that $B = A^T A$ is a block tridiagonal matrix of the form

$$B = \begin{bmatrix} B_{11} & B_{21}^T & & & & \\ B_{21} & B_{22} & B_{32}^T & & & \\ & B_{32} & \cdot & \cdot & & \\ & & \cdot & \cdot & \cdot & \\ & & & \cdot & \cdot & \cdot \\ & & & & B_{N+1,N+1} & B_{N+2,N+1}^T \\ & & & & B_{N+2,N+1} & B_{N+2,N+2} \end{bmatrix}$$

where the blocks are of the same dimension as the corresponding blocks of G. This implies that

$$G_\nu = G + \nu A^T A,$$

for a given value of ν, is a block tridiagonal matrix of the same type. If $Z^T G^{(k)} Z$ is positive definite the modified Newton method in $M(A)$ requires the solution of the equations

$$Z^T G^{(k)} Z\omega^{(k)} = -Z^T g^{(k)} \tag{9}$$

from which the direction of search is obtained by setting

$$p^{(k)} = Z\omega^{(k)}.$$

An efficient method of obtaining a close approximation to $p^{(k)}$ is to compute the vector $ZZ^T y$, where y is the solution of the set of equations

$$(G^{(k)} + \nu A^T A)y = -g^{(k)}.$$

This follows from Lemma 2 since in the limit, both directions are identical. In general $Z^T G^{(k)} Z$ is not positive definite at points outside some neighbourhood of a strong local minimum. The algorithm given by Gill and Murray [7] uses the modified Cholesky algorithm to factorize the matrix $Z^T G^{(k)} Z$. The factorization obtained is of a matrix $Z^T G^{(k)} Z + E^{(k)}$ where $E^{(k)}$ is a diagonal matrix and $Z^T G^{(k)} Z + E^{(k)}$ is positive definite. The direction of search is then found by solving (using the factorization obtained)

$$(Z^T G^{(k)} Z + E^{(k)}) w^{(k)} = -Z^T g \tag{10}$$

and setting

$$p^{(k)} = Z w^{(k)}. \tag{11}$$

For the present application we instead factorize the matrix $G_v^{(k)}$ by the modified Cholesky algorithm, which gives the factors

$$L_v^{(k)} D_v^{(k)} L_v^{(k)T} = G_v^{(k)} + E_v^{(k)}.$$

The direction of search is then found by solving

$$L_v^{(k)} D_v^{(k)} L_v^{(k)T} y = -g^{(k)}$$

and computing

$$p^{(k)} = Z Z^T y.$$

The triangular factors have exactly the same zero/non-zero structure as those in the unconstrained algorithm and the vector $p^{(k)}$ is such that

$$F(x^{(k)} + \alpha^{(k)} p^{(k)}) < F^{(k)}.$$

Note that when $Z^T G_v^{(k)} Z$ is indefinite $p^{(k)}$ is not the same as that obtained from (10) and (11). The convergence proofs still hold since the significant properties of $p^{(k)}$ are unaltered.

3.1 *A basic iteration for the solution of the discretized optimal control problem*

The kth step of the basic iteration for performing linearly constrained optimization is as follows. It is assumed that initially a value of $x^{(0)}$ is given such that $Ax^{(0)} = b$.

Step 1. Calculate the Hessian matrix, $G^{(k)}$, of $F(x)$ at $x^{(k)}$.

Step 2. Calculate the matrix $G_v^{(k)}$ where

$$G_v^{(k)} = G^{(k)} + v_k A^T A.$$

Step 3. Form the modified Cholesky factorization of $G_v^{(k)}$ which gives

$$\bar{G}_v^{(k)} = L_v^{(k)} D_v^{(k)} L_v^{(k)T} = G_v^{(k)} + E_v^{(k)}.$$

Step 4

Case 1. $\|Z^T g^{(k)}\| < \varepsilon$. If $\|E_\nu^{(k)}\| = 0$ then $x^{(k)}$ is an adequate approximation to a strong local minimum of DLCP1 and the iteration is terminated. Assume that $\|E_\nu^{(k)}\| > 0$. The vector $p^{(k)}$ is determined using an alternative search procedure as follows. Determine the vector y by solving the set of equations

$$L_\nu^{(k)T} y = e_j.$$

Set

$$p^{(k)} = \begin{cases} -\text{sign}\,(y^T g^{(k)})\, ZZ^T y, & \|Z^T g^{(k)}\| > 0 \\ ZZ^T y, & \|Z^T g^{(k)}\| = 0 \end{cases}$$

where the index j is such that

$$\zeta_j \leqslant \zeta_i, \qquad i = 1, 2, \ldots, \bar{n}$$

ζ_i being the ith diagonal element of $D_\nu^{(k)} - E_\nu^{(k)}$.

Case 2. $\|Z^T g^{(k)}\| > \varepsilon$. Solve for $p^{(k)}$ via the equations

$$L_\nu^{(k)} D_\nu^{(k)} L_\nu^{(k)T} y = -g^{(k)}$$

$$p^{(k)} = ZZ^T y.$$

Step 5. Compute the step length $\alpha^{(k)}$ as in the basic iteration for the unconstrained problem. If an $\alpha^{(k)}$ cannot be found satisfying the relevant conditions put $\nu_{k+1} = 2\nu_k$ and continue at Step 2 with $G^{(k+1)} = G^{(k)}$ and $g^{(k+1)} = g^{(k)}$. Otherwise set $\nu_{k+1} = \nu_k$ and $x^{(k+1)} = x^{(k)} + \alpha^{(k)} p^{(k)}$.

Step 6. If it has not been found in the process of finding $\alpha^{(k)}$ compute $g^{(k+1)}$, set $k = k + 1$ and continue at Step 1.

3.2 *Discussion*

The Hessian matrix $G^{(k)}$ can be approximated using the techniques described in Section 2.4. In practice the value of ν_0 can be set at a fixed value (say 10^3) and it will only be altered if the alternative search procedure fails to locate a downhill direction. If ν is large enough the alternative search procedure is guaranteed to find such a direction. This is proved in the following theorem.

THEOREM. *The vector p generated by the alternative search procedure is such that*

$$p^T G p < 0.$$

Proof. For convenience we will drop the superfix k since $G^{(k)}$ and $p^{(k)}$ remain constant during this part of the algorithm. From the definition of p we have

$$p^T Gp = y(v)^T ZZ^T GZZ^T y(v). \tag{12}$$

Note the addition of the argument v to the vector y in order to emphasise its dependence upon v. It can be shown that Eqn (12) can be written as

$$p^T Gp = y(v)^T (I - A^T(AA^T)^{-1}A)\, G(I - A^T(AA^T)^{-1}A)\, y(v). \tag{13}$$

If Lemma 2 is applied to the matrix

$$H = G + E_v$$

then

$$H_v^{-1} = Z(Z^T HZ)^{-1} Z^T + O\left(\frac{1}{v}\right).$$

The triangular factors of H_v are known and consequently

$$L_v^{-T} D_v^{-1} L_v^{-1} = H_v^{-1}.$$

From this last equation it is clear that

$$\lim_{v \to \infty} L_v^{-T} \in N(A), \text{ the nullspace of } A$$

which in turn implies that $\overset{*}{y} = \lim\limits_{v \to \infty} y(v) \in N(A)$,

$$\|\overset{*}{y} - y(v)\| = O\left(\frac{1}{v}\right)$$

and

$$(I - A^T(AA^T)^{-1}A)\, y(v) = y(v) + O\left(\frac{1}{v}\right).$$

Eqn(13) can now be written as

$$\begin{aligned} p^T Gp &= y(v)^T G y(v) + O\left(\frac{1}{v}\right) \\ &= y(v)^T (H_v - vA^T A - E_v) y(v) + O\left(\frac{1}{v}\right) \\ &= d_j - E_j - v y(v)^T A^T A y(v) - \sum_{r=j+1}^{n+\bar{n}} E_r y_r(v)^2 + O\left(\frac{1}{v}\right) \end{aligned}$$

where d_j and E_j are the jth diagonal elements of D_ν and E_ν respectively. It can be shown [6] that $d_j - E_j < 0$ which implies that there exists ν_0 such that

$$p^T G p < 0 \qquad \text{for all} \qquad \nu > \nu_0.$$

This is the required result.

Instead of computing $p^{(k)} = ZZ^T y$, it is possible to define an alternative basic iteration in which $p^{(k)}$ is computed as

$$\begin{aligned} p^{(k)} &= (I - A^T(AA^T)^{-1}A)y \\ &= (I - A^T(L_A L_A{}^T)^{-1}A)y. \end{aligned} \tag{14}$$

Since the lower triangular factor L_A of A is sparse and independent of any column ordering in A computing $p^{(k)}$ by (14) may be a preferable strategy when the algorithm is extended to include inequality constraints on the control variables. In such an algorithm it would be necessary to store A and L rather than Q and L.

Conclusions

Two algorithms have been presented for the solution of a problem in the Calculus of Variations, the first requiring knowledge of the first and second variation and the second requiring only the first variation. These algorithms have several properties:

(i) The algorithms require a very small amount of storage and utilize the most powerful methods for unconstrained optimization that are currently available

(ii) The algorithms can be shown to be convergent to a strong local minimum of a wide class of functions

(iii) The initial approximation to the solution can be arbitrary, providing the objective function is twice continuously differentiable.

Although there have been algorithms which minimize storage requirements [4, 13] they have not been linked with a powerful optimization procedure and have not satisfied (i) and (ii) above. Finally, an outline of a method for more difficult problems in optimal control has been given which satisfies similar properties to those described above.

Acknowledgements

The authors wish to thank Miss Susan Picken, Mr. E. L. Albasiny and Dr. M. A. Saunders for a number of helpful suggestions and Mrs. Margaret Thomson for her help in obtaining the computational results.

References

1. Akhiezer, N. I., "The Calculus of Variations". Blaisdell Publishing Company (1962).
2. Ciarlet, P. G., Schultz, M. H. and Varga, R. S., Nonlinear boundary value problems I. *Numer. Math.* **9** (1967), 394–430.
3. Curtis, A. R., Powell, M. J. D. and Reid, J. K., "On the Estimation of Sparse Jacobian Matrices". A.E.R.E. Report TP.476.
4. Elkin, R., "Convergence Theorems for Gauss–Seidel and Other Minimization Algorithms". Ph.D. Dissertation, University of Maryland (1968).
5. Gill, P. E., Murray, W. and Pitfield, R. A., "The Implementation of Two Revised Quasi-Newton Algorithms for Unconstrained Optimization". National Physical Laboratory DNAC 11 (1972).
6. Gill, P. E., Murray, W. and Picken, S. M., "The Implementation of Two Modified Newton Methods for Unconstrained Optimization" National Physical Laboratory DNAC 24 (1972).
7. Gill, P. E. and Murray, W., "Two Methods for the Solution of Linearly Constrained and Unconstrained Optimization Problems". National Physical Laboratory DNAC 25 (1972).
8. Herbold, R. J., Schultz, M. H. and Varga, R. S., The effect of quadrature errors in the numerical solution of boundary value problems by variational techniques. *Aequationes Math.* **3** (1969), 247–270.
9. Hestenes, M. R., "Calculus of Variations and Optimal Control Theory". John Wiley, New York (1966).
10. Martin, R. S. and Wilkinson, J. H., Symmetric decomposition of a positive definite band matrix. *In* "Handbook for Automatic Computation", Vol. II, Springer–Verlag (1971).
11. Murray, W., Analytical expressions for the eigenvalues andeigenvectors of the Hessian matrices of barrier and penalty functions. *J. Opt. Th. Applics.* **7** (1971), 189–196.
12. Murray, W., Second derivative methods. *In* W. Murray (ed.) "Numerical Methods for Unconstrained Optimization", Academic Press, London and New York (1972).
13. Schechter, S., Minimization of a convex function by relaxation. *In* J. Abadie (ed.) "Integer and and Nonlinear Programming". North Holland, Amsterdam (1970).
14. Simpson, R. B., The Rayleigh–Ritz process for the simplest problem in the calculus of variations. *SIAM J. Numer. Anal.* **6** (1969), 258–271.
15. Wilkinson, J. H., "The Algebraic Eigenvalue Problem". Oxford University Press, London (1965).

Discussion

FREEMAN (*Sunderland Polytechnic*). In the solution of the linear quadratic cost problem based on an orthogonal decomposition method it has been found that Chebyshev polynomials give more accurate sub-optimal controls with fewer terms in the approximation (and hence more quickly). This arises from the fact that these polynomials not only minimise a weighted mean square error function but also minimise the maximum error in the approximation. I wonder therefore if you have attempted to use Chebyshev polynomials and if so with what success.

GILL. Yes, we initially used Chebyshev polynomials because of their good truncation properties. Unfortunately if you attempt to solve problems using Chebyshev polynomials, the Hessian matrices are dense. Assuming twenty coefficients per variable, a ten time-dependent variable problem becomes a 200 variable finite-dimensional problem and the storage of a 200 × 200 Hessian matrix could cause some difficulty on a small computer, This is why one must give up the slightly superior truncation properties of Chebyshev polynomials in order to obtain the considerable benefits resulting from the sparse matrix properties of piecewise polynomials.

8. Singular Perturbations of Linear Time-optimal Control Problems

W. D. COLLINS

Department of Mathematics, University of Strathclyde, Glasgow, Scotland

1. Introduction

In this paper we consider some aspects of the time-optimal control problem for a linear system with constant coefficients which depends on a small parameter ε in such a way that the order of the system drops when ε tends to zero. The control process considered is

$$\frac{dx}{dt} = A_1x + A_2z + b_1u$$
$$\varepsilon\frac{dz}{dt} = A_3x + A_4z + b_2u \tag{1}$$

on the interval $0 \leqslant t \leqslant T(\varepsilon)$, where $x(t, \varepsilon) \in \mathfrak{R}^n$, $z(t, \varepsilon) \in \mathfrak{R}^m$, $u(t, \varepsilon) \in \mathfrak{R}^1$ and ε is a positive parameter. The matrices A_1, A_2, A_3 and A_4 and the vectors b_1 and b_2 are constant and supposed independent of ε though this latter requirement is not necessary for most of the results obtained. The system is to be moved from a given initial state $(x(0), z(0))$ to the origin $(x(T), z(T)) = (0,0)$ in minimum time $T(\varepsilon)$ using the scalar control $u(t, \varepsilon)$ which is subject to the constraint $|u(t, \varepsilon)| \leqslant 1$. In physical systems the parameter ε could represent small time constants, masses, moments of inertia, etc. The presence of the parameter makes the determination of the optimal control more difficult in that it raises the order of the system and, since ε multiplies derivatives, the differential equations are accordingly more difficult to handle numerically.

If however we set $\varepsilon = 0$ and suppose A_4 is non-singular, the second equation gives

$$z = -A_4{}^{-1}A_3x - A_4{}^{-1}b_2u$$

and, on substituting this expression for z in the first equation, we obtain

$$\frac{dx}{dt} = A_0 x + b_0 u \tag{2}$$

where

$$A_0 = A_1 - A_2 A_4^{-1} A_3, \quad b_0 = b_1 - A_2 A_4^{-1} b_2$$

which we term the reduced system. The question arises as to whether the time-optimal control for this control process can be used as an approximation to the optimal control for the full process and, if so, how the control for the full process differs from it when ε is small. The problem is one in the theory of singular perturbations of differential equations. We cannot, for instance, satisfy all the conditions at $t = 0$ and $t = T$ with a solution of the reduced equation.

Singular perturbation techniques have been used in optimal control by Kokotović and his co-workers, amongst them Hadlock, Sannuti and Yackel, at the University of Illinois in a series of papers over the last five years [1–7], by Kelley [8, 9, 10] and by O'Malley [11,12]. A survey of recent developments is given in [17]. One problem considered by both O'Malley and the Illinois group is the linear state regulator problem for the control process (1) with $u \in \mathfrak{R}^r$ and unbounded, in which it is required to minimize the performance index

$$J(\varepsilon) = \tfrac{1}{2} y'(T) F y(T) + \frac{1}{2} \int_0^T [y' D y + u' G u] dt$$

where $y = C_1 x + C_2 z$, $y \in \mathfrak{R}^s$ and F, D, G, C_1 and C_2 are constant matrices, the time T and the initial state $(x(0), z(0))$ being given. Conditions under which the solution of the full system tends to that of the reduced system are established and the nature of the solution for small ε examined. Kokotović and Yackel [7], in particular, obtain results on the Riccati matrix and the system it satisfies. The Illinois group has also applied singular perturbation methods to some non-linear problems. Accounts of singular perturbation methods for ordinary differential equations are given by O'Malley [13] and Wasow [14].

2. A Simple Example

For the control processes (1) and (2) the time-optimal controls if they exist are bang-bang, so that a knowledge of the times at which the controls switch and their values on the initial intervals is sufficient to determine them. Before examining the time-optimal control problem for the process (1) we consider a simple example for which an exact solution can be found in order to bring out some features of the problem. We consider the control process

$$\begin{aligned} \frac{dx}{dt} &= -x + z \\ \varepsilon \frac{dz}{dt} &= -z + u \end{aligned} \tag{3}$$

$0 \leqslant t \leqslant T(\varepsilon)$, where x, z and u are scalars. We take the initial conditions as $(x(0), z(0)) = (-1, -1)$, the terminal ones being $(x(T), z(T)) = (0, 0)$.

The system (3) can be written as

$$\varepsilon \frac{dy}{dt} = Ay + bu \tag{4}$$

where

$$y = \begin{pmatrix} x \\ z \end{pmatrix}, \quad A = \begin{pmatrix} -\varepsilon, & \varepsilon \\ 0 & -1 \end{pmatrix}, \quad b = \begin{pmatrix} 0 \\ 1 \end{pmatrix}.$$

Since the process is controllable and the eigenvalues of A, which are -1 and $-1/\varepsilon$, are real, the number of switches in u is one less than the order of the system [15, p.143], so that there is only one switch in u. Denoting the time at which this switch occurs as $t = T_1(\varepsilon)$, we have that the solution to the problem is

$$\left.\begin{aligned} x(t) &= 1 + \frac{2}{(1-\varepsilon)}[\varepsilon e^{-t/\varepsilon} - e^{-t}] \\ z(t) &= 1 - 2e^{-t/\varepsilon}, \qquad u(t) = 1 \end{aligned}\right\} 0 \leqslant t < T_1(\varepsilon)$$

and

$$\left.\begin{aligned} x(t) &= -1 - \frac{1}{(1-\varepsilon)}[\varepsilon e^{(T-t)/\varepsilon} - e^{(T-t)}] \\ z(t) &= -1 + e^{(T-t)/\varepsilon}, \qquad u(t) = -1 \end{aligned}\right\} T_1(\varepsilon) < t \leqslant T(\varepsilon)$$

where $T(\varepsilon)$ and $T_1(\varepsilon)$ satisfy

$$2(1-\varepsilon) + 2\varepsilon e^{-T_1/\varepsilon} - 2e^{-T_1} = -\varepsilon e^{T_2/\varepsilon} + e^{T_2}$$

$$2 = 2e^{-T_1/\varepsilon} + e^{T_2/\varepsilon}$$

$$T_2 = T - T_1.$$

For small ε the solution of these equations is

$$T(\varepsilon) = (1 + 2\varepsilon)\ln 2 + O(\varepsilon^2)$$

$$T_1(\varepsilon) = (1 + \varepsilon)\ln 2 + O(\varepsilon^2).$$

The reduced system obtained by putting $\varepsilon = 0$ in (3) is

$$\frac{dx^{(0)}}{dt} = -x^{(0)} + u^{(0)}$$

whose solution subject to the initial condition $x^{(0)}(0) = -1$ and the terminal condition $x^{(0)}(T^{(0)}) = 0$ is

$$x^{(0)}(t) = 1 - 2e^{-t}, \qquad u^{(0)}(t) = 1, \qquad 0 \leqslant t \leqslant T^{(0)}$$

where $T^{(0)} = \ln 2$.

We can note the following features of the solutions of the full and the reduced systems. The additional switch required in the optimal control for the full process takes place at a time within $O(\varepsilon)$ of the terminal time. The terminal time $T(\varepsilon)$ for the full control process tends to that of the reduced process. The solution $x(t, \varepsilon)$ in the interval $0 \leqslant t \leqslant T(\varepsilon)$ tends to the solution $x^{(0)}(t)$ for the reduced process since $T_1(\varepsilon)$ also tends to $T^{(0)}$ as ε tends to zero. The solution for $z(t, \varepsilon)$ does not however tend to the solution $z^{(0)}(t) = u^{(0)}(t) = 1$ for the reduced process over the whole interval $0 \leqslant t \leqslant T(\varepsilon)$ but only for times away from the ends of the interval. This reflects the loss of the initial and terminal conditions on z when $\varepsilon = 0$. The solution for z adjusts to the correct initial and terminal values over 'boundary layers' at $t = 0$ and $t = T(\varepsilon)$, which are intervals of $O(\varepsilon)$. The term 'boundary layer' comes from singular perturbation theory. There is a switch in the boundary layer at $t = T(\varepsilon)$ but not one in the layer at $t = 0$.

This example reflects the main features of the solution for the control process (1) as ε tends to zero. Under certain controllability conditions on the A's and the b's and conditions on the eigenvalues of A_4 the solution of the full system for $x(t, \varepsilon)$ tends to the solution $x^{(0)}(t)$ of the reduced system while the solution for $z(t, \varepsilon)$ tends to $z^{(0)}(t)$ only for times not within $O(\varepsilon)$ of $t = 0$ and $t = T(\varepsilon)$. The optimal control $u(t, \varepsilon)$ tends to the optimal control $u^{(0)}(t)$ for the reduced problem in the sense that

$$\text{measure} \quad U(\varepsilon) \to 0 \quad \text{as} \quad \varepsilon \to 0$$

where

$$U(\varepsilon) = \{t' \mid u(T(\varepsilon)t', \varepsilon) \neq u^{(0)}(T^{(0)}t')\}.$$

Additional switches to those in the optimal control for the reduced system occur in the boundary layers at $t = 0$ and $t = T(\varepsilon)$. It is possible to obtain results on the number of additional switches and in which boundary layer they occur.

3. The Time-optimal Control Problem

To establish the desired results we suppose the process (1) is controllable and replace it by a canonical form which is more convenient for our purposes. If we set

$$y = \begin{pmatrix} x \\ z \end{pmatrix}, \qquad A = \begin{pmatrix} \varepsilon A_1 & \varepsilon A_2 \\ A_3 & A_4 \end{pmatrix}, \qquad b = \begin{pmatrix} \varepsilon b_1 \\ b_2 \end{pmatrix},$$

the system (1) can be written as

$$\varepsilon \frac{dy}{dt} = Ay + bu. \tag{5}$$

The process (5) is controllable if

$$\operatorname{rank} [A^{n+m-1} b, A^{n+m-2} b, \ldots, Ab, b] = n + m \tag{6}$$

sufficient conditions for this to hold for all positive ε being

$$\operatorname{rank} [A_0{}^{n-1} b_0, A_0{}^{n-2} b_0, \ldots, A_0 b_0, b_0] = n \tag{7}$$

and

$$\operatorname{rank} [A_4{}^{m-1} b_2, A_4{}^{m-2} b_2, \ldots, A_4 b_2, b_2] = m \tag{8}$$

the first of these being the condition for the reduced system (2) to be controllable. Kokotović and Yackel [7] term the system (5) boundary layer controllable when condition (8) is satisfied.

When (6) holds, the process (5) is linearly equivalent [15, p.90] to the control process

$$\varepsilon \frac{dw}{dt} = Cw + \tilde{b}u \tag{9}$$

where C, the companion matrix to A, and $\tilde{b}$ are given by

$$C = \begin{bmatrix} 0 & 1 & 0 & \cdots\cdots & 0 \\ 0 & 0 & 1 & & 0 \\ 0 & 0 & 0 & & 0 \\ \cdot & \cdot & \cdot & & \cdot \\ \cdot & \cdot & \cdot & & \cdot \\ \cdot & \cdot & \cdot & & \cdot \\ 0 & 0 & 0 & & 1 \\ \alpha_{n+m} & \alpha_{n+m-1} & \alpha_{n+m-2} & & \alpha_1 \end{bmatrix}$$

$\tilde{b} = e_{n+m}$

the coefficients $\alpha_1(\varepsilon), \ldots \alpha_{n+m}(\varepsilon)$ being uniquely determined from

$$A^{n+m}b = \alpha_1 A^{n+m-1} b + \ldots + \alpha_{n+m} b.$$

Since the characteristic polynomial of A is also its minimal polynomial, A being non-derogatory [16, p.48], the coefficients $\alpha_1, \ldots \alpha_{n+m}$ are uniquely determined from the characteristic equation of A,

$$\det |\rho I_{n+m} - A| = \rho^{n+m} - \alpha_1 \rho^{n+m-1} - \ldots - \alpha_{n+m} = 0$$

and are thus independent of b. The vectors $y(t)$ and $w(t)$ are related by

$$w(t) = \tilde{P} P^{-1} y(t) \tag{10}$$

where the time-independent non-singular matrices P and $\tilde{P}$ are given by

$$P = [A^{n+m-1} b, A^{n+m-2} b, \ldots, Ab, b]$$
$$\tilde{P} = [C^{n+m-1} \tilde{b}, C^{n+m-2} \tilde{b}, \ldots, C\tilde{b}, \tilde{b}] . \tag{11}$$

Since the matrices A_i are independent of ε, the α_i's are polynomials in ε and for small ε are given by

$$\alpha_j(\varepsilon) = \alpha_j(0) + O(\varepsilon), \qquad 1 \leqslant j \leqslant m \tag{12}$$

where

$$A_4{}^m b_2 = \alpha_1(0) A_4{}^{m-1} b_2 + \ldots + \alpha_m(0)\, b_2$$

with $\alpha_m(0) = (-1)^{m+1} |A_4|$. Since A_4 is assumed non-singular, $\alpha_m(0)$ is non-zero. Further, for $1 \leqslant j \leqslant n$ we have

$$\alpha_{m+j}(\varepsilon) = -\alpha_m(0)\, \gamma_j\, \varepsilon^j + O(\varepsilon^{j+1}) \tag{13}$$

where the γ_i's are independent of ε and are determined from

$$A_0{}^n b_0 = \gamma_1 A_0{}^{n-1} b_0 + \ldots + \gamma_n b_0$$

and hence from the characteristic equation for A_0, A_0 being non-derogatory.

We can also transform the reduced system (2) to a canonical form similar to (9), the vector $w^{(0)}(t)$ corresponding to $w(t)$ being given by

$$w^{(0)}(t) = \tilde{P}_0 P_0{}^{-1} x(t)$$

where P_0 and $\tilde{P}_0$ are the time-independent, non-singular matrices for the reduced system (2) corresponding to the matrices P and $\tilde{P}$ given by (11).

We then have that for small ε

$$\varepsilon^{n-r+1}\alpha_m(0)w_r(t) = -w_r^{(0)}(t) + O(\varepsilon), \qquad 1 \leqslant r \leqslant n \tag{14}$$

and so to a first approximation $w_r(t)$ depends only on $x(t)$ for $1 \leqslant r \leqslant n$. For $n+1 \leqslant r \leqslant n+m$ the first approximation to $w_r(t)$ however depends on $z(t)$ as well as $x(t)$.

The solution of Eqn (9) can be written in terms of the transition matrix $\Phi(t, s)$ as

$$\varepsilon w(t) = \varepsilon\Phi(t, 0)w(0) + \int_0^t \Phi(t, s)\, e_{n+m}\, u(s)\, ds \tag{15}$$

where $\Phi(t, s)$ satisfies

$$\varepsilon\frac{d}{dt}\Phi(t, s) = C\Phi(t, s),$$

$$\Phi(s, s) = I_{n+m}.$$

Each component $\Phi_i(t, s)$ of $\Phi(t, s)$ can be expressed as a linear combination of terms of the form

$$B_{i,\lambda}\psi_\lambda e^{\lambda(t-s)}, \qquad B_{i,2\mu-1}\operatorname{Re}(\psi_\mu e^{\mu(t-s)}), \qquad B_{i,2\mu}\operatorname{Im}(\psi_\mu e^{\mu(t-s)})$$

where λ is a real and μ a complex root of the characteristic equation for (9)

$$\det|\varepsilon\lambda I_{n+m} - C| = \varepsilon^{n+m}\lambda^{n+m} - \alpha_1(\varepsilon)\varepsilon^{n+m-1}\lambda^{n+m-1} \ldots - \alpha_{n+m}(\varepsilon) = 0 \tag{16}$$

so that $\varepsilon\lambda$ and $\varepsilon\mu$ are eigenvalues of C, while ψ_λ and ψ_μ are the corresponding eigenvectors and are constant. We suppose $2q$ eigenvalues $\sigma_k \pm i\tau_k$, $k = 1, \ldots q$, are complex, the remaining $n + m - 2q$ eigenvalues λ_k, $k = 2q + 1, \ldots n + m$, being real. We assume the eigenvalues of C to be distinct, so that the corresponding eigenvectors form a linearly independent set. The coefficients $B_{i,j}$ in $\Phi_i(t, s)$ are determined from the condition $\Phi(s, s) = I$, the transformation of (5) to the canonical form (9) facilitating the calculation of these coefficients.

The initial and terminal conditions on $x(t)$ and $z(t)$ lead to initial and terminal conditions on $w(t)$, so that from (10) we have $w(0)$ given and $w(T) = 0$. Hence, when $t = T$, Eqn (15) becomes

$$0 = \varepsilon\Phi(T, 0)w(0) + \int_0^T \Phi(T, s)\, e_{n+m}\, u(s)\, ds.$$

Equating the coefficients of the linearly independent eigenvectors in this equation to zero then gives

$$\varepsilon \sum_{j=1}^{n+m} w_j(0)\, B_{j,2k-1} + \int_0^T u(s)\,[B_{n+m,2k-1} \cos(\tau_k s) - B_{n+m,2k} \sin(\tau_k s)]\, e^{-\sigma_k s}\, ds = 0$$

$$\varepsilon \sum_{j=1}^{n+m} w_j(0)\, B_{j,2k} + \int_0^T u(s)\,[B_{n+m,2k-1} \sin(\tau_k s) + B_{n+m,2k} \cos(\tau_k s)]\, e^{-\sigma_k s}\, ds = 0, \qquad k = 1, \ldots q$$

and

$$\varepsilon \sum_{j=1}^{n+m} w_j(0)\, B_{j,k} + B_{n+m,k} \int_0^T u(s)\, e^{-\lambda_k s}\, ds = 0, \qquad k = 2q+1, \ldots n+m. \tag{17}$$

Since the optimal control $u(t, \varepsilon)$ is bang-bang and thus takes only the values ± 1, it is determined when we know the number $N(\varepsilon)$ of switches it makes, the times $T_1, \ldots T_N$ at which it switches, where

$$0 < T_1 < T_2 < \ldots < T_N < T$$

and its value Δ, equal to either $+1$ or -1, on the first interval $0 < t < T_1$. Then

$$u(t) = (-1)^r \Delta, \quad T_r < t < T_{r+1}, \quad r = 0, 1, \ldots N$$

with $T_0 = 0$, $T_{N+1} = 1$. Substitution for $u(t)$ in (17) leads to the equations

$$1 + (-1)^{N+1} e^{-\sigma_k T} \cos \tau_k T + 2 \sum_{r=1}^{N} (-1)^r e^{-\sigma_k T_r} \cos \tau_k T_r$$

$$= -\varepsilon \Delta \left[C_{n+m,2k-1} \sum_{j=1}^{n+m} w_j(0) B_{j,2k-1} + C_{n+m,2k} \sum_{j=1}^{n+m} w_j(0) B_{j,2k} \right],$$

$$(-1)^{N+1} e^{-\sigma_k T} \sin \tau_k T + 2 \sum_{r=1}^{N} (-1)^r e^{-\sigma_k T_r} \sin \tau_k T_r$$

$$= -\varepsilon \Delta \left[C_{n+m,2k-1} \sum_{j=1}^{n+m} w_j(0) B_{j,2k} - C_{n+m,2k} \sum_{j=1}^{n+m} w_j(0) B_{j,2k-1} \right],$$

$$k = 1, \ldots q$$

where

$$C_{n+m,2k-1} = \frac{\sigma_k B_{n+m,2k-1} - \tau_k B_{n+m,2k}}{B^2_{n+m,2k-1} + B^2_{n+m,2k}}$$

$$C_{n+m,2k} = \frac{\tau_k B_{n+m,2k-1} + \sigma_k B_{n+m,2k}}{B^2_{n+m,2k-1} + B^2_{n+m,2k}}$$

and

$$1 + (-1)^{N+1} e^{-\lambda_k T} + 2 \sum_{r=1}^{N} (-1)^r e^{-\lambda_k T_r}$$

$$= - \varepsilon \Delta \lambda_k \sum_{j=1}^{n+m} w_j(0) \frac{B_{j,k}}{B_{n+m,k}}, \qquad k = 2q+1, \ldots n+m. \tag{18}$$

From the maximum principle [15, p.135] the switching times are the zeros in the interval $[0, T]$ of the function

$$p_{n+m}(t, \varepsilon) = \sum_{k=1}^{q} e^{-\sigma_k t} (P_k \cos \tau_k t + Q_k \sin \tau_k t) + \sum_{k=2q+1}^{n+m} R_k e^{-\lambda_k t}, \tag{19}$$

where the coefficients P_k, Q_k and R_k are independent of t. The function $p_{n+m}(t,\varepsilon)$ is the $(n+m)$th component of the vector $p(t, \varepsilon) \in \Re^{n+m}$, which satisfies the equation

$$\varepsilon \frac{dp}{dt} = - C'p$$

C' being the transpose of C. The control $u(t, \varepsilon)$ is then given as

$$u(t,\varepsilon) = \operatorname{sgn} p_{n+m}(t, \varepsilon).$$

The switching times $T_r(\varepsilon)$ thus satisfy

$$p_{n+m}(T_r, \varepsilon) = 0, \quad r = 1, \ldots N \tag{20}$$

with $p_{n+m}(t, \varepsilon) \neq 0$ at all other times in the interval $0 < t \leqslant T(\varepsilon)$. That $p_{n+m}(T, \varepsilon) \neq 0$ is a consequence of the Hamiltonian being constant along the optimal trajectory. Equations (18) and (20) form a set of $N+n+m$ equations for the $N+1+n+m$ unknowns $T_1, T_2, \ldots T_N, T, P_1, \ldots P_q, Q_1, \ldots Q_q, R_{2q+1}, \ldots R_{n+m}$ together with N and Δ which are also unknown. Since $p_{n+m}(t, \varepsilon)$ cannot be identically zero in $0 \leqslant t \leqslant T(\varepsilon)$, at least one of the coefficients in $p_{n+m}(t, \varepsilon)$ is non-zero. This can be taken as unity without affecting the switching times so that there are only $N+n+m$ unknowns. The equations are to be solved subject to the condition $p_{n+m}(t, \varepsilon) \neq 0$ if $t \neq T_r$, $t \in (0, T]$. We want to determine how the number of switches $N(\varepsilon)$, the switching times $T_r(\varepsilon)$ and $T(\varepsilon)$ vary as ε tends to zero and in particular their relations to the corresponding quantities N_0, $T_r^{(0)}$ and $T^{(0)}$ for the reduced process (2).

Using the results (12) and (13) for the coefficients $\alpha_j(\varepsilon)$ occurring in equation (16), we note that, when ε is small, n roots of Eqn (16) are $O(1)$ and to a first approximation satisfy

$$(\lambda^{(0)})^n - \gamma_1(\lambda^{(0)})^{n-1} - \ldots - \gamma_n = 0 \tag{21}$$

which is the characteristic equation for A_0, so that $\lambda^{(0)}$ is an eigenvalue of A_0. The remaining m roots of (16) are $O(1/\varepsilon)$ and to a first approximation are given by $\varepsilon\lambda = \lambda^{(1)}$, where $\lambda^{(1)}$ satisfies

$$(\lambda^{(1)})^m - \alpha_1(0)\,(\lambda^{(1)})^{m-1} - \ldots - \alpha_{m-1}(0)\,\lambda^{(1)} - \alpha_m(0) = 0 \tag{22}$$

which is the characteristic equation of A_4 so that $\lambda^{(1)}$ is an eigenvalue of A_4. Since $\alpha_m(0) \neq 0$, all the roots of this equation are non-zero. We suppose Eqn (22) has r pairs of complex conjugate roots, where $r \leqslant q$, $2r \leqslant m$, the remaining $m - 2r$ roots being real.

The basic results obtained are that, provided none of the roots of (22) are pure imaginary, the switching times $T_r(\varepsilon), r = 1, \ldots N$ and $T(\varepsilon)$ are such that

(1) $T(\varepsilon)$ tends to $T^{(0)}$ as ε tends to zero

(2) there are N_1 switching times $T_1(\varepsilon), \ldots T_{N_1}(\varepsilon)$, such that $T_r(\varepsilon)$ is $O(\varepsilon)$ for small ε (N_1 may be zero)

(3) there are N_0 switching times $T_{N_1+1}(\varepsilon), \ldots T_{N_1+N_0}(\varepsilon)$, such that $T_{N_1+r}(\varepsilon) \to T_r^{(0)}$ as $\varepsilon \to 0, r = 1, \ldots N_0$, where $T_r^{(0)}, r = 1, \ldots N_0$, are the switching times for the reduced process

(4) there are N_2 switching times $T_{N_1+N_0+1}(\varepsilon), \ldots T_N(\varepsilon), N_1 + N_0 + N_2 = N$, such that $\big(T(\varepsilon) - T_{N_1+N_0+r}(\varepsilon)\big)$ is $O(\varepsilon)$ for small ε (N_2 may be zero).

The switches additional to those corresponding to the ones for the reduced system thus occur in the boundary layers at $t = 0$ and $t = T(\varepsilon)$. That cases when Eqn (22) has pure imaginary roots need to be excluded might be anticipated from work by Wasow [14, pp.226–234] on boundary value problems for scalar linear differential equations involving a small parameter.

To obtain these results we divide the switching times $T_1, \ldots T_N$, into three groups as follows. We suppose there are N_1 times $T_1, \ldots T_{N_1}$ satisfying

$$T_r(\varepsilon) < m\varepsilon, \quad r = 1, \ldots N_1$$

and N_2 times $T_{N_1+\tilde{N}+1}, \ldots T_{N_1+\tilde{N}+N_2}$ $(N_1 + \tilde{N} + N_2 = N)$ satisfying

$$T(\varepsilon) - T_{N_1+\tilde{N}+r}(\varepsilon) < M\varepsilon, \quad r = 1, \ldots N_2$$

where m and M are constants independent of ε. The remaining times $T_{N_1+1}, \dots T_{N_1+\tilde{N}}$, satisfy neither of these inequalities. We assume $N_1(\varepsilon)$ and $N_2(\varepsilon)$ are bounded as ε tends to zero. This has been proved under certain conditions on the roots of the characteristic equation (16) but not so far in general. Equations (20) are used to express the coefficients P_k, Q_k and R_k in (19) in terms of the switching times. It is then possible to estimate $p_{n+m}(t, \varepsilon)$ in the boundary layers, that is, for times t of $O(\varepsilon)$ for the initial layer and $T - t$ of $O(\varepsilon)$ for the terminal layer as well as outside these layers. The forms of Eqns (18) can also be found as ε tends to zero. Provided there are no pure imaginary roots of (22), then outside the boundary layers those Eqns (18) for which k takes values corresponding to the roots of (16) of $O(1)$ tend to the corresponding equations for the reduced system with $\tilde{N}$ in place of N_0. The switching function also tends to the switching function for the reduced process. Since the optimal control for the reduced process is unique, it follows that outside the boundary layers the switching times tend to those for the reduced process and that $T(\varepsilon)$ tends to $T^{(0)}$.

The equations for the switching times in the boundary layers are obtained from those Eqns (18) for which k takes values corresponding to the roots of (16) of $O(1/\varepsilon)$.

Setting

$$T_r(\varepsilon) = \varepsilon\, U_r(\varepsilon), \qquad r = 1, \dots N_1$$

$$T(\varepsilon) - T_{N_1+\tilde{N}+r}(\varepsilon) = \varepsilon\, V_r(\varepsilon), \qquad r = 1, \dots N_2$$

and

$$\sigma_k + i\,\tau_k = \varepsilon^{-1}(\sigma_k{}^{(1)} + i\,\tau_k{}^{(1)}) + O(1)$$

$$\lambda_k = \varepsilon^{-1}\,\lambda_k{}^{(1)} + O(1)$$

so that $\sigma_k{}^{(1)} + i\,\tau_k{}^{(1)}$ and $\lambda_k{}^{(1)}$ are eigenvalues of A_4 and satisfy Eqn (22), we find

$$1 + 2\sum_{r=1}^{N_1} (-1)^r \exp[-\sigma_k{}^{(1)}\, U_r] \cos[\tau_k{}^{(1)}\, U_r]$$
$$= -\,\Delta\, \mathrm{Re}\,[(l_{2k-1}^m\, v(0))/(l_{2k-1}^m\, b_2)], \tag{23}$$

$$2\sum_{r=1}^{N_1} (-1)^r \exp[-\,\sigma_k{}^{(1)}\, U_r] \sin[\tau_k{}^{(1)}\, U_r]$$
$$= \Delta\, \mathrm{Im}\,[(l_{2k-1}^m\, v(0))/(l_{2k-1}^m\, b_2)] \tag{24}$$

for those k such that $\sigma_k{}^{(1)} > 0$ and

$$1 + 2\sum_{r=1}^{N_1} (-1)^r \exp[-\,\lambda_k{}^{(1)}\, U_r] = -\,\Delta(l_k{}^m\, v(0))/(l_k{}^m\, b_2) \tag{25}$$

for those k such that $\lambda_k^{(1)} > 0$, where $v(0) = A_3 x(0) + A_4 z(0), l_{2k-1}^m$ is the left eigenvector of A_4 corresponding to the eigenvalue $\sigma_k^{(1)} + i\tau_k^{(1)}$ and l_k^m the left eigenvector corresponding to the eigenvalue $\lambda_k^{(1)}$. Also we have

$$(-1)^{N_2+1} + 2\sum_{r=1}^{N_2} (-1)^r \exp\,[\sigma_k^{(1)} V_r] \cos\,[\tau_k^{(1)} V_r] = 0 \tag{26}$$

$$2\sum_{r=1}^{N_2} (-1)^r \exp\,[\sigma_k^{(1)} V_r] \sin\,[\tau_k^{(1)} V_r] = 0 \tag{27}$$

for those k such that $\sigma_k^{(1)} < 0$ and

$$(-1)^{N_2+1} + 2\sum_{r=1}^{N_2} (-1)^r \exp\,[\lambda_k^{(1)} V_r] = 0 \tag{28}$$

for those k such that $\lambda_k^{(1)} < 0$. The equations for the U_r's do not involve switching times after T_{N_1} since terms such as $\exp\,[-\sigma_k^{(1)} T_r/\varepsilon]$ tend to zero exponentially for $r > N_1$ and $\sigma_k^{(1)} > 0$. For a similar reason the equations for the V_r's do not involve other switching times.

Setting $t = \varepsilon t'$, we find the switching function in the initial boundary layer at $t = 0$ has the form

$$\sum_{\substack{k \\ \sigma_k^{(1)} > 0}} \exp\,[-\sigma_k^{(1)} t']\,[P_k \cos\,(\tau_k^{(1)} t') + Q_k \sin\,(\tau_k^{(1)} t')] + \sum_{\substack{k \\ \lambda_k^{(1)} > 0}} R_k \exp\,(-\lambda_k^{(1)} t') + p_n^{(0)}(0) \tag{29}$$

where $p_n^{(0)}(t)$ is the switching function for the reduced problem. The switching function in the terminal boundary layer at $t = T$ has a similar form, the sums being over those k corresponding to eigenvalues with negative real parts.

The modified switching times U_r satisfy Eqns (23), (24) and (25) and are zeros of the switching function (29) which has no other zeros. We see that switching in the initial boundary layer is associated with those roots of Eqn (22) having positive real parts and in the terminal boundary layer with those roots having negative real parts. If there are no roots with positive real parts there are no switches in the initial boundary layer and similarily for the terminal layer if there are no roots with negative real parts.

If all the roots of (22) are real and there are m_+ with positive real parts and m_- with negative real parts so that $m_+ + m_- = m$, it can be shown that there are at most m_+ switches in the initial layer and m_- in the terminal layer. For example, for the system

$$\frac{dx}{dt} = 2x + z_1, \qquad \varepsilon\frac{dz_1}{dt} = z_1 + z_2, \qquad \varepsilon\frac{dz_2}{dt} = x - z_2 + u$$

Eqn (16) is

$$\varepsilon^3\lambda^3 - 2\varepsilon^3\lambda^2 - \varepsilon\lambda + \varepsilon = 0$$

and to a first approximation has roots $\pm 1/\varepsilon$ and 1. The optimal control has at most one switch in the initial layer and exactly one in the terminal layer, there being no switches outside these layers.

The solution of the equations for the U_r's leads to restrictions on the initial values $x(0)$ and $z(0)$. Such restrictions are to be expected since, for instance, if Eqn (16) has positive real roots, not all initial states can be steered to the origin by bounded controls. In the example just given the reduced system

$$\frac{dx^{(0)}}{dt} = x^{(0)} - u^{(0)}$$

has the solution

$$x^{(0)}(t) = \Delta_0 + (x^{(0)}(0) - \Delta_0)\, e^t, \qquad 0 \leqslant t \leqslant T^{(0)}$$

where

$$\begin{aligned} \Delta_0 &= 1 \quad \text{for} \quad 0 < x^{(0)}(0) < 1 \\ &= -1 \quad \text{for} \quad -1 < x^{(0)}(0) < 0 \end{aligned}$$

and

$$T^{(0)} = \ln |\Delta_0/(\Delta_0 - x^{(0)}(0))|.$$

An optimal control does not exist for $|x^{(0)}(0)| \geqslant 1$. For the full system there is one switch in each boundary layer. For the initial layer U_1 satisfies

$$2e^{-U_1} = 1 - \Delta_0\,[x(0) + 2z_1(0) + z_2(0)]$$

and, since $U_1 > 0$, a solution of this equation is only possible if

$$|x(0) + 2z_1(0) + z_2(0)| < 1$$

which imposes a restriction on $x(0)$, $z_1(0)$ and $z_2(0)$ additional to the one on $x(0)$, that is $|x(0)| < 1$. The modified switching time V_1 is determined independently of the initial conditions and is given by $V_1 = \ln 2$. The corresponding switching time $T_2(\varepsilon) = T(\varepsilon) - \varepsilon V_1(\varepsilon)$ is known to $O(\varepsilon)$,h owever, only when $T(\varepsilon)$ is known to $O(\varepsilon)$ and is thus not independent of the initial conditions.

References

1. Kokotović, P. V. and Sannuti, P., Singular perturbation method for reducing the model order in optimal control design. *I.E.E.E. Trans. autom. Control* **13** (1968), 377–384.

2. Sannuti, P. and Kokotović, P. V., Near optimum design of linear systems by a singular perturbation method. *I.E.E.E. Trans. autom. Control* **14** (1969), 15–21.
3. Sannuti, P., Continuity and differentiability properties of optimal control with respect to singular perturbations. *I.E.E.E. Trans. autom. Control* **14** (1969), 762–763.
4. Sannuti, P. and Kokotović, P. V., Singular perturbation method for near optimum design of high-order non-linear systems. *Automatica* **5** (1969), 773–779.
5. Hadlock, C., Jamshidi, M. and Kokotović, P. V., Near optimal design of three-time scale systems. *In* "Proc. Fourth Ann. Princeton Conf. Information Sciences and Systems" (1970), 118–122.
6. Haddad, A. H. and Kokotović, P. V., Note on singular perturbation of linear state regulators. *I.E.E.E. Trans. autom. Control* **16** (1971), 279–281.
7. Kokotović, P. V. and Yackel, R. A., Singular perturbation of linear regulators: Basic theorems. *I.E.E.E. Trans. autom. Control* **17** (1972), 29–37.
8. Kelley, H. J., Singular perturbations for a Mayer variational problem. *AIAA Journal* **8** (1970), 1177–1178.
9. Kelley, H. J., Flight path optimization with multiple time scales. *Journal of Aircraft* **8** (1971), 238–240.
10. Kelley, H. J., Aircraft maneuver optimization by reduced-order approximation. *Advances in Control Systems* **9** (1972).
11. O'Malley Jr., R. E., Singular perturbation of the time-invariant linear state regulator problem. *J. Differential Equations* **12** (1972), 117–128.
12. O'Malley, Jr., R. E., The singularly perturbed linear state regulator problem. *SIAM J. Control* **10** (1972), 399–413.
13. O'Malley Jr., R. E., Topics in singular perturbations. *In* "Lectures on Ordinary Differential Equations", R. McKelvey (ed.). Academic Press, London and New York (1970).
14. Wasow, W., "Asymptotic Expansions for Ordinary Differential Equations". Interscience (1965).
15. Lee, E. B. and Markus, L., "Foundations of Optimal Control Theory". John Wiley, 1967.
16. Barnett, S. and Storey, C., "Matrix Methods in Stability Theory". Nelson, 1970.
17. American Automatic Control Council, Symposium on Singular Perturbations: Order Reduction in Control System Design. American Society of Mechanical Engineers (1972).

Discussion

Fuller (*University of Cambridge*). Your work has been concerned with open loop control. What happens if you have feedback control? Presumably the actual system will be unstable if operated with a controller designed on the basis of a reduced system.

Collins. I don't know. This would really arise in the work of Kokotović and his group. They require the matrix A to be stable and put on rather more stringent conditions than I do here. Without looking further into it I am not able to answer your question.

9. Trajectory Sensitivity Reduction in the Optimal Linear Regulator

P. J. FLEMING AND M. M. NEWMANN

Department of Engineering Mathematics
The Queen's University, Belfast, N. Ireland

The optimal linear regulator provides a convenient approach to the design of deterministic control systems with predictable trajectories. Such a design procedure requires knowledge of the plant parameters. However, in practice these may either be unknown or varying so that the system designed to be optimal for one set of parameter values may no longer be optimal for a different set of values. Hence any given trajectory may deviate considerably from the optimal trajectory with the same initial conditions (trajectory sensitivity).

Various authors have attempted to reduce the trajectory sensitivity to parameter variations of the optimal linear regulator by using design algorithms based on the idea of including a quadratic trajectory sensitivity term in the integrand of the cost functional, and specifying a linear feedback control comprised of state and trajectory sensitivity terms. Most of these methods neglect a second-order derivative term so that the trajectory sensitivity is only approximately modelled throughout the design algorithm and, in particular, the cost functional does not include an accurate measure of the trajectory sensitivity.

In this paper the actual trajectory sensitivity vector is included in the cost functional although of necessity the control uses the approximate sensitivity vector. The method also requires the introduction of two new state variable vectors so that the resulting augmented state equation is four times as large as the original state equation. In order to make the design of high-order systems computationally efficient the numerical procedures required for optimizing the augmented problem have been considerably refined.

The method is demonstrated using a first-order system and the resulting trajectory sensitivity is compared with that obtained by two other leading methods.

1. Introduction

The paper discusses modifications to the time-invariant infinite-duration optimal linear regulator problem designed to give reduced sensitivity to variations in a system parameter.

The straight-forward linear regulator problem is to determine the optimal feedback control vector, $\mathbf{u}^*(t)$, for the nth-order system

$$\dot{\mathbf{x}}(t) = A\mathbf{x}(t) + B\mathbf{u}(t), \qquad \mathbf{x}(0) = \mathbf{x}_0 \tag{1}$$

which minimizes the cost functional

$$J = \int_0^\infty (\mathbf{x}'Q\mathbf{x} + \mathbf{u}'R\mathbf{u})\, dt \tag{2}$$

where Q is an $n \times n$ symmetric, positive semi-definite matrix and R is an $r \times r$ positive definite matrix. A suitable choice of Q and R will lead to an optimal control, $\mathbf{u}^*(t)$, with the property of keeping the state vector $\mathbf{x}(t)$ near zero without using an excessive amount of control energy.

Consider the case when the matrices A and B are functions of a scalar time-invariant parameter α. The feedback control chosen to minimize the cost functional (2) when α takes a nominal value, α_{nom}, may no longer be optimal when α takes a different value and an increased cost might occur (cost sensitivity), and also any given trajectory may deviate considerably from the optimal trajectory with the same initial conditions (trajectory sensitivity). If the optimal regulator problem is used as a convenient design procedure for obtaining a system with predictable trajectories then it is more important to reduce trajectory sensitivity than cost sensitivity. This is the point of view taken in this paper.

One approach to the problem of reducing trajectory sensitivity is to choose a feedback control $\mathbf{u}(t)$ to minimize a modified cost functional which includes a sensitivity term. Two important algorithms based on this approach will be described and compared with a new one.

2. Basic Theory

Partially differentiating (1) with respect to α gives

$$\dot{\boldsymbol{\sigma}} = A_\alpha\mathbf{x} + A\boldsymbol{\sigma} + B_\alpha\mathbf{u} + B\frac{\partial \mathbf{u}}{\partial \alpha}, \qquad \boldsymbol{\sigma}(0) = 0 \tag{3}$$

where $A_\alpha = \partial A/\partial\alpha$, $B_\alpha = \partial B/\partial\alpha$ and $\boldsymbol{\sigma}$ denotes the trajectory sensitivity vector $\partial\mathbf{x}/\partial\alpha$.

Anticipating a linear feedback control law

$$\mathbf{u} = K_1\mathbf{x} + K_2\boldsymbol{\sigma} \tag{4}$$

so that

$$\frac{\partial \mathbf{u}}{\partial \alpha} = K_1\boldsymbol{\sigma} + K_2\frac{\partial \boldsymbol{\sigma}}{\partial \alpha}$$

Eqn (3) can be written

$$\dot{\boldsymbol{\sigma}} = (A_\alpha + B_\alpha K_1)\mathbf{x} + (A + BK_1 + B_\alpha K_2)\boldsymbol{\sigma} + BK_2\frac{\partial \boldsymbol{\sigma}}{\partial \alpha}, \qquad \boldsymbol{\sigma}(0) = 0. \tag{5}$$

Neglecting the $\partial\boldsymbol{\sigma}/\partial\alpha$ term (which cannot be found except by differentiating (5) with respect to α, so introducing an unknown $\partial^2\boldsymbol{\sigma}/\partial\alpha^2$ term and so on, *ad infinitum*) when solving (5) will lead to an approximate trajectory sensitivity vector which we denote by $\boldsymbol{\rho}(t)$. Thus we are led to the idea of using a linear feedback control

$$\mathbf{u} = K_1\mathbf{x} + K_2\boldsymbol{\rho} \tag{6}$$

so that on substituting (6) into (1) we have

$$\dot{\mathbf{x}} = (A + BK_1)\mathbf{x} + BK_2\boldsymbol{\rho}, \quad \mathbf{x}(0) = \mathbf{x}_0 \tag{7}$$

where $\boldsymbol{\rho}(t)$ is obtained from the differential equation

$$\dot{\boldsymbol{\rho}} = (A_\alpha + B_\alpha K_1)\mathbf{x} + (A + BK_1 + B_\alpha K_2)\boldsymbol{\rho}, \quad \boldsymbol{\rho}(0) = 0. \tag{8}$$

The solution of Eqn (8) will only approximate to the trajectory sensitivity vector, $\boldsymbol{\sigma}(t)$, which is given by

$$\dot{\boldsymbol{\sigma}} = (A_\alpha + B_\alpha K_1)\mathbf{x} + B_\alpha K_2\boldsymbol{\rho} + (A + BK_1)\boldsymbol{\sigma} + BK_2\frac{\partial \boldsymbol{\rho}}{\partial \alpha}, \qquad \boldsymbol{\sigma}(0) = 0 \tag{9}$$

where

$$\frac{\partial \dot{\boldsymbol{\rho}}}{\partial \alpha} = (A_\alpha + B_\alpha K_1)\boldsymbol{\sigma} + (A + BK_1 + B_\alpha K_2)\frac{\partial \boldsymbol{\rho}}{\partial \alpha}, \qquad \frac{\partial \boldsymbol{\rho}}{\partial \alpha}(0) = 0. \tag{10}$$

A closer agreement between $\boldsymbol{\rho}(t)$ and $\boldsymbol{\sigma}(t)$ can be obtained, if so desired, by modelling $\boldsymbol{\rho}(t)$ by the equation

$$\dot{\boldsymbol{\rho}} = (A_\alpha + B_\alpha \tilde{K}_1)\mathbf{x} + (A + B\tilde{K}_1 + B_\alpha \tilde{K}_2)\boldsymbol{\rho}, \qquad \boldsymbol{\rho}(0) = 0 \tag{11}$$

so that

$$\frac{\partial \dot{\boldsymbol{\rho}}}{\partial \alpha} = (A_\alpha + B_\alpha \tilde{K}_1)\boldsymbol{\sigma} + (A + B\tilde{K}_1 + B_\alpha \tilde{K}_2)\frac{\partial \boldsymbol{\rho}}{\partial \alpha}, \qquad \frac{\partial \boldsymbol{\rho}}{\partial \alpha}(0) = 0 \tag{12}$$

where $\tilde{K}_1$ and $\tilde{K}_2$ can be chosen to make $\boldsymbol{\sigma}(t)$ and $\boldsymbol{\rho}(t)$ as close as desired.

Equations (7), (8), (9), (10), (11) and (12) form the basis of all of the three methods of trajectory sensitivity reduction to be considered.

3. Kreindler's Method [4]

This method is based upon Eqns (7) and (8) which may be written in the augmented form (of order 2n)

$$\dot{\mathbf{v}} = \begin{bmatrix} A & 0 \\ A_\alpha & A + BK_1 \end{bmatrix} \mathbf{v} + \begin{bmatrix} B \\ B_\alpha \end{bmatrix} [K_1 \; K_2] \, \mathbf{v}$$

or

$$\dot{\mathbf{v}} = A_1 \mathbf{v} + B_1 \mathbf{u}, \quad \mathbf{v}(0) = \mathbf{v}_0 \tag{13}$$

where

$$\mathbf{v} = [\mathbf{x}', \boldsymbol{\rho}']', \qquad A_1 = \begin{bmatrix} A & 0 \\ A_\alpha & A + BK_1 \end{bmatrix} \quad \text{and} \quad B_1 = \begin{bmatrix} B \\ B_\alpha \end{bmatrix}$$

since

$$\mathbf{u} = [K_1 \; K_2] \mathbf{v}.$$

The cost functional (2) is modified to include a trajectory sensitivity term, thus

$$J_1 = \int_0^\infty (\mathbf{x}' Q_1 \mathbf{x} + \boldsymbol{\rho}' Q_2 \boldsymbol{\rho} + \mathbf{u}' R \mathbf{u}) \, dt$$

where Q_1 and Q_2 are $n \times n$ symmetric positive semi-definite matrices, and may be written

$$J_1 = \int_0^\infty (\mathbf{v}' \bar{Q} \mathbf{v} + \mathbf{u}' R \mathbf{u}) \, dt \tag{14}$$

where

$$\bar{Q} = \begin{bmatrix} Q_1 & 0 \\ 0 & Q_2 \end{bmatrix}.$$

Therefore we have the specific optimal control problem of choosing K_1 and K_2 in Eqn (6) so as to minimize the cost functional (14), when $\alpha = \alpha_{\text{nom}}$, subject to the differential constraint (13). This problem is no longer in the standard linear regulator form because of the dependence of A_1 upon K_1 and Newmann [5] has shown that it is impossible to obtain constant matrices K_1 and K_2 which will enable $\mathbf{u}(t)$ to minimize J_1 for any initial conditions on $\mathbf{x}(t)$.

If the dependence of A_1 upon K_1 was absent the control which minimizes the cost functional (14) for any initial conditions would, by standard linear regulator theory, be given by

$$\mathbf{u} = K\mathbf{v} = [K_1\ K_2]\mathbf{v}$$

where

$$K = -R^{-1}B_1{}'P \tag{15}$$

and P is the symmetric positive definite solution of

$$PA_1 + A_1{}'P - PB_1R^{-1}B_1{}'P + \bar{Q} = 0. \tag{16}$$

Motivated by this approach Kreindler arbitrarily chooses for his feedback matrices K_1 and K_2 those values which satisfy (15) and (16), in spite of the fact that, because A_1 contains K_1, such a choice of K_1 and K_2 will not, in general, minimize J_1. Nevertheless, in practice the method seems quite successful in reducing trajectory sensitivity and, moreover, has the advantage that the design is independent of initial conditions of $\mathbf{x}(t)$.

4. Rao and Soudack's Method [8]

If minimization of the cost functional (14) is to ensure reduction of trajectory sensitivity, $\boldsymbol{\rho}(t)$ should closely approximate $\boldsymbol{\sigma}(t)$. A possible modification of the above method would be to use Eqn (11) instead of Eqn (8) to model trajectory sensitivity, with a suitable choice of $\tilde{K}_1$ and $\tilde{K}_2$. Kreindler [3] suggested a procedure along these lines but left many design details unresolved.

Rao and Soudack have considered a generalisation of the above approach and have devised a detailed design algorithm. Their method replaces the trajectory sensitivity model Eqn (11) by the new model equation

$$\dot{\boldsymbol{\rho}} = C\mathbf{x} + D\boldsymbol{\rho}, \qquad \boldsymbol{\rho}(0) = 0 \tag{17}$$

so that (12) is replaced by

$$\frac{\partial \dot{\boldsymbol{\rho}}}{\partial \alpha} = C\boldsymbol{\sigma} + D\frac{\partial \boldsymbol{\rho}}{\partial \alpha}, \qquad \frac{\partial \boldsymbol{\rho}}{\partial \alpha}(0) = 0 \tag{18}$$

where C and D are matrices containing as many arbitrary parameters as desired. Hence the $2n$th order augmented system becomes

$$\dot{\mathbf{v}} = A_1\mathbf{v} + B_1\mathbf{u}, \quad \mathbf{v}(0) = \mathbf{v}_0$$

where now

$$A_1 = \begin{bmatrix} A & 0 \\ C & D \end{bmatrix} \quad \text{and} \quad B_1 = \begin{bmatrix} B \\ 0 \end{bmatrix}.$$

Then for some initial choice of the parameters of C and D the cost functional

(14) is minimized by the K which is found by solving (15) and (16), where A_1 is given by this new expression.

For this optimal K the parameters in C and D are now adjusted using a gradient technique to decrease the scalar J_2, where

$$J_2 = \int_0^\infty \|\boldsymbol{\rho}(t) - \boldsymbol{\sigma}(t)\|^2 \, dt$$

and $\|(.)\|$ refers to the Euclidean norm. To do this it is necessary to integrate the simultaneous differential Eqns (7), (9), (17) and (18) for some particular initial conditions. All the above calculations are made with $\alpha = \alpha_{\text{nom}}$.

Using the new values of C and D in A_1 the optimal K can be recalculated and the procedure repeated until J_2 is reduced to some acceptable level. Clearly this method could be implemented using Eqns (11) and (12) in place of (17) and (18) as suggested by Kreindler.

The algorithm has the disadvantage that the iterative procedure need not necessarily converge and also that the design is dependent on the initial conditions of $\mathbf{x}(t)$.

5. A new method

We propose that Eqns (7), (8), (9) and (10) be used in a different way to represent a 4nth order dynamical system in which K_1 and K_2 are chosen to minimize the cost functional.

$$J_3 = \int_0^\infty (\mathbf{x}'Q_1\mathbf{x} + \boldsymbol{\sigma}'Q_2\boldsymbol{\sigma} + \mathbf{u}'R\mathbf{u}) \, dt \tag{19}$$

when $\alpha = \alpha_{\text{nom}}$. Here the exact trajectory sensitivity, $\boldsymbol{\sigma}(t)$, is included in the cost functional but, for practical reasons, the linearly modelled trajectory sensitivity vector $\boldsymbol{\rho}(t)$ is retained in the feedback control which is given by Eqn (6).

Clearly Eqns (8) and (10) can be replaced in the design procedure by the alternative Eqns (11) and (12) or (17) and (18) so that more design parameters ($\tilde{K}_1$, $\tilde{K}_2$ or C, D) are available for optimal adjustment along with K_1 and K_2. However, due to the presence of $\boldsymbol{\sigma}(t)$ in the cost functional (19) it is not as essential to constrain $\boldsymbol{\rho}(t)$ and $\boldsymbol{\sigma}(t)$ to near equality as it is in the method of Rao and Soudack, therefore Eqns (8) and (10) can usually be used to model $\boldsymbol{\rho}(t)$ and $\partial\boldsymbol{\rho}(t)/\partial\alpha$.

Nevertheless, $\boldsymbol{\rho}(t)$ enters J_3 through $\mathbf{u}(t)$ and should $\partial\boldsymbol{\rho}(t)/\partial\alpha$ be large enough for $\boldsymbol{\rho}(t)$ and $\boldsymbol{\sigma}(t)$ to be appreciably different it is possible that optimal choice of K_1 and K_2 would have insufficient effect in reducing sensitivity. In this case, the introduction of extra design parameters in the modelling of

$\boldsymbol{\rho}(t)$ and $\partial\boldsymbol{\rho}(t)/\partial\alpha$ by using the alternative equations should lead to a better system design.

Experience has shown, however, that when $\boldsymbol{\rho}(t)$ and $\boldsymbol{\sigma}(t)$ are not appreciably different the extra computational labour involved in this more general approach is not justified by the additional reduction of sensitivity.

Equations (7), (8), (9) and (10) can be written in augmented form

$$\dot{\ } = \bar{A}\mathbf{v}, \quad \mathbf{v}(0) = \mathbf{v}_0 \tag{20}$$

where

$$\mathbf{v} = \left[\mathbf{x}' \ \boldsymbol{\rho}' \ \boldsymbol{\sigma}' \ \frac{\partial\boldsymbol{\rho}'}{\partial\alpha}\right]'$$

and

$$\bar{A} = \begin{bmatrix} A + BK_1 & BK_2 & 0 & 0 \\ A_\alpha + B_\alpha K_1 & A + BK_1 + B_\alpha K_2 & 0 & 0 \\ A_\alpha + B_\alpha K_1 & B_\alpha K_2 & A + BK_1 & BK_2 \\ 0 & 0 & A_\alpha + B_\alpha K_1 & A + BK_1 + B_\alpha K_2 \end{bmatrix}$$

and the cost functional (19) can be rewritten

$$J_3 = \int_0^\infty \mathbf{v}'(\bar{Q} + E_1'K'RKE_1)\mathbf{v}\, dt \tag{21}$$

$$\text{where } \bar{Q} = \begin{bmatrix} Q_1 & 0 & 0 & 0 \\ 0 & 0 & 0 & 0 \\ 0 & 0 & Q_2 & 0 \\ 0 & 0 & 0 & 0 \end{bmatrix}, \quad E_1 = [I_{2n}\ 0_{2n}] \quad \text{and}$$

$K = [K_1\ K_2]$. The procedure is to choose K to minimize the cost functional (21) subject to the differential constraint (20).

This problem is not of the standard linear regulator form and various computational methods have been devised for obtaining the optimal matrix K. These will be described in the next section.

6. Computational Procedure

It is well known that, provided K is chosen so that the system (20) is asymptotically stable, the cost functional (21) can be written

$$J_3 = \mathbf{v}_0'P\mathbf{v}_0 = \mathrm{tr}\{P\mathbf{v}_0\mathbf{v}_0'\} = \mathrm{tr}\{PV_0\} \tag{22}$$

where P is the symmetric positive semi-definite matrix solution of the Lyapunov matrix equation

$$P\bar{A} + \bar{A}'P + \bar{Q} + E_1'K'RKE_1 = 0. \tag{23}$$

Since $\boldsymbol{\rho}(0) = \boldsymbol{\sigma}(0) = \partial\boldsymbol{\rho}(0)/\partial\alpha = 0$ is the obvious choice for the initial conditions

$$V_0 = \begin{bmatrix} X_0 & 0 & 0 & 0 \\ 0 & 0 & 0 & 0 \\ 0 & 0 & 0 & 0 \\ 0 & 0 & 0 & 0 \end{bmatrix}$$

where

(i) $X_0 = \mathbf{x}_0\mathbf{x}_0'$, if $\mathbf{x}_0$ is known, or

(ii) $X_0 = E\{\mathbf{x}_0\mathbf{x}_0'\}$, where $E\{\mathbf{x}\}$ is the expected value of the random variable $\mathbf{x}$, if there is statistical knowledge of $\mathbf{x}_0$, or

(iii) $X_0 = I$; this is a special case of (ii) which assumes that $\mathbf{x}_0$ is a random variable uniformly distributed over the surface of an n-dimensional unit sphere and is used when there is no *a priori* knowledge of $\mathbf{x}_0$.

Thus the problem can be restated as that of choosing the feedback matrix K to minimize the cost function (22) subject to the algebraic constraint (23).

Procedure 1

This is the most direct approach to the numerical solution of the above problem. Through Eqn (23) P is implicitly a function of K so that the cost function J_3 (22) can be regarded as a function of K. Therefore Powell's conjugate direction method [7] for minimizing a function of several variables can be used to find the optimum K.

A numerical procedure which will solve the Lyapunov matrix equation (23) for P and hence evaluate J_3 for any given value of K has to be incorporated into the program. Investigation has shown that the series solution method devised by Smith [9] is one of the best available for solving large-order Lyapunov matrix equations. In fact, for systems of up to fourth-order further simplification is possible because of the special structure of $\bar{A}$, namely,

$$\bar{A} = \begin{bmatrix} M & 0 \\ N & M \end{bmatrix},$$

where

$$M = \begin{bmatrix} A + BK_1 & BK_2 \\ A_\alpha + B_\alpha K_1 & A + BK_1 + B_\alpha K_2 \end{bmatrix}$$

and

$$N = \begin{bmatrix} A_\alpha + B_\alpha K_1 & B_\alpha K_2 \\ 0 & 0 \end{bmatrix}.$$

Since K must be chosen so that the system (20) is asymptotically stable, the eigenvalues of the matrix $\bar{A}$, corresponding to each trial choice of K in the minimization procedure, must have negative real parts. If this condition is violated the choice of K can be penalized by assigning a large value to the corresponding J_3 so that the search for an optimal K continues in a new direction. Again the procedure may be simplified because owing to the special structure of $\bar{A}$ it is only necessary to investigate the eigenvalues of M each time since they are the same as for $\bar{A}$.

This numerical procedure has worked well on all the problems with which it has been tested.

Procedure 2

An alternative procedure for solving the above type of minimization problem, which has been documented in [2], is based upon the Lagrange multiplier approach to constrained minimization problems. Following the method described in [6], it can be shown that necessary conditions for K and P to minimize the cost function (22) are given by the equations

$$P\bar{A} + \bar{A}'P + \bar{Q} + E_1'K'RKE_1 = 0 \tag{24}$$

$$\Lambda\bar{A}' + \bar{A}\Lambda + V_0 = 0 \tag{25}$$

and

$$RKE_1\Lambda E_1' + B_1'P\Lambda E_1' + B_2'P\Lambda E_2' + B_3'P\Lambda E_3' + B_4'P\Lambda E_4' = 0 \tag{26}$$

where Λ is a $4n \times 4n$ matrix of constant Lagrange multipliers and

$$B_1 = [B' \; B_\alpha' \; B_\alpha' \; 0]', \quad B_2 = [0 \; 0 \; B' \; B_\alpha']', \quad B_3 = [0 \; B' \; 0 \; 0]', \quad B_4 = [0 \; 0 \; 0 \; B']',$$

$$E_2 = [0_{2n} \; I_{2n}], \quad E_3 = \begin{bmatrix} 0 & I_n & 0 & 0 \\ 0 & 0 & 0 & 0 \end{bmatrix} \quad \text{and} \quad E_4 = \begin{bmatrix} 0 & 0 & 0 & I_n \\ 0 & 0 & 0 & 0 \end{bmatrix}.$$

These equations can be solved for stationary values of K and P by the procedures described in [2]. A modification of the procedure in which Smith's series method is used to solve the two Lyapunov matrix Eqns (24) and (25) in place of Bingulac's approach [1] increases the efficiency of numerical solution for large-order Lyapunov matrix equations. This procedure, however, is less direct than procedure 1 and requires more computing time.

Procedure 3

An attractive iterative algorithm follows from Eqns (24)–(26). Equation (26) can be written

$$K_{i+1} = -R^{-1}(B_1'P_i\Lambda_i E_1' + B_2'P_i\Lambda_i E_2' + B_3'P_i\Lambda_i E_3' + B_4'P_i\Lambda_i E_4')(E_1\Lambda_i E_1')^{-1};$$

P_i and Λ_i are found from

$$P_i\bar{A} + \bar{A}'P_i + \bar{Q} + E_1'K_i'RK_iE_1 = 0$$

$$\Lambda_i\bar{A}' + \bar{A}\Lambda_i + V_0 = 0$$

where $\bar{A} = A_0 + B_1K_iE_1 + B_2K_iE_2 + B_3K_iE_3 + B_4K_iE_4$ and

$$A_0 = \begin{bmatrix} A & 0 & 0 & 0 \\ A_\alpha & A & 0 & 0 \\ A_\alpha & 0 & A & 0 \\ 0 & 0 & A_\alpha & A \end{bmatrix}.$$

Convergence of this method is not guaranteed in every case but in two examples tested, one first-order and one second-order system, K was found to the required accuracy after about eight iterations.

7. Example

(i) The straight-forward optimal regulator design (algorithm 1) is compared with the new design (algorithm 2).

Consider the following first-order system:

$$\dot{x} = \alpha^2 x + u$$

where $x(0) = 1$ and $\alpha_{nom} = 1$.

Algorithm 1: The control which minimizes the cost functional

$$J = \int_0^\infty (x^2 + u^2)\, dt$$

is $u = -2{\cdot}414x$ when $\alpha = 1$. Fig. 1 shows the system response using this control when $\alpha = 0{\cdot}5$, $1{\cdot}0$ and $1{\cdot}5$. The trajectory sensitivity, σ_1, is shown in Fig. 2 and is evaluated for $\alpha = 1$.

Algorithm 2: Applying the method proposed in this paper, the augmented system equations are

$$\dot{x} = (\alpha^2 + k_1)x + k_2\rho, \qquad x(0) = 1$$

$$\dot{\rho} = 2x + (1 + k_1)\rho, \qquad \rho(0) = 0$$

$$\dot{\sigma} = 2\alpha x + (\alpha^2 + k_1)\sigma + k_2\frac{\partial\rho}{\partial\alpha}, \quad \sigma(0) = 0$$

and

$$\frac{\partial \dot{\rho}}{\partial \alpha} = 2\sigma + (1 + k_1)\frac{\partial \rho}{\partial \alpha}, \quad \frac{\partial \rho}{\partial \alpha}(0) = 0.$$

Then the control which minimizes the cost functional

$$J_3 = \int_0^\infty (x^2 + s\sigma^2 + u^2)\, dt$$

for $s = 1$, $\alpha = 1$ is $u = -2{\cdot}777x - 0{\cdot}214\rho$. Figure 1 shows the system response using this control when $\alpha = 0{\cdot}5$, $1{\cdot}0$ and $1{\cdot}5$. The trajectory sensitivity, σ_2, evaluated for $\alpha = 1$, is shown along with the modelled sensitivity, ρ_2, in Fig. 2.

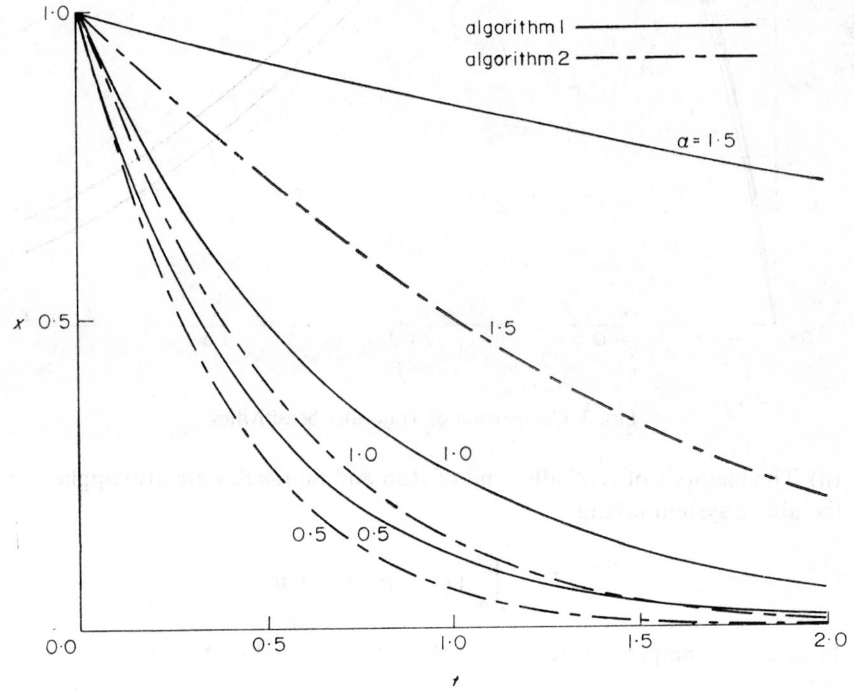

FIG. 1. System Response (Algorithm 1 and Algorithm 2)

Comparing the two sets of curves in Fig. 1, it is found that the system responses for $\alpha = 0{\cdot}5$ and $1{\cdot}5$ deviate less from the nominal trajectory ($\alpha_{nom} = 1$) for algorithm 2. Figure 2 shows a considerable reduction in trajectory sensitivity when algorithm 2 is employed. Trajectory sensitivity is reduced even more if larger values of s are taken in the cost functional J_3.

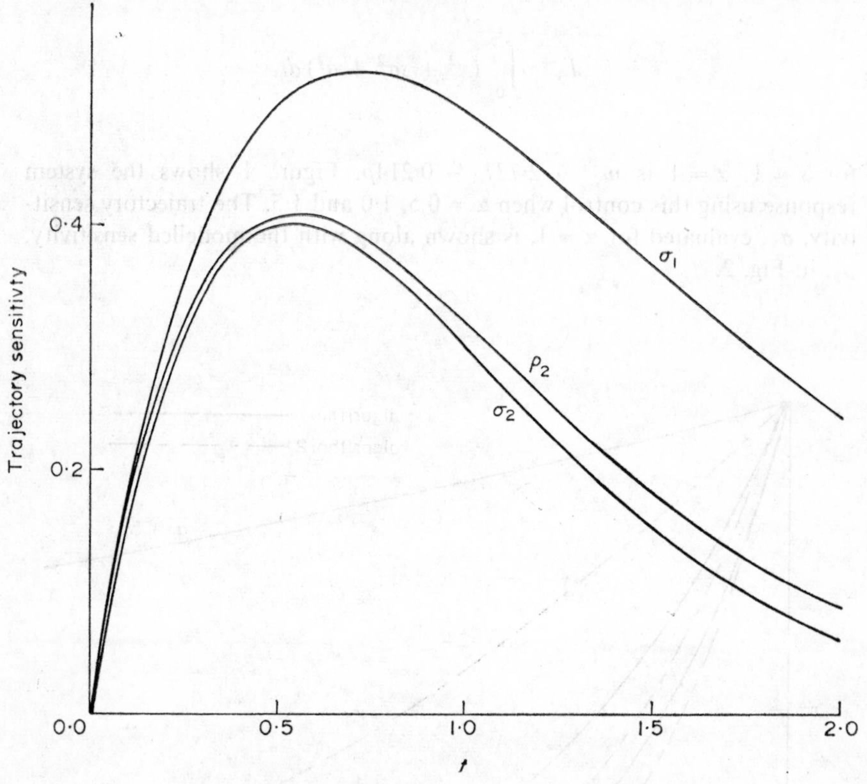

FIG. 2. Comparison of Trajectory Sensitivities

(ii) The methods of Kreindler and of Rao and Soudack were also applied to the above system taking

$$J_1 = \int_0^\infty (x^2 + \rho^2 + u^2)\, dt.$$

In order to compare all four algorithms the original cost

$$J = \int_0^\infty (x^2 + u^2)\, dt$$

was evaluated along with the sensitivity integral

$$S = \int_0^\infty \sigma^2 \, dt$$

for each controller design. S is used as a measure of overall trajectory sensitivity. Results are shown in the table.

	Algorithm 1	Algorithm 2	Kreindler	Rao and Soudack
J	2·414	2·478	2·450	2·443
S	0·354	0·140	0·176	0·189

It can be seen that all the sensitivity reduction algorithms work and that any decrease in sensitivity results in a slight increase in the original cost, J. Surprisingly, for the example treated, Rao and Soudack's algorithm is not as effective as that of Kreindler. This is mainly due to numerical difficulties arising from their method and perhaps also to the choice of example. Algorithm 2 gives the best result with a 60% decrease in S for a 2·7% increase in J. Curves, similar to those given in Figs. 1 and 2, were also obtained for these other algorithms and these show the same trends as indicated by the table.

8. Conclusions

The algorithms have been applied to a second-order problem and trajectory sensitivity was reduced in each case. Kreindler's method and the algorithm proposed in this paper are simpler to apply than Rao and Soudack's method and gave better results for the examples considered. The new method, although more dependent on initial conditions than Kreindler's method, gave the best overall sensitivity reduction and has the additional feature that it can be easily modified for the case when some of the states are not measurable. This is not possible for the other methods since their dependence on the matrix Riccati equation presupposes that all the states are present.

References

1. Bingulac, S. P., *I.E.E.E. Trans. autom. Control* **AC-15** (1970), 135–137.
2. Fleming, P. J. and Newmann, M. M., *Electronics Lett.* **8** (1972), 50–52.
3. Kreindler, E., *Int. J. Control* **8** (1968), 89–96.
4. Kreindler, E., *I.E.E.E. Trans. autom. Control* **AC-14** (1969), 206–207.
5. Newmann, M. M., *Int. J. Control* **11** (1970), 1079–1084.
6. Newmann, M. M., *Electronics Lett.* **6** (1970), 360–362.
7. Powell, M. J. D., *Comput. J.* (1965), 303–307.

8. Rao, S. G. and Soudack, A. C., *I.E.E.E. Trans. autom. Control* **AC-16** (1971), 194–196.
9. Smith, R. A., *SIAM J. Appl. Math.* **16** (1968), 198–201.

Discussion

HEALEY (*Cardiff*). In your example, I see from Fig. 1 that the system response using algorithm 2 is better than the response for the optimal control case (algorithm 1) when α takes its nominal value (i.e. $\alpha = 1$). Can you explain this?

FLEMING. The reason is that, for the particular cost function chosen, using algorithm 2 to design the control leads to a higher feedback gain.

HEALEY. In other words you are getting more control out of it. Surely the comparison you must make is between algorithms which give rise to the same control effort.

FLEMING. There are many possible ways of comparing these two procedures and we have just demonstrated that trajectory sensitivity can be reduced by a systematic approach. You would like to place more emphasis on the relative control magnitudes.

HEALEY. I think this is the most important basis for comparison. Have you made a comparison on these lines?

FLEMING. Yes. We checked the response with a control designed by algorithm 2 in which the maximum control magnitude was equal to the maximum control magnitude resulting from the standard optimal case (algorithm 1) and we still got an improved sensitivity response. However, because the gain was lower the sensitivity reduction was not so great.

HEALEY. Have you tried your method on higher order systems?

FLEMING. Only up to 3rd-order systems.

BARNETT (*University of Bradford*). Many workers in sensitivity reduction attemp to demonstrate the superiority of their particular method by applying it to some specific example. It is possible that a different choice of example could lead to the opposite conclusion. How many examples have you tried? Is there a class of examples for which your algorithm doesn't offer an improvement?

FLEMING. We haven't found such a class of problems. We have tried the algorithm on one first-, three second-, and one third-order system and in each case an improvement in sensitivity has resulted.

NEWMANN. We have tended to treat examples that other people have treated. Obviously such an approach cannot be a proof that an algorithm always works. All we can say is that here is an alternative design procedure, we've tried it and it seems to work. As has been pointed out, it depends on the criterion we use just how good it is. The choice of criterion depends on what you are trying to do.

FLEMING. Referring to our method, further improvement can be made by including the matrices C and D in the sensitivity model equation. But obviously the size of the problem gets a bit out of hand because of the number of extra parameters in C and D.

NEWMANN. Of course it will be more expensive to use methods like this because you must have a dynamic feedback. In any particular case it will have to be decided whether the decrease in sensitivity is worth the extra expense.

10. Domains of Controllability

M. J. DAVIES

Department of Applied Mathematics,
University College of Wales,
Aberystwyth, Wales

1. Introduction

This paper is a light-hearted pictorial presentation of some results procured during a numerical investigation of certain features associated with the Duffing oscillator,

$$\ddot{y} + y + \varepsilon y^3 = u(t), \qquad |u| \leqslant 1, \tag{1}$$

in the case of time optimal, origin seeking problems.

In the state space of (x_1, x_2), where

$$x_1 = y, \qquad x_2 = \dot{y}, \tag{2}$$

the system equations are

$$\dot{x}_1 = x_2, \qquad \dot{x}_2 = u - x_1 - \varepsilon x_1{}^3 \tag{3}$$

and one need labour no longer the usual procedure in these problems. Two overall features of the solution are of interest:

(A) *The Switching Curve.* This is the line in the state space which defines the synthesis or control policy. It is the union of the locus of switching points together with the lines of indifference [1] or separatrices [2], and is more precisely defined (since this will include features as yet unnoticed) as the locus of points at which optimal trajectories fail to have derivatives [3].

(B) *T-Controllable Curves.* The set of points in the state space from which the minimum time to the origin (along an optimal path) is exactly *T*, is the *T*-controllable curve. Intuitively, these will be closed curves surrounding the origin; their interiors, the *T*-controllable domains, being the set of states controllable to the origin in some lesser time. They may also be regarded as the level surfaces of minimal time, or as the solutions to the Bellman equation for such a problem. These curves are also called isochrones in the interests of brevity. When the isochrone develops a re-entrant spur, the whole curve is called a pseudo–isochrone, its outer boundary retaining the name isochrone.

Attention is mainly directed at the way in which the two curves A and B interact with each other when one blindly examines *all* time extremal trajectories to the origin, secure in the knowledge that the global minimum, if it exists, is somewhere amongst them.

2. The Results

The switching curve and isochrones for the system

$$\ddot{y} = u, \qquad |u| \leqslant 1 \tag{4}$$

and the simple harmonic oscillator

$$\ddot{y} + y = u, \qquad |u| \leqslant 1 \tag{5}$$

are probably familiar. In both these cases, the features are well-behaved objects and satisfy one's intuitive ideas. Figures 1 and 2 show five lobes of

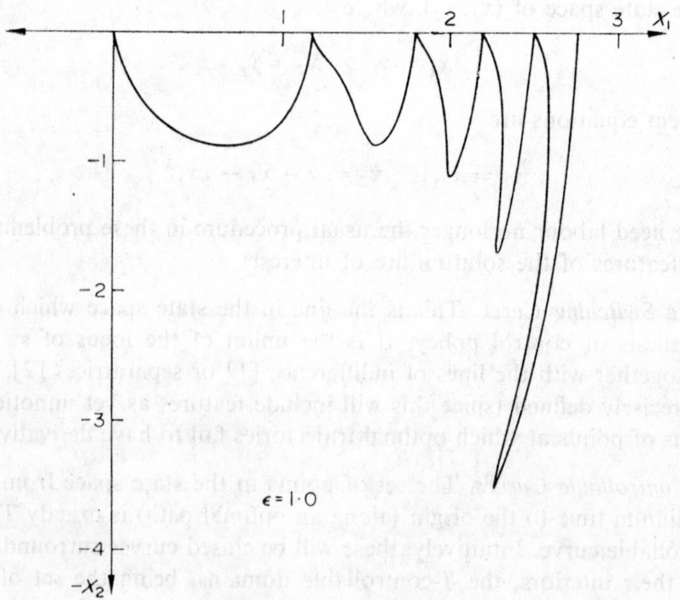

FIG. 1. Five lobes of the switching curve for the Duffing oscillator, $\varepsilon = 1{\cdot}0$.

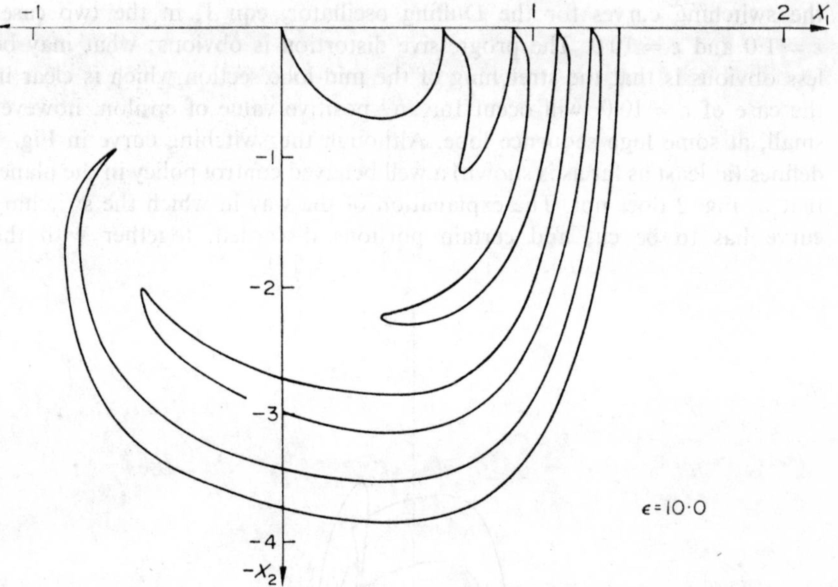

FIG. 2. Five lobes of the switching curve for the Duffing oscillator, $\varepsilon = 10{\cdot}0$.

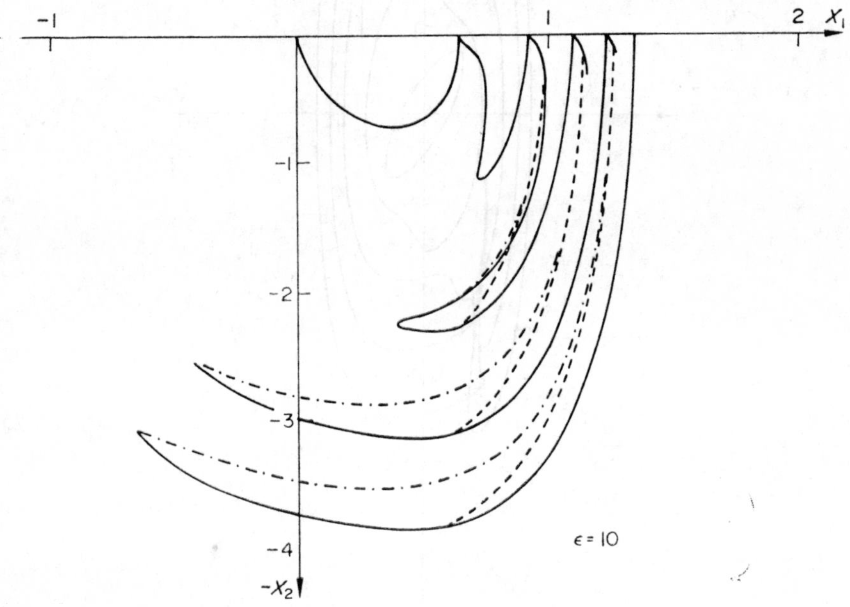

FIG. 3. Swept lobes, $\varepsilon = 10{\cdot}0$

the switching curves for the Duffing oscillator, eqn 1, in the two cases $\varepsilon = 1{\cdot}0$ and $\varepsilon = 10{\cdot}0$. The progressive distortion is obvious; what may be less obvious is that the stretching of the mid-lobe section which is clear in the case of $\varepsilon = 10{\cdot}0$, will occur for any positive value of epsilon, however small, at some high sequence lobe. Although the switching curve in Fig. 1 defines (at least as far as is shown) a well behaved control policy in the plane, that in Fig. 2 does not. The explanation of the way in which the switching curve has to be cut and certain portions discarded, together with the

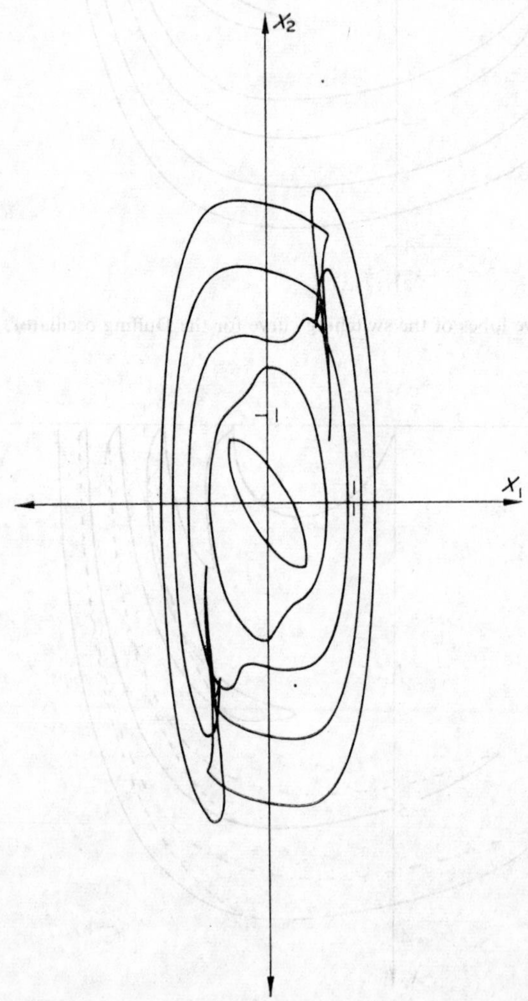

FIG. 4. Isochrones for the Duffing oscillator, $\varepsilon = 10{\cdot}0$, for $T = 1(1)5$.

construction of separatrices, has been given elsewhere [2, 4] and connects closely with the work presented in this volume by Dr. Mirica in his paper on stratified sets. Figure 3 shows the switching curve for $\varepsilon = 10$ after this process, known as lobe sweeping, has clarified the situation. However it is of interest to retain the whole switching curve and to follow the effect this retention has on the isochrones. The level surfaces of extremal time are then shown in the figures.

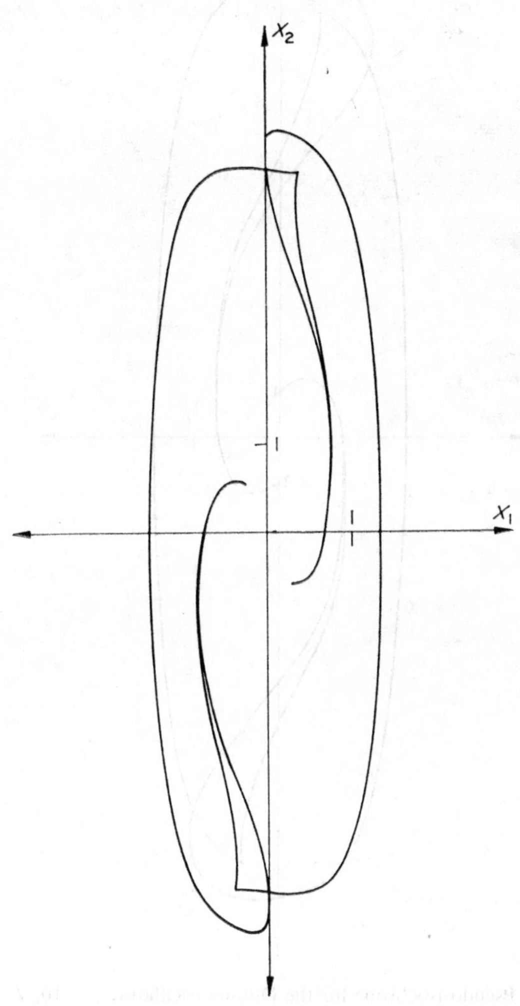

FIG. 5. Pseudo-isochrone for the Duffing oscillator, $\varepsilon = 10$, $T = 6$.

Figure 4 shows the isochrones for $T = 1$ time unit, by steps of 1 until 5 time units, while Figs 5, 6, and 7 show isolated pseudo isochrones for $T = 6$, 7 and 8 respectively.

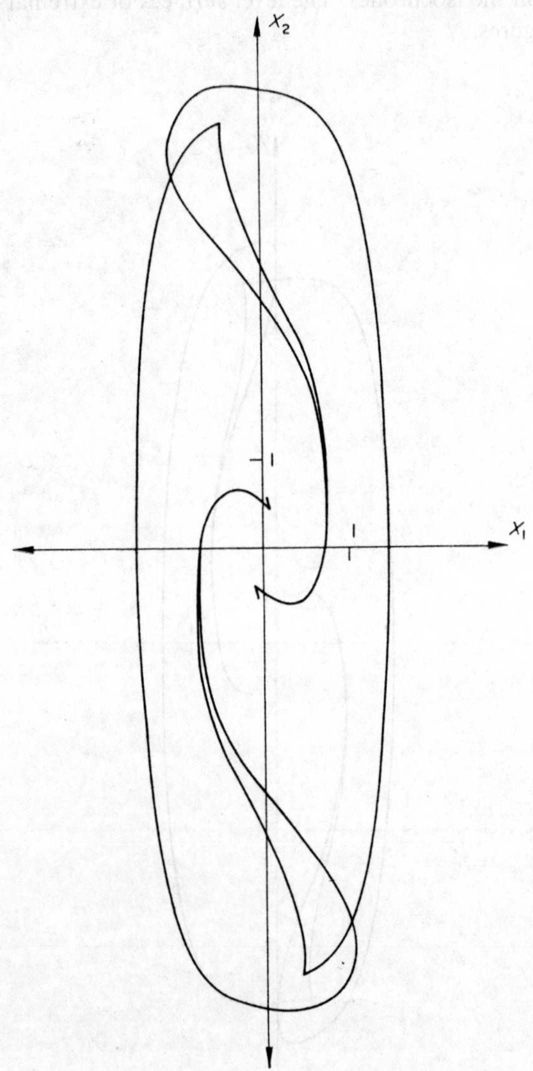

FIG. 6. Pseudo-isochrone for the Duffing oscillator, $\varepsilon = 10$, $T = 7$.

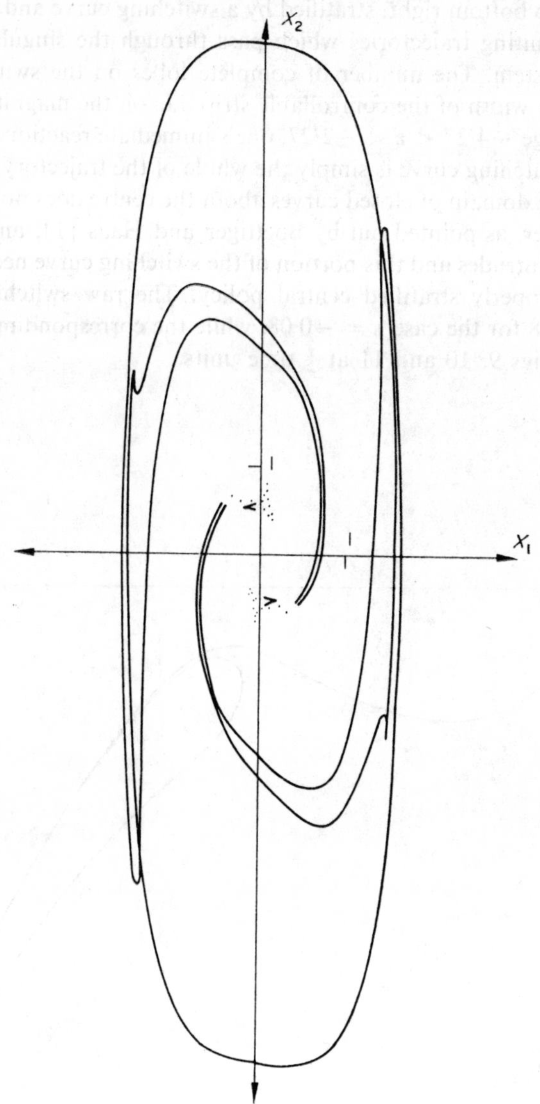

FIG. 7. Pseudo-isochrone for the Duffing oscillator, $\varepsilon = 10$, $T = 8$.

In the case of the Soft Duffing oscillator, epsilon negative, the situation is generally much simpler. The controllable domain—that subset of the state space which can be controlled to the origin for *some* T—is a strip running from top left to bottom right, stratified by a switching curve and bounded by appropriate limiting trajectories which pass through the singular points of the Duffing system. The number of complete lobes on the switching curve depends on the width of the controllable strip, i.e. on the magnitude of epsilon. In the range $-4/27 < \varepsilon < -2/27$, one's immediate reaction would be to say that the switching curve is simply the whole of the trajectory through the origin since the domain of closed curves about the centre does not include the origin. However, as pointed out by Boettiger and Haas [1], an anomalous second switch intrudes and this portion of the switching curve needs to be cut to define a properly stratified central policy. The raw switching curve is shown in Fig. 8 for the case $\varepsilon = -0{\cdot}08$, while the corresponding isochrones are given in Figs 9, 10 and 11 at $\frac{1}{2}$ time units.

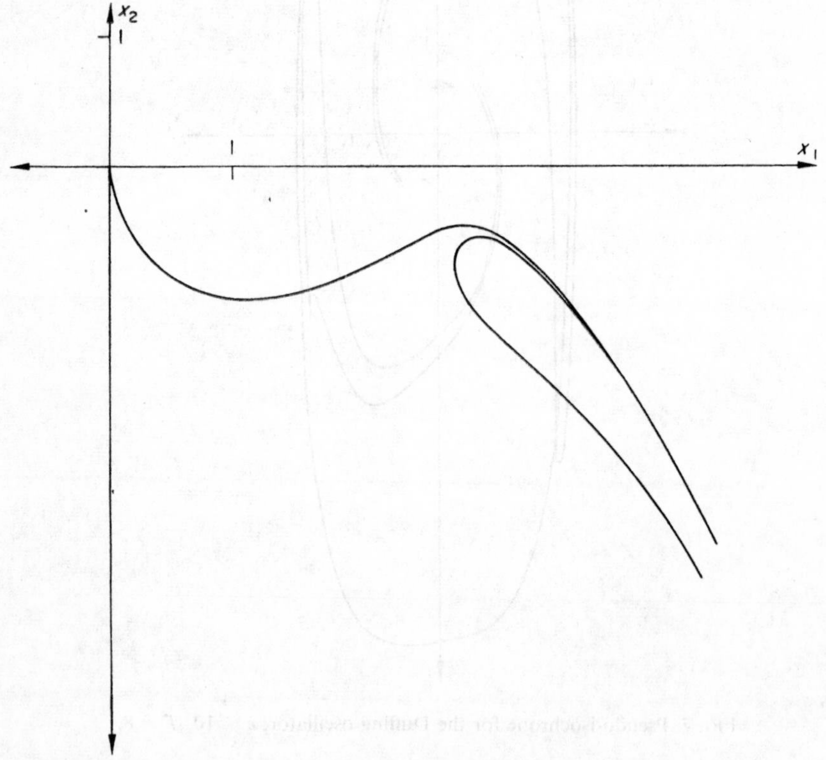

FIG. 8. Switching curve for the Duffing oscillator, $\varepsilon = -0{\cdot}08$.

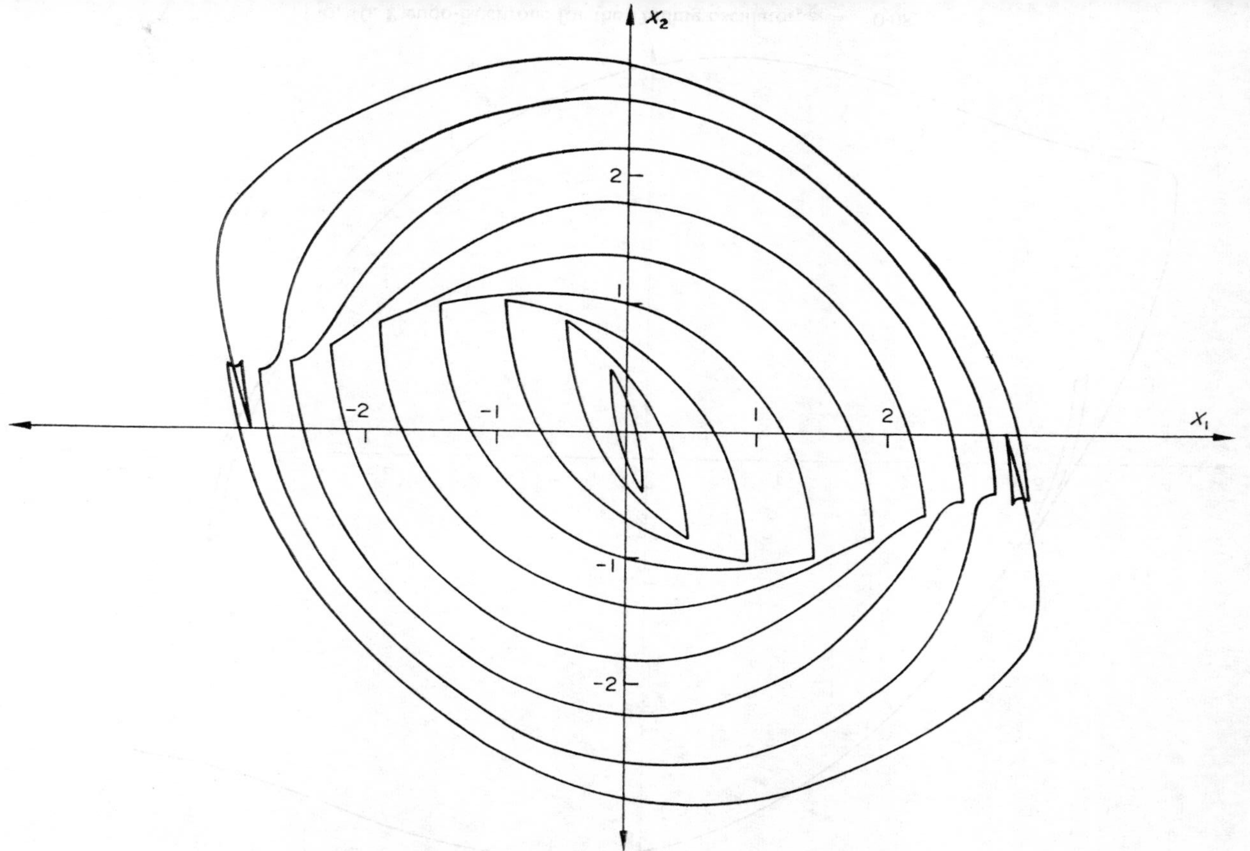

FIG. 9. Isochrones for the Duffing oscillator, $\varepsilon = -0{\cdot}08$, $T = \frac{1}{2}(\frac{1}{2})4\frac{1}{2}$.

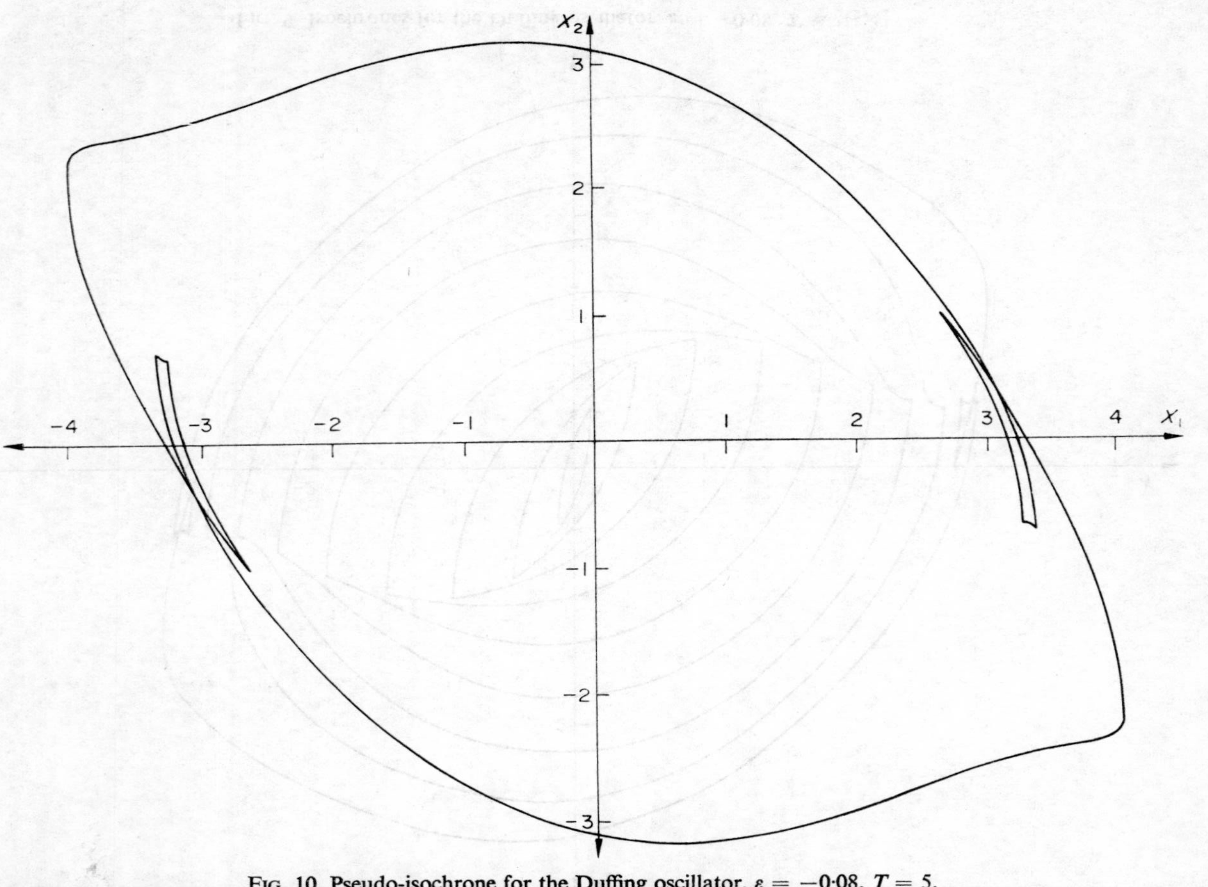

FIG. 10. Pseudo-isochrone for the Duffing oscillator, $\varepsilon = -0{\cdot}08$, $T = 5$.

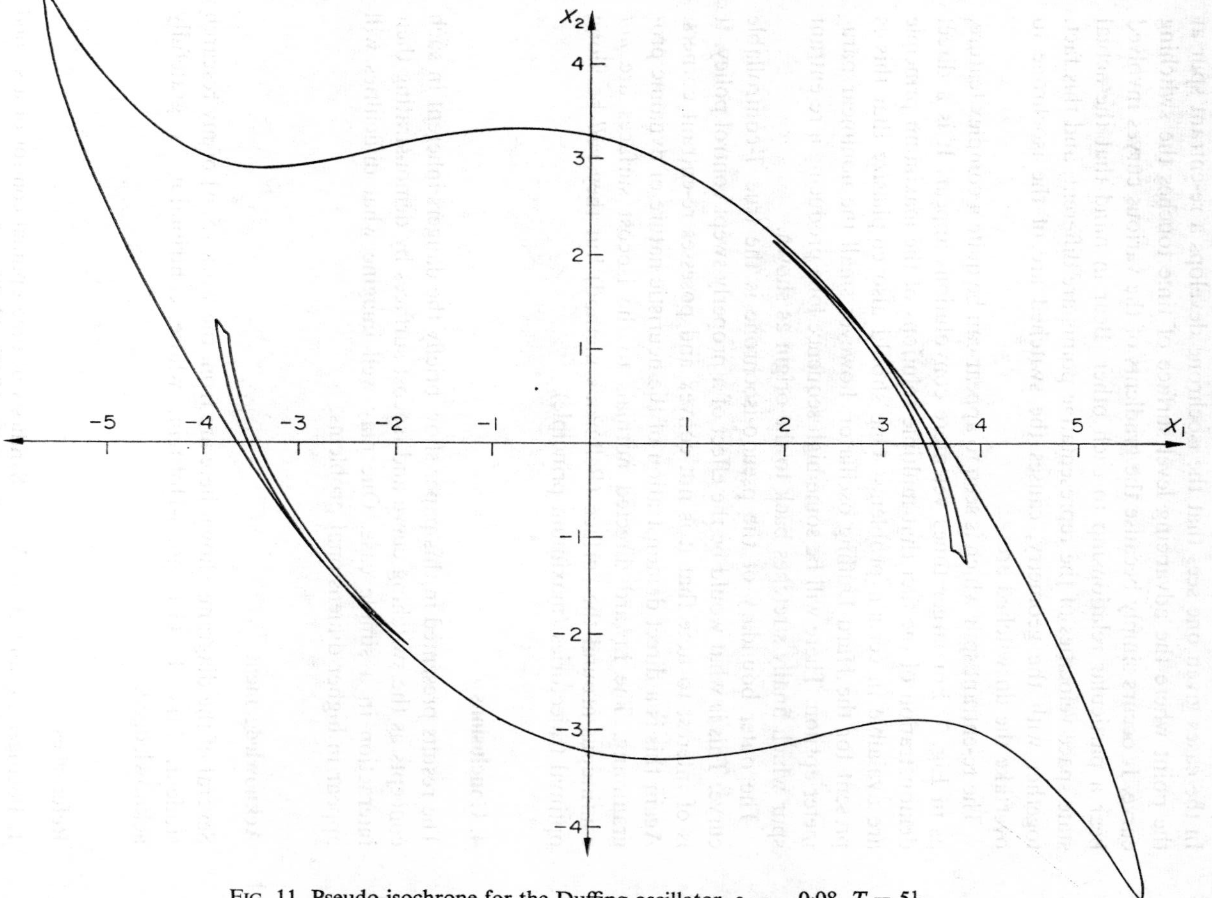

FIG. 11. Pseudo-isochrone for the Duffing oscillator, $\varepsilon = -0{\cdot}08$, $T = 5\frac{1}{2}$.

3. Discussion

In the cases given one sees that the isochrone develops a re-entrant spur at the point where the advancing level surface of time touches the switching curve. It occurs simply because the gradients of the various curves involved bear a particular relationship to each other. Bear in mind that the actual state space velocities of the representative points are different, and this fact, together with the geometry, causes the switched arc of the isochrone to overtake the unswitched arc.

The re-entrant spur which is seen to appear can be quite a complex feature, as in Fig. 7. For larger times yet more convolutions appear. It is a direct demonstration of the fact that multiple solutions of the maximum principle are avaialble in certain problems. One should also emphasize that this is present for the Hard Duffing oscillator, however small the nonlinear parameter epsilon. There will be some high sequence lobe producing a re-entrant spur which finally stretches back to the origin as shown.

The outer boundary of the pseudo-isochrone is the true T-controllable curve. This is what would be the effect of a properly swept control policy. It is of interest to note that it is not convex and posesses re-entrant corners. Again this is a direct demonstration of the heuristic nature of dynamic programming. The inward directed normals to the isocost surfaces are *not* continuous around the surface (Bellman's method) but they *will* be along optimal trajectories (maximum principle).

4. Conclusions

The results presented in this paper show briefly the dangers inherent in such concepts as the switching curve and isocost surfaces by demonstrating their interaction in a simple case. One may well imagine what difficulties will appear in higher dimensional problems.

Acknowledgement

Several of the diagrams shown here are from the work [5, 6] of my research student, Mrs. I. Grisogono-Holstein, whose contribution is gratefully acknowledged.

References

1. Boettiger, A. and Haas, V. B., Synthesis of time optimal control of a second order nonlinear process. *J.O.T.A.* **4** (1) (1969), 22–39.
2. Davies, M. J. Discontinuities of switching curves. *Int. J. Control* **14** (1) (1971), 175–181.

3. Lee, E. B. and Markus, L., "Foundations of Optimal Control Theory". Wiley, New York (1967).
4. Davies, M. J., Time optimal control and the Duffing oscillator. *J. Inst. Maths. Applics.* **9** (1972), 357–369.
5. Davies, M. J. and Grisogono, I., The evolution of controllable domains. *Int. J. Control* **15** (3) (1972), 523–527.
6. Grisogono, I., Master's thesis presented in the University of Wales (1971).

Discussion

PARKS (*University of Warwick*). These sets are rather interesting because they are not convex.

DAVIES. They are not convex in this case, or in any problem where the lobesweeping feature is present. The re-entrant spurs are to be eliminated, in the same way as the lobes are swept, and this whole concept is considered in detail in Dr. Mirica's paper.†

PARKS. A lot of the conventional wisdom on this topic depends on the assumption that the sets are convex.

DAVIES. They are not convex in many cases, even with such simple problems as I have considered.

STOREY (*University of Loughborough*). It is the nonlinearities which cause this.

DAVIES. I think the truth is that every nonlinear system is a law unto itself and a new problem.

† Page 167 of these Proceedings.

11. The Optimal Synthesis of a Control Problem Defines a Stratification of the Phase Space

ŞTEFAN MIRICĂ

University of Warwick, Coventry, England†

The study of stratified sets was initiated in 1957 by Whitney [5] in the description of the structure of analytic varieties. Later on, R. Thom [4] and J. Mather, [2], used the notion of stratification in the study of topological stability of differentiable mappings.

We recall first that a set S is said to be *a stratified set* (or an abstract stratified set) if the following axioms, I–IX, are satisfied:

I. S is a Haussdorf, locally compact topological space with a countable basis for its topology;

II. S is the union of pairwise disjoint subsets $A \subset S$ called the strata of S; denote by $\mathscr{S}$ the family of all strata of S;

III. Each stratum of S is a differentiable manifold (in the induced topology) and as a subset of S is locally closed;

IV. The family $\mathscr{S}$ of strata is locally finite;

V. Any two strata of S satisfy the condition of the frontier: $A \cap \bar{B} \neq \emptyset$ implies $A \subset \bar{B}$; if $A \subset \bar{B}$ and $A \neq B$ (so $A \subset \{\bar{B} - B\}$) then we write $A < B$;

VI. For any stratum A of S a triple (T_A, Π_A, ρ_A) is given where $T_A \subset S$ is an open neighbourhood of A in S, $\Pi_A: T_A \to A$ is a continuous retraction and $\rho_A: T_A \to [0, \infty)$ is a continuous function; by abuse of language the triple (T_A, Π_A, ρ_A) is called the tubular neighbourhood of A and the family $\mathscr{T}$ of all such triples is called the set of *control data* on S;

VII. For any stratum A of S, $A = \rho_A^{-1}(0)$;

VIII. If we denote $T_{A,B} = T_A \cap B, \Pi_{A,B} = \Pi_A | T_{A,B}, \rho_{A,B} = \rho_A | T_{A,B}$ for any two strata A, B of S for which $T_A \cap B \neq \emptyset$, then the map $(\Pi_{A,B}, \rho_{A,B})$: $T_{A,B} \to A \times [0, \infty)$ is a smooth submersion;

IX. For any strata $A, B, C \in \mathscr{S}$ we have $\Pi_{A,B} \circ \Pi_{B,C} = \Pi_{A,C}, \rho_{A,B} \circ \Pi_{B,C} = \rho_{A,C}$ (on the sets where the compositions are defined).

Present address: Dept. of Mathematics, University of Bucharest, Romania.

Given a stratified set S, *a controlled vector field on* S is a collection $\eta = \{\eta_A | A \in \mathscr{S}\}$ where $\eta_A : A \to T(A)$ is a vector field such that for any stratum $B > A$ (i.e. $A \subset \{\bar{B} - B\}$) we have:

$$\eta_B(\rho_{A,B}) = 0, \qquad (\Pi_{A,B})_* \circ \eta_B = \eta_A \circ \Pi_{A,B} \tag{1}$$

that is, the map $\Pi_{A,B}: T_{A,B}(\subset B) \to A$ projects the trajectories of η_B onto trajectories of η_A and the function ρ_A is constant on any trajectory of η_B. Roughly speaking the integral curves of η_B are, in a certain sense, "parallel" to any stratum which is included in the frontier of B.

We remark that the word "control" which has appeared so far in the definition of the stratified set, has nothing to do with what is known as "control theory".

However, the optimal synthesis (or "the closed-loop" optimal control or the optimal control as a function of state variable) of a control problem, defines a partition of the phase space into differentiable submanifolds called "the cells" of the synthesis such that on every cell the optimal control is a smooth function of the state variable.

One can expect that such a partition of the phase space defines a stratification in the natural topology but the optimal trajectories cannot define a controlled vector field because there are cells which are intersected "transversally" by the optimal trajectories and in a stratified set the controlled vector field is tangent to any stratum.

We show that one can define a stratification in which the optimal trajectories are integral curves of a controlled vector field in the sense of Mather.

For simplicity we will construct this stratification on an example.

Let us consider the time-optimal control problems to the origin $0 \in R^2$ for the system:

$$\dot{x}_1 = x_2, \qquad \dot{x}_2 = u, \qquad u \in [-1, +1]. \tag{2}$$

It is well known that the optimal control as a function of the state variable $x = (x_1, x_2)$ is given by:

$$v(x) = \begin{cases} +1 & \text{for} \quad x \in c_1 \cup c_3 \\ -1 & \text{for} \quad x \in c_2 \cup c_4 \end{cases} \tag{3}$$

where c_1, c_2, c_3, c_4 and $\{0\}$ are the cells of the synthesis (Fig. 1):

$$c_1 = \{x \mid x_1 = \tfrac{1}{2}(x_2)^2, x_2 < 0\},$$
$$c_2 = \{x \mid x_1 = -\tfrac{1}{2}(x_2)^2, x_2 > 0\},$$
$$c_3 = \{x \mid x_1 = \tfrac{1}{2}(x_2)^2 - a^2, x_2 < a, a > 0\},$$
$$c_4 = \{x \mid x_1 = -\tfrac{1}{2}(x_2)^2 + b^2, x_2 > b, b < 0\}$$

Therefore the optimal trajectories are integral curves of the discontinuous differential system:

$$\begin{aligned} \dot{x}_1 &= x_2 \\ \dot{x}_2 &= v(x_1, x_2) \end{aligned} \tag{4}$$

The sets c_1, c_2, c_3, c_4 and $\{0\}$ define obviously a stratification of R^2 and it is easy to construct a set of control data which satisfies the axioms I–IX but the vector field (4) defined by the optimal trajectories cannot be a controlled vector field for any set of control data.

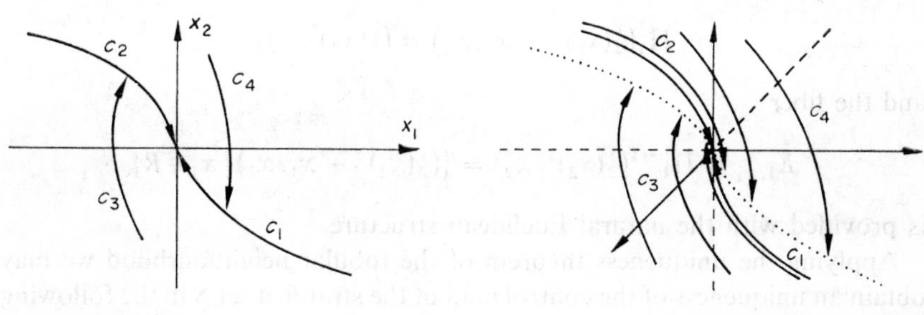

FIG. 1. FIG. 2.

In order to obtain a stratified set such that the optimal field is a controlled vector field we define the set S to be the topological sum of the sets $c_1 \cup c_3$ and $c_2 \cup c_4$ (each of them provided with the natural topology) therefore as a set S is just $R^2 - \{0\}$ and as a topological space has two distinct connected components. Intuitively one can imagine S imbedded in the space R^3 obtained by cutting a plane along $c_1 \cup \{0\} \cup c_2$ and rotate a little one of the half-planes (Fig. 2).

The stratified set S has now four strata, c_1, c_2, c_3, c_4, and we may choose a set of control data (i.e. a tubular neighbourhood for each stratum) such that the optimal field (4) is a controlled vector field.

Obviously the strata c_3 and c_4 are open subsets of S; therefore we may choose: $T_{c_3} = c_3, \Pi_{c_3} = id_{c_3}, \rho_{c_3} \equiv 0$ and analogously for c_4.

For c_1 we may choose T_{c_1} to be an open subset of $c_1 \cup c_3$, namely: $T_{c_1} = \{x | \, x_1 = \frac{1}{2}(x_2)^2 - a^2, x_2 < 0, a \in [0, \varepsilon)\}$ for an $\varepsilon > 0$ and $\Pi_{c_1}(x_1, x_2) = (\frac{1}{2}(x_2)^2, x_2), \rho_{c_1}(x_1, x_2) = \{\frac{1}{2}(x_2)^2 - x_1\}^2$ and analogously for c_2. It is easy to verify that these control data satisfy the axioms of a stratified set and the optimal field (4) is a controlled vector field.

Moreover, $T_{c_1}(T_{c_2})$ is the restriction to $c_1 \cup c_3$ (respectively to $c_2 \cup c_4$) of an ordinary tubular neighbourhood of $c_1(c_2)$ in R^2; that is, there exists an inner product bundle $\Pi_1 : E_1 \to c_1$ and a diffeomorphism ϕ_1 of a neighbourhood B_δ of c_1 in E_1 onto T_{c_1} which commutes with the zero section of E_1 and such that

$$\Pi_{c_1} = \Pi_1 \circ \phi_1^{-1}, \qquad \rho_{c_1} = \rho_1 \circ \phi^{-1}$$

where $\rho_1(e) = \|e\|_1^2$ for $e \in E_1$.

In our case

$$E_1 = \{(\tfrac{1}{2}(x_2)^2 - x_1, x_2) \mid x_1 \in R, \qquad x_2 < 0\},$$

$\Pi_1 : E_1 \to c_1$ is given by:

$$\Pi_1(\tfrac{1}{2}(x_2)^2 - x_1, x_2) = (\tfrac{1}{2}(x_2)^2, x_2)$$

and the fiber

$$E_{1,x_2} = \Pi_1^{-1}(\tfrac{1}{2}(x_2)^2, x_2) = \{(\tfrac{1}{2}(x_2)^2 - x_1, x_2) \mid x_1 \in R\}$$

is provided with the natural Euclidean structure.

Applying the uniqueness theorem of the tubular neighbourhood we may obtain an uniqueness of the control data of the stratified set S in the following sense: for any other set

$$\mathscr{T}' = \{(T_A', \Pi_A', \rho_A') \mid A \in \mathscr{S}\}$$

of control data on S there exists a homeomorphism $h: S \to S$ which preserves the strata and for any stratum A of S there exists a neighbourhood T_A'' of A in T_A' such that

$$(\Pi_A \circ h^{-1}) \mid T_A'' = (h^{-1} \circ \Pi_A') \mid T_A'', \qquad (\rho_A \circ h^{-1}) \mid T_A'' = \rho_A' \mid T_A''$$

The construction can be carried out in the general case using the rigourous definition and some properties of the synthesis [1, 3].

Besides the academic interest of this connection between the optimal synthesis and the stratified sets, it is likely that the associated stratification contains the essential topological-differential properties of the synthesis and therefore may serve to a deeper study of this object.

For example, it is intuitively clear that the equivalence of two syntheses defined via the associated stratifications (i.e. the syntheses are equivalent if the associated stratifications are homeomorphic) is less restrictive that the equivalence defined by the existence of a homeomorphism of the whole phase space.

Further on, a "good" definition of the equivalence of two syntheses may be a significant step towards the study of the structural stability of the optimal synthesis; that is, the study of the variations of the "global picture" of the optimal trajectories for small variations of the initial control problem.

References

1. Boltyanskii, V. G., "Mathematical Methods of Optimal Control". Holt, Reinhart and Winston (1971).
2. Mather, J., "Lectures on Topological Stability". Harvard (1970).
3. Mirica, S., *R.I.R.O.* **1** (1971), 73–104.
4. Thom, R., *L'Enseignement Mathematique* **8** (1962), 24–33.
5. Whitney, H., *Ann. of Math.* **66** (1957), 545–556.

syntheses, that is, the study of the variations of the "global picture" of the optimal trajectories for small variations of the initial control problem.

References

1. Boltyanskii, V. G., "Mathematical Methods of Optimal Control". Holt, Rinehart and Winston (1971).
2. Mather, J., "Lectures on Topological Stability", Harvard (1970).
3. Mirica, S., *R.I.R.O.* 1 (1971), 73–104.
4. Thom, R., *Enseignement Mathématique* 8 (1962), 24–31.
5. Whitney, H., *Ann. of Math.* 66 (1957), 545–556.

12. Improving the Conditioning of Optimal Control Problems Using Simple Models

J. C. ALLWRIGHT

Department of Computing and Control

Imperial College of Science and Technology, London, England

1. Introduction

Control function optimisation problems of the following type are considered here: minimise with respect to the control function u the performance index

$$V(u) = \int_{\mathbf{T}} L(y(t), u(t); t)\, dt \tag{1.1}$$

for the dynamical system

$$S \begin{cases} \dot{x}(t) = f(x(t), u(t); t) : x(t_s) = x_s \\ y(t) = C(x(t), t) \end{cases} \tag{1.2}$$

where $x(t) \in \mathbf{R}^n$; $y(t) \in \mathbf{R}^r$; $u(t) \in \mathbf{R}^m$; $t \in \mathbf{T} = [t_s, t_f]$, $t_f - t_s < \infty$.

Often a state-space model $\boldsymbol{M}$ of the system $\boldsymbol{S}$ will be available which is simple in the sense that it is easier (hopefully much easier) to compute the optimal control function for the model than it is to find the optimal control for the system $\boldsymbol{S}$. A basic question is: how can a given model be used to help speed control optimisation for the system $\boldsymbol{S}$?

The obvious way to use the model would be to minimise $V(u)$ of (1.1) for it and then to hope that the resulting optimal control for the model is an adequate approximation to the optimal control for the system. Lower-bound results [1] for the minimal performance index value can be used in some cases (mainly for linear systems with quadratic performance indices) to check if an adequate approximation to the optimal control for the system has in fact been obtained. In general, an adequate approximation will not be obtained and then the obvious technique for using the model will have failed.

In this paper, the model is employed to help find the actual optimal control for the system using the following ideas:

(i) an approximation to the second-derivative operator for the optimisation problem for the system can be associated with a model of the system.

(ii) algorithms can use the approximate second-derivative operator to help speed the solution of the optimisation problem for the system by improving its conditioning.

(iii) in such algorithms, the solution of each equation involving the 'infinite-dimensional' approximate second-derivative operator can be carried out without inverting it by just solving an optimisation problem for the model.

(iv) because the model is simple, each optimisation for it can be carried out rather easily.

The basic structure of algorithms using linear models for the case of continuous-time linear systems with quadratic performance indices is treated for both continuous- and discrete-time models, in Sections 3 and 4.

The algorithm used for the linear system, quadratic performance index case is a generalisation (first described by Hestenes [2]) of the usual conjugate-gradient algorithm and is called the conjugate-gradient ($\boldsymbol{K}^{-1}$) algorithm here. This algorithm and its connections with some other algorithms which could be used for optimisation using models are discussed first, in Section 2.

For more general optimisation problems, with non-linear systems, etc., a more complicated algorithm seems to be needed. Such an algorithm is outlined in Section 5.

2. Approximate Second-derivative Operators and Contraction-mapping, Davidon and Conjugate-gradient Algorithms

Consider the problem of minimising the quadratic

$$V(\boldsymbol{u}) = a + \langle \boldsymbol{b}, \boldsymbol{u} \rangle + \tfrac{1}{2}\langle \boldsymbol{u}, G\boldsymbol{u} \rangle \tag{2.1}$$

where $a \in \mathbf{R}$, $\boldsymbol{b}$ and $\boldsymbol{u}$ belong to a real Hilbert space $\mathbf{U}$ and where $G : \mathbf{U} \to \mathbf{U}$ is a positive-definite, self-adjoint linear operator. This optimisation problem is considered because the optimal control problem for a linear system with a quadratic performance index function can be posed in this way. It is easy to see that the optimising $\boldsymbol{u}$ is

$$\boldsymbol{u}^0 = -G^{-1}\boldsymbol{b}. \tag{2.2}$$

In the optimal control case, $\boldsymbol{G}$ is an operator which is not particularly easy to invert. It turns out that a self-adjoint, positive-definite, approximation $\boldsymbol{K}$ to $\boldsymbol{G}$ which is easier to invert than $\boldsymbol{G}$ can be associated with a model of the system. In fact the inverses of the operators $\boldsymbol{G}$ and $\boldsymbol{K}$ are not actually needed, only the solutions to equations like $\boldsymbol{Gu} = \boldsymbol{d}$ and $\boldsymbol{Ku} = \boldsymbol{d}$ are needed. For descriptive convenience, however, $\boldsymbol{K}$ will be said to be easier to invert (in some sense) than $\boldsymbol{G}$ if $\boldsymbol{Ku} = \boldsymbol{d}$ is easier to solve for $\boldsymbol{u}$ than $\boldsymbol{Gu} = \boldsymbol{d}$. The purpose of this section is to discuss how an approximation $\boldsymbol{K}$ to $\boldsymbol{G}$ can be used in iterative algorithms for finding $\boldsymbol{u}^0$.

Since $\boldsymbol{K}$ is an approximation to $\boldsymbol{G}$,

$$\boldsymbol{G} = \boldsymbol{K} + \boldsymbol{E} \tag{2.3}$$

where $\boldsymbol{E}$ is the error between $\boldsymbol{G}$ and its approximation $\boldsymbol{K}$. From (2.2), $\boldsymbol{u}^0$ satisfies

$$\boldsymbol{G}\boldsymbol{u}^0 = -\boldsymbol{b}$$

i.e.

$$\boldsymbol{K}\boldsymbol{u}^0 = -\boldsymbol{b} - \boldsymbol{E}\boldsymbol{u}^0.$$

Since $\boldsymbol{K}$ is easy to invert, premultiply this by its inverse to give

$$\boldsymbol{u}^0 = -\boldsymbol{K}^{-1}\boldsymbol{b} - \boldsymbol{K}^{-1}\boldsymbol{E}\boldsymbol{u}^0$$

which suggests the following iterative algorithm for finding $\boldsymbol{u}^0$:

$$\boldsymbol{u}_{j+1} = -\boldsymbol{K}^{-1}\boldsymbol{b} - \boldsymbol{K}^{-1}\boldsymbol{E}\boldsymbol{u}_j : \boldsymbol{u}_0 \in \mathrm{U}. \tag{2.4}$$

THEOREM 2.1 *The algorithm of* (2.4) *is a contraction-mapping algorithm and generates a sequence* $\{\boldsymbol{u}_j\}$ *convergent to* $\boldsymbol{u}^0$ *if* $\boldsymbol{K}$ *is a sufficiently good approximation to* $\boldsymbol{G}$ *in that* $\|\boldsymbol{K}^{-1}\boldsymbol{E}\| < 1$.

The proof involves an application of standard results for contractions, is very simple and can be found in [3].

In general the condition $\|\boldsymbol{K}^{-1}\boldsymbol{E}\| < 1$ will not be satisfied and so convergence is not guaranteed.

Another way to use $\boldsymbol{K}^{-1}$ would be to use it as the initial estimate $\boldsymbol{H}_0$ of the inverse second-derivative operator of the Davidon algorithm [4], as formulated by Fletcher and Powell:

The Davidon algorithm

Choose $\tilde{\boldsymbol{u}}_0 \in \mathrm{U}$, a guess at $\boldsymbol{u}^0$. Set $j = 0$, $\boldsymbol{H}_0 = \boldsymbol{K}^{-1}$

(1) Set $\tilde{\boldsymbol{s}}_j = -\boldsymbol{H}_j\tilde{\boldsymbol{g}}_j$

where $\tilde{\boldsymbol{g}}_j$ is the gradient of $V(\boldsymbol{u})$ with respect to $\boldsymbol{u}$, evaluated at $\tilde{\boldsymbol{u}}_j$, i.e. where $\tilde{\boldsymbol{g}}_j = \boldsymbol{b} + \boldsymbol{G}\tilde{\boldsymbol{u}}_j$.

$$\text{Set } \tilde{w}_j = \arg\min_{w \in \mathbf{R}} V(\tilde{\boldsymbol{u}}_j + w\tilde{\boldsymbol{s}}_j)$$

$$\tilde{\boldsymbol{u}}_{j+1} = \tilde{\boldsymbol{u}}_j + \tilde{w}_j\tilde{\boldsymbol{s}}_j$$

$$\Delta\tilde{\boldsymbol{u}}_j = \tilde{\boldsymbol{u}}_{j+1} - \tilde{\boldsymbol{u}}_j$$

$$\Delta\tilde{\boldsymbol{g}}_j = \tilde{\boldsymbol{g}}_{j+1} - \tilde{\boldsymbol{g}}_j$$

$$\boldsymbol{H}_{j+1} = \boldsymbol{H}_j + \frac{\Delta\tilde{\boldsymbol{u}}_j\rangle\langle\Delta\tilde{\boldsymbol{u}}_j}{\langle\Delta\tilde{\boldsymbol{u}}_j, \Delta\tilde{\boldsymbol{g}}_j\rangle} - \frac{\boldsymbol{H}_j\Delta\tilde{\boldsymbol{g}}_j\rangle\langle\boldsymbol{H}_j\Delta\tilde{\boldsymbol{g}}_j}{\langle\Delta\tilde{\boldsymbol{g}}_j, \boldsymbol{H}_j\Delta\tilde{\boldsymbol{g}}_j\rangle}$$

and go to (1).

The disadvantage of this algorithm is the amount of storage needed for the operators $\boldsymbol{H}_j$ (or for some approximation by a finite-dimensional matrix). It may be noted [5] that the algorithm actually only needs $\boldsymbol{H}_j$ so as to be able to evaluate $\boldsymbol{H}_j\boldsymbol{q}$ for elements $\boldsymbol{q}$, so if $\boldsymbol{H}_0\boldsymbol{q}$ can be evaluated without needing a matrix approximation to $\boldsymbol{H}_0$, then $\boldsymbol{H}_j\boldsymbol{q}$ can be evaluated without use of a matrix approximation to $\boldsymbol{H}_j$ if the terms $\Delta\tilde{\boldsymbol{u}}_i$, $\boldsymbol{H}_i\Delta\tilde{\boldsymbol{g}}_i$ are stored. For then $\boldsymbol{H}_j\boldsymbol{q}$ can be evaluated using

$$\boldsymbol{H}_j\boldsymbol{q} = \boldsymbol{H}_0\boldsymbol{q} + \sum_{i=0}^{j-1} \Delta\tilde{\boldsymbol{u}}_i \frac{\langle\Delta\tilde{\boldsymbol{u}}_i, \boldsymbol{q}\rangle}{\langle\Delta\tilde{\boldsymbol{u}}_i, \Delta\tilde{\boldsymbol{g}}_i\rangle} - \sum_{i=0}^{j-1} \boldsymbol{H}_i\Delta\tilde{\boldsymbol{g}}_i \frac{\langle\boldsymbol{H}_i\Delta\tilde{\boldsymbol{g}}_i, \boldsymbol{q}\rangle}{\langle\Delta\tilde{\boldsymbol{g}}_i, \boldsymbol{H}_i\Delta\tilde{\boldsymbol{g}}_i\rangle}.$$

The storage needed for the terms $\Delta\tilde{\boldsymbol{u}}_i$, $\boldsymbol{H}_i\Delta\tilde{\boldsymbol{g}}_i$ will tend to be much less than that needed for a matrix approximation to $\boldsymbol{H}_j$. However, unless j is very small, a considerable amount of storage will *still* be needed.

It is well-known [6] that the usual conjugate-gradient algorithm is equivalent to the Davidon algorithm for a quadratic performance index if $\boldsymbol{H}_0 = \boldsymbol{I}$, in the sense that if both start with the same $\boldsymbol{u}_0$, then the sequence $\{\boldsymbol{u}_j\}$ generated by the conjugate-gradient algorithm is identical to the sequence generated by the Davidon algorithm. An important advantage of the conjugate-gradient algorithm over the Davidon algorithm is that it requires little storage. A version of the conjugate-gradient algorithm which is equivalent to the Davidon algorithm for any given $\boldsymbol{H}_0$ would therefore be valuable. Such an algorithm is the conjugate-gradient ($\boldsymbol{K}^{-1}$) algorithm of Hestenes, which is stated next.

The conjugate-gradient ($\boldsymbol{K}^{-1}$) *algorithm*

$$\text{Choose } \hat{\boldsymbol{u}}_0 \in \mathbf{U}.$$

$$\text{Set } \hat{\boldsymbol{s}}_0 = -\boldsymbol{K}^{-1}\hat{\boldsymbol{g}}_0, j = 0$$

where $\hat{\boldsymbol{g}}_j$ is the gradient of V with respect to $\boldsymbol{u}$, evaluated at $\hat{\boldsymbol{u}}_j$

$$\text{(1)} \quad \text{Set } \hat{w}_j = \arg\min_{w \in \mathbf{R}} V(\hat{\boldsymbol{u}}_j + w\hat{\boldsymbol{s}}_j)$$

$$\hat{\boldsymbol{u}}_{j+1} = \hat{\boldsymbol{u}}_j + \hat{w}_j \hat{\boldsymbol{s}}_j$$

$$\hat{\beta}_{j+1} = \langle \hat{\boldsymbol{g}}_{j+1}, \boldsymbol{K}^{-1}\hat{\boldsymbol{g}}_{j+1} \rangle / \langle \hat{\boldsymbol{g}}_j, \boldsymbol{K}^{-1}\hat{\boldsymbol{g}}_j \rangle$$

$$\hat{\boldsymbol{s}}_{j+1} = -\boldsymbol{K}^{-1}\hat{\boldsymbol{g}}_{j+1} + \hat{\beta}_{j+1}\hat{\boldsymbol{s}}_j$$

$$j = j + 1$$

Go to (1).

Obviously the usual conjugate-gradient algorithm is the conjugate-gradient ($\boldsymbol{I}$) algorithm.

THEOREM 2.2 *If* $\boldsymbol{H}_0 = \boldsymbol{K}^{-1}$ *and* $\tilde{\boldsymbol{u}}_0 = \hat{\boldsymbol{u}}_0$, *the sequences* $\{\tilde{\boldsymbol{u}}_j\}$ *generated by the Davidon algorithm and* $\{\hat{\boldsymbol{u}}_j\}$ *generated by the conjugate-gradient* ($\boldsymbol{K}^{-1}$) *algorithm are identical.*

This result may be proved using the following two Lemmas, where $\boldsymbol{K}^{\frac{1}{2}}$ denotes the symmetric square root of $\boldsymbol{K}$.

LEMMA 2.1 *Suppose* $\{\hat{\boldsymbol{u}}_j\}$ *is the sequence generated by the conjugate-gradient* ($\boldsymbol{K}^{-1}$) *algorithm for minimising* $V(\boldsymbol{u})$. *Let* $\{\hat{\bar{\boldsymbol{u}}}_j\}$ *denote the sequence obtained by using the usual conjugate-gradient algorithm for minimising* $V(\boldsymbol{K}^{-\frac{1}{2}}\bar{\boldsymbol{u}})$, *starting with* $\hat{\bar{\boldsymbol{u}}}_0 = \boldsymbol{K}^{\frac{1}{2}}\hat{\boldsymbol{u}}_0$. *Then the sequences* $\{\hat{\bar{\boldsymbol{u}}}_j\}$ *and* $\{\hat{\boldsymbol{u}}_j\}$ *are related by*

$$\hat{\boldsymbol{u}}_j = \boldsymbol{K}^{-\frac{1}{2}}\hat{\bar{\boldsymbol{u}}}_j, \forall j \geqslant 0.$$

LEMMA 2.2 *Suppose* $\{\tilde{\boldsymbol{u}}_j\}$ *is the sequence generated by the Davidon algorithm with* $\boldsymbol{H}_0 = \boldsymbol{K}^{-1}$ *for minimising* $V(\boldsymbol{u})$ *and that* $\{\boldsymbol{u}_j\}$ *is the sequence generated by the Davidon algorithm with* $\boldsymbol{H}_0 = \boldsymbol{I}$ *for minimising* $V(\boldsymbol{K}^{-\frac{1}{2}}\bar{\boldsymbol{u}})$, *starting with* $\tilde{\bar{\boldsymbol{u}}}_0 = \boldsymbol{K}^{\frac{1}{2}}\tilde{\boldsymbol{u}}_0$, *Then the sequences* $\{\tilde{\boldsymbol{u}}_j\}$ *and* $\{\tilde{\bar{\boldsymbol{u}}}_j\}$ *are related by*

$$\tilde{\boldsymbol{u}}_j = \boldsymbol{K}^{-\frac{1}{2}}\tilde{\bar{\boldsymbol{u}}}_j, \forall j \geqslant 0.$$

Also, if the jth estimate of the inverse second derivative operator for the first case is denoted by $\boldsymbol{H}_j$ *and by* $\tilde{\boldsymbol{H}}_j$ *for the second case, then*

$$\boldsymbol{H}_j = \boldsymbol{K}^{-\frac{1}{2}}\tilde{\boldsymbol{H}}_j\boldsymbol{K}^{-\frac{1}{2}}, \forall j \geqslant 0.$$

Proof of Theorem 2.2 Myers [6] has shown that the Davidon algorithm with $\boldsymbol{H}_0 = \boldsymbol{I}$ and the usual conjugate-gradient algorithm are equivalent. so $\hat{\bar{\boldsymbol{u}}}_j = \tilde{\bar{\boldsymbol{u}}}_j, \forall j \geqslant 0$, for the sequences $\{\hat{\bar{\boldsymbol{u}}}_j\}$ and $\{\tilde{\bar{\boldsymbol{u}}}_j\}$ of Lemmas 2.1 and 2.2, The results of Lemmas 2.1 and 2.2 therefore give $\tilde{\boldsymbol{u}}_j = \hat{\boldsymbol{u}}_j, \forall j \geqslant 0$, as required [7].

It follows from Lemma 2.1 that the convergence properties of the conjugate-gradient (K^{-1}) algorithm for minimising $V(u)$ are essentially the same as those of the usual conjugate-gradient algorithm for minimising

$$V(K^{-\frac{1}{2}}u) = a + \langle K^{-\frac{1}{2}}b, u\rangle + \tfrac{1}{2}\langle u, K^{-\frac{1}{2}}GK^{-\frac{1}{2}}u\rangle.$$

It is well known that the convergence rate of the usual conjugate-gradient algorithm can be bounded in terms of the ratio of the upper- and lower-bounds for the second-devivative operator. Therefore, were the ratio smaller for $K^{-\frac{1}{2}}GK^{-\frac{1}{2}}$ than for G, the conjugate-gradient (K^{-1}) algorithm would give potentially faster convergence than the usual conjugate-gradient algorithm.

Suppose that for all $u \in U$

$$m_G\|u\|^2 \leqslant \langle u, Gu\rangle \leqslant M_G\|u\|^2, \qquad m_K\|u\|^2 \leqslant \langle u, Ku\rangle$$

with $m_G > 0$ and $m_K > 0$ (since $G > 0$ and $K > 0$).

Recall that E is the difference between G and K, so that

$$K^{-\frac{1}{2}}GK^{-\frac{1}{2}} = I + K^{-\frac{1}{2}}EK^{-\frac{1}{2}}$$

Hence

$$(1 - \|E\|/m_K)\|u\|^2 \leqslant \langle u, K^{-\frac{1}{2}}GK^{-\frac{1}{2}}u\rangle \leqslant (1 + \|E\|/m_K)\|u\|^2$$

so that the ratio $(m_K + \|E\|)/(m_K - \|E\|)$ of the bounds for $K^{-\frac{1}{2}}GK^{-\frac{1}{2}}$ is smaller than the ratio M_G/m_G for G if K is a good approximation to G.

Thus if K is a good approximation to G, the conjugate-gradient (K^{-1}) algorithm would be expected to give more rapid convergence than the usual conjugate-gradient algorithm. This is the justification for considering using it instead of the usual conjugate-gradient algorithm.

Finally, it can be shown [3] that the conjugate-gradient (K^{-1}) algorithm always gives better convergence than the contraction-mapping algorithm of (2.4), in the sense of

THEOREM 2.3 *If the contraction-mapping algorithm of* (2.4) *generates the sequence* $\{u_j\}$ *and the conjugate-gradient* (K^{-1}) *algorithm generates the sequence* $\{\hat{u}_j\}$ *and if both start at the same point* (*i.e. if* $u_0 = \hat{u}_0$), *then*

$$V(\hat{u}_j) < V(u_j), \forall j > 0$$

unless the contraction-mapping algorithm achieves the minimum in its first iteration, in which case the conjugate-gradient (K^{-1}) *algorithm will also achieve the minimum in the first iteration.*

It is not surprising that the contraction-mapping algorithm is less effective than the conjugate-gradient (K^{-1}) algorithm because it is easy to see that it is 'isomorphic' to a steepest-descent algorithm which has fixed step-length and so does not even optimise along successive search directions.

In view of the advantages of the conjugate-gradient (K^{-1}) algorithm over the contraction-mapping and Davidon algorithms, it is used in Sections 3 and 4 for control optimisation using models.

3. Optimisation of Continuous-time Systems with Quadratic Performance Indices using Continuous-time Models

Consider the problem of finding the control function u which minimises the performance index

$$V(u) = \tfrac{1}{2}\int_{\mathbf{T}} \{\langle [y(t) - y_d(t)], Q(t)[y(t) - y_d(t)]\rangle + \langle u(t), R(t)u(t)\rangle\}\, dt \tag{3.1}$$

for the continuous-time dynamical system

$$\boldsymbol{S}\begin{cases} \dot{x}(t) = A(t)x(t) + B(t)u(t) : x(t_s) = x_s & (3.2)\\ y(t) = C(t)x(t) & (3.3)\end{cases}$$

where $x(t) \in \mathbf{R}^n$; $y_d(t), y(t) \in \mathbf{R}^r$; $u(t) \in \mathbf{R}^m$; $R(t) > 0$ and $Q(t) \geqslant 0$ are symmetric matrices; y_d, A, B, C, Q, R are continuous functions of time and where $\mathbf{T} = [t_s, t_f], t_f - t_s < \infty$.

The optimisation problem will be considered in a Hilbert space context, so let $\mathbf{U}$ be the Hilbert space of equivalence classes $\boldsymbol{u}$ of square integrable control functions $u : \mathbf{T} \to \mathbf{R}^m$. Similarly let $\mathbf{Y}$ be the Hilbert space for square-integrable functions $y : \mathbf{T} \to \mathbf{R}^r$.

It will first be shown that the above optimal control problem can be posed as the optimisation of a quadratic of the kind considered in Section 2.

The response y of the system $\boldsymbol{S}$ can be written as

$$y = y_s + \boldsymbol{W}\boldsymbol{u} \tag{3.4}$$

where

$$(y_s)(t) = \phi(t, t_s)\, x_s$$

$$\boldsymbol{W} : \mathbf{U} \to \mathbf{Y} : (\boldsymbol{W}\boldsymbol{u})(t) = \int_{t_s}^{t} W(t, \tau)u(\tau)\, d\tau.$$

Here ϕ is the state-transition matrix for the system of (3.2) and $\boldsymbol{W}$ is the associated convolution operator (or input-output map).

The performance index of (3.1) can be written as

$$V(\boldsymbol{u}) = \tfrac{1}{2}\langle(\boldsymbol{y} - \boldsymbol{y}_d), \boldsymbol{Q}(\boldsymbol{y} - \boldsymbol{y}_d)\rangle + \tfrac{1}{2}\langle\boldsymbol{u}, \boldsymbol{R}\boldsymbol{u}\rangle \tag{3.5}$$

where

$$\boldsymbol{Q} : \mathbf{Y} \to \mathbf{Y} : (\boldsymbol{Q}\boldsymbol{y})(t) = Q(t)y(t)$$

$$\boldsymbol{R} : \mathbf{U} \to \mathbf{U} : (\boldsymbol{R}\boldsymbol{u})(t) = R(t)u(t) \tag{3.6}$$

$$\langle\boldsymbol{y}, \bar{\boldsymbol{y}}\rangle = \int_{\mathbf{T}} \langle y(t), \bar{y}(t)\rangle\, dt.$$

Since $Q(t)$ and $R(t)$ are symmetric and since $Q(t) \geqslant 0$, $R(t) > 0$, the operators $\boldsymbol{Q}$ and $\boldsymbol{R}$ are self-adjoint with $\boldsymbol{Q} \geqslant 0$, $\boldsymbol{R} > 0$.

Substitution of (3.4) into (3.5) reveals that

$$V(\boldsymbol{u}) = \tfrac{1}{2}\langle\boldsymbol{y}_d, \boldsymbol{Q}\boldsymbol{y}_d\rangle + \langle\boldsymbol{u}, \boldsymbol{W}^*\boldsymbol{Q}(\boldsymbol{y}_s - \boldsymbol{y}_d)\rangle + \tfrac{1}{2}\langle\boldsymbol{u}, (\boldsymbol{R} + \boldsymbol{W}^*\boldsymbol{Q}\boldsymbol{W})\boldsymbol{u}\rangle \tag{3.7}$$

where $\boldsymbol{W}^*$ is the Hilbert space adjoint of $\boldsymbol{W}$. Hence $V(\boldsymbol{u})$ of (3.7) can be identified with $V(\boldsymbol{u})$ of (2.1) if $\boldsymbol{G} = \boldsymbol{R} + \boldsymbol{W}^*\boldsymbol{Q}\boldsymbol{W}$, etc.

Here the operator $\boldsymbol{R} + \boldsymbol{W}^*\boldsymbol{Q}\boldsymbol{W}$ cannot be inverted using ordinary matrix inversion techniques, because it maps the infinite-dimensional space $\mathbf{U}$ into itself and does not have, in general, a sparse structure. Approximations to $\boldsymbol{R} + \boldsymbol{W}^*\boldsymbol{Q}\boldsymbol{W}$ which can be inverted easily will be considered next.

A simple, positive-definite, self-adjoint, approximation $\boldsymbol{K}$ to $\boldsymbol{R} + \boldsymbol{W}^*\boldsymbol{Q}\boldsymbol{W}$ is $\boldsymbol{K} = \boldsymbol{R}$. The operator $\boldsymbol{R}$ is easy to invert because it follows from (3.6) that $\boldsymbol{R}^{-1}$ is defined by

$$(\boldsymbol{R}^{-1}\boldsymbol{u})(t) = R(t)^{-1}u(t).$$

The contraction-mapping algorithm described by Freeman in [8] is the algorithm of (2.4) with $\boldsymbol{K} = \boldsymbol{R}$. In general the condition $\|\boldsymbol{R}^{-1}\boldsymbol{W}^*\boldsymbol{Q}\boldsymbol{W}\| < 1$ will not be satisfied and so the algorithm does not necessarily yield convergence. Indeed, it may well diverge [9]. In order to overcome this, the equally simple approximation $\boldsymbol{K} = \boldsymbol{R} + \theta\boldsymbol{I}, \theta \in \mathbf{R}$, was suggested in [9], where it was shown that convergence is guaranteed for $\theta \geqslant \|\boldsymbol{W}^*\boldsymbol{Q}\boldsymbol{W}\|/2$ for all positive-definite R. The aim was to approximate $\boldsymbol{W}^*\boldsymbol{Q}\boldsymbol{W}$ by $\theta\boldsymbol{I}$. Freeman [10] later considered the effect of the size of θ on a bound for the convergence rate of the algorithm of [9], but the *a-priori* determination of the value of θ which optimises the actual convergence rate for a particular case seems to be impractical. Probably the best scheme for choosing θ is the adaptive one (suggested in [9]) of adjusting it while the algorithm is operating to give the best observed convergence rate.

The above approximation $\boldsymbol{R} + \theta\boldsymbol{I}$ to $\boldsymbol{R} + \boldsymbol{W}^*\boldsymbol{Q}\boldsymbol{W}$ is not very satisfactory for any value of θ because $\boldsymbol{R} + \theta\boldsymbol{I}$ does not contain a term giving a good approximation to the dynamic contribution W^*QW to $R+W^*QW$. In [11] Freeman considers the construction of a finite-rank approximation to $R^{-1}W^*QW$ and the use of such an approximation in a contraction-mapping algorithm for iterative control function optimisation.

A more natural approximation to $R+W^*QW$ would be $K=R+\overline{W}^*Q\overline{W}$ where $\overline{W}$ is an approximation to W which is simpler than W in the sense that its canonical state space realisation is easier to optimise than that of W. This type of approximation is considered here. The reason why $R+\overline{W}^*Q\overline{W}$ is then potentially easier to invert than $R+W^*QW$ will become apparent later.

The above approximation $\overline{W}$ could be the convolution-operator for a simple continuous-time model of the system S of the following form:

$$M_C\begin{cases}\dot{\bar{x}}(t)=\bar{A}(t)\bar{x}(t)+\bar{B}(t)u(t)\\ y(t)=\bar{C}(t)\bar{x}(t)\end{cases}$$

with $\bar{x}(t)\in\mathbf{R}^{\bar{n}}$ and u and y as before. For M_C to be a simple model of S, $\bar{n}<$ (preferably $\ll$) n.

In the rest of this section, optimisation of the system S is considered using the conjugate-gradient (K^{-1}) algorithm with $K=R+\overline{W}^*Q\overline{W}$ and $\overline{W}$ the convolution-operator for the above model M_C.

The only use which is made of K in the conjugate-gradient (K^{-1}) algorithm is for finding functions q of the form

$$q=K^{-1}z$$

for given z. The function q is actually found by solving the equation

$$Kq=z$$

for q, i.e. by solving the equation

$$(R+\overline{W}^*Q\overline{W})q=z. \tag{3.7}$$

Now (3.7) is just the optimality condition for the problem

$$\min_{q\in\mathbf{U}}-\langle q,z\rangle+\tfrac{1}{2}\langle y,Qy\rangle+\tfrac{1}{2}\langle q,Rq\rangle$$

when

$$y=\overline{W}q.$$

Hence the function q of (3.7) is just that function $q\in\mathbf{U}$ which minimises the performance index

$$V^z(q)\triangleq\int_{\mathbf{T}}\{-\langle q(t),z(t)\rangle+\tfrac{1}{2}\langle y(t),Q(t)y(t)\rangle+\tfrac{1}{2}\langle q(t),R(t)q(t)\rangle\}dt \tag{3.8}$$

for the dynamical system (actually the model M_C)

$$\dot{\bar{x}}(t) = \bar{A}(t)\bar{x}(t) + \bar{B}(t)q(t) : \bar{x}(t_s) = 0 \tag{3.9}$$

$$y(t) = \bar{C}(t)\bar{x}(t) \tag{3.10}$$

The minimisation of $V^z(\boldsymbol{q})$ of (3.8) for the model of (3.9), (3.10) requires just a slight modification of the usual state-space techniques for solving the optimal regulator problem. An important point is that the associated Riccati equation is independent of z and so need only be solved once, at the start, even though the conjugate-gradient ($\boldsymbol{K}^{-1}$) algorithm needs $V^z(\boldsymbol{q})$ to be minimised for more than one z. The minimising $\boldsymbol{q}$ for each z can then be found using that Riccati solution and the solution of a linear differential equation. Such minimisations are potentially easier than the minimisation of V of (3.1) for the system $\boldsymbol{S}$ of (3.2, 3) because the dimension of the state of the model is smaller than that of the system.

For clarity, the application of the conjugate-gradient ($\boldsymbol{K}^{-1}$) algorithm to minimising $V(\boldsymbol{u})$ for the system $\boldsymbol{S}$ using a model is summarised next.

The conjugate-gradient ($\boldsymbol{K}^{-1}$) *algorithm for system optimisation using a model.*

Choose an initial estimate $\boldsymbol{u}_0 \in \mathrm{U}$ of $\boldsymbol{u}^0$ (the optimal control for the system).

Set $\quad \boldsymbol{K}^{-1}\boldsymbol{g}_0 = \arg\min_{\boldsymbol{q}\in \mathrm{U}} (\text{for the model})\, V^{\boldsymbol{g}_0}(\boldsymbol{q})$

$$\boldsymbol{s}_0 = -\boldsymbol{K}^{-1}\boldsymbol{g}_0, j = 0$$

(1) Set $\quad w_j = \arg\min_{w\in \mathrm{R}} (\text{for the system})\, V(\boldsymbol{u}_j + w\boldsymbol{s}_j)$

$$\boldsymbol{u}_{j+1} = \boldsymbol{u}_j + w_j\boldsymbol{s}_j$$

$$\boldsymbol{K}^{-1}\boldsymbol{g}_{j+1} = \arg\min_{\mathrm{q}\in \mathrm{U}} (\text{for the model})\, V^{\boldsymbol{g}_{j+1}}(\boldsymbol{q})$$

$$\beta_{j+1} = \langle \boldsymbol{g}_{j+1}, \boldsymbol{K}^{-1}\boldsymbol{g}_{j+1}\rangle / \langle \boldsymbol{g}_j, \boldsymbol{K}^{-1}\boldsymbol{g}_j\rangle$$

$$\boldsymbol{s}_{j+1} = -\boldsymbol{K}^{-1}\boldsymbol{g}_{j+1} + \beta_{j+1}\boldsymbol{s}_j$$

$$j = j + 1$$

Go to (1).

In the above algorithm, $\{\boldsymbol{u}_j\}$ converges to the optimal control for the system and $\boldsymbol{g}_j$ denotes the gradient of the performance index $V(\boldsymbol{u})$ of (3.1) with respect to the control for the system $\boldsymbol{S}$ of (3.2), (3.3), evaluated for $\boldsymbol{u} = \boldsymbol{u}_j$. This can be found in the following (well-known) way:

Solve

$$\dot{x}_j(t) = A(t)x_j(t) + B(t)u_j(t) : x(t_s) = x_s \quad \text{for } x_j.$$

Next solve

$$-\dot{\lambda}_j(t) = A(t)^T\lambda_j(t) + Q(t)[x_j(t) - x_d(t)] : \lambda_j(t_f) = 0 \qquad \text{for } \lambda_j.$$

Then

$$g_j(t) = R(t)u_j(t) + B(t)^T\lambda_j(t)$$

4. Optimisation of Continuous-time Systems with Quadratic Performance Indices using Discrete-time Models

The performance index V and system $\boldsymbol{S}$ considered here are those of Section 3 (i.e. of (3.1, 2, 3)).

Recall that an approximation of the form $\boldsymbol{R} + \overline{\boldsymbol{W}}^*\boldsymbol{Q}\,\overline{\boldsymbol{W}}$ to $\boldsymbol{R} + \boldsymbol{W}^*\boldsymbol{Q}\boldsymbol{W}$ was used in Section 3 where $\overline{\boldsymbol{W}}$ is the convolution-operator for a continuous-time model $\boldsymbol{M}_C$ of $\boldsymbol{S}$. This resulted in the optimisation of the continuous-time model being needed for a sequence of performance indices when the conjugate-gradient $(\boldsymbol{K}^{-1})$ algorithm was used for optimising $\boldsymbol{S}$ with $\boldsymbol{K} = \boldsymbol{R} + \overline{\boldsymbol{W}}^*\boldsymbol{Q}\overline{\boldsymbol{W}}$. The numerical solution of optimisation problems with discrete-time performance indices and discrete-time systems is relatively simple compared to the optimisation of continuous-time performance indices for continuous-time models because essentially all that is required is the solution of a discrete-time Riccati equation. This suggests that there might be potential advantages in using a discrete-time model to optimise a given continuous-time system. Each optimisation of the discrete-time model which is needed is not a standard optimisation for discrete-time problems because the performance index required to be minimised is continuous-time and because the control function space is not discretised. The use of discrete-time models in this context therefore deserves further study.

An approximation $\overline{\boldsymbol{W}}$ to $\boldsymbol{W}$ is needed which is associated with a discrete-time model but which, like $\boldsymbol{W}$, maps the (continuous-time) control function space $\mathbf{U}$ into the (continuous-time) output function space $\mathbf{Y}$. This suggests the following factorisation of $\overline{\boldsymbol{W}}$:

$$\overline{\boldsymbol{W}} = \boldsymbol{V}_1\tilde{\boldsymbol{W}}\boldsymbol{V}_2$$

where $\boldsymbol{V}_2$ maps $\mathbf{U}$ into a finite-dimensional space $\mathbf{U}_D$ of discretized controls, $\tilde{\boldsymbol{W}}$ is the (discrete) convolution operator for a discrete-time system which maps the space $\mathbf{U}_D$ of discretized controls into a space $\mathbf{Y}_D$ of discretised outputs and where V_1 maps $\mathbf{Y}_D$ into the continuous-time output function space $\mathbf{Y}$.

Perhaps the simplest case is that of the discrete-time model

$$M_D \begin{cases} y(t) = C_{k+1}x_{k+1},\ t \in \mathbf{T}_{k+1} & (4.1) \\ x_{k+1} = A_k x_k + B_k \displaystyle\int_{\mathbf{T}_k} u(\tau)d\tau & (4.2) \end{cases}$$

where $\mathbf{T}$ is partitioned as $\{\mathbf{T}_1, \mathbf{T}_2, \ldots, \mathbf{T}_N\}$ with

$$\mathbf{T}_k = [t_k, t_{k+1}], \qquad k = 1, 2, \ldots, N-1$$

$$\mathbf{T}_N = [t_N, t_{N+1}]$$

and where

$$t_s \triangleq t_1 < t_2 < t_3 \ldots < t_N < t_{N+1} \triangleq t_f$$

Here also

$$x_k \Delta x(t_k) \in \mathbf{R}^{\bar{n}}.$$

This model can easily be generalised to allow, say, $y(t), t \in \mathbf{T}_k$, to depend on both x_k and x_{k+1} (perhaps using linear interpolation) and to allow x_{k+1} to depend in a more complicated way on the restriction of u to $\mathbf{T}_k$. Such generalisations have been considered by D. G. Chapman of Imperial College. The above case is sufficiently general to demonstrate the approach, however, and so is the only case considered here.

For the choice of $\boldsymbol{W}$ induced by M_D, the conjugate-gradient ($\boldsymbol{K}^{-1}$) algorithm can be applied to optimise the control for the system in the same way as in Section 3 for the case of the continuous-time model M_C. The only difference occurs in the way the subsidiary optimisations

$$\min_{\mathbf{q} \in \mathbf{U}} V^z(\boldsymbol{q}) \tag{4.3}$$

for the model are carried out. A dynamic programming algorithm for performing such optimisations is developed in the rest of this section. The structure of the derivation is standard but features which are not standard arise due to the performance index and control function space being for continuous-time while the model is discrete-time.

Recall from (3.8) that

$$V^z(q) = \int_{\mathbf{T}} \{-\langle q(t), z(t)\rangle + \tfrac{1}{2}\langle y(t), Q(t)y(t)\rangle + \tfrac{1}{2}\langle q(t), R(t)q(t)\rangle\}dt. \tag{4.4}$$

Let

$$V^0_{k+1}(x_{k+1}) = \min_{q} \int_{t_{k+1}}^{t_f} \{-\langle q(t), z(t)\rangle + \tfrac{1}{2}\langle y(t), Q(t)y(t)\rangle + \tfrac{1}{2}\langle q(t), R(t)q(t)\rangle\}dt \tag{4.5}$$

given that the state of M_D at time t_{k+1} is x_{k+1}.

In the usual way, assume that $V^0_{k+1}(x_{k+1})$ can be written as

$$V^0_{k+1}(x_{k+1}) = a_{k+1} + \langle b_{k+1}, x_{k+1}\rangle + \tfrac{1}{2}\langle x_{k+1}, P_{k+1}x_{k+1}\rangle. \tag{4.6}$$

Then the principle of optimality gives

$$V^0_k(x_k) = \min_{q_k} \int_{\mathbf{T}_k} \{-\langle z(t), q(t)\rangle + \tfrac{1}{2}\langle y(t), Q(t)y(t)\rangle + \tfrac{1}{2}\langle q(t), R(t)q(t)\rangle\}dt + V^0_{k+1}(A_k x_k + B_k \mathbf{1}_k \mathbf{q}_k) \tag{4.7}$$

where $\mathbf{q}_k$ denotes the restriction of $\mathbf{q}$ to $\mathbf{T}_k$ and

$$\mathbf{1}_k \mathbf{q}_k \triangleq \int_{\mathbf{T}_k} q_k(\tau)d\tau.$$

The unminimised right-hand side of (4.7) can be written as

$$\begin{aligned} a_{k+1} &+ \langle x_k, A_k{}^T b_{k+1}\rangle + \langle \mathbf{q}_k, \mathbf{1}_k{}^* B_k{}^T(b_{k+1} + P_{k+1}A_k x_k) - z_k\rangle_k \\ &+ \tfrac{1}{2}\langle x_k, (Q_k + A_k{}^T P_{k+1} A_k)x_k\rangle \\ &+ \tfrac{1}{2}\langle \mathbf{q}_k, (\mathbf{R}_k + \mathbf{1}_k{}^* B_k{}^T P_{k+1} B_k \mathbf{1}_k)\mathbf{q}_k\rangle_k \end{aligned} \tag{4.8}$$

where

$$Q_k \triangleq \int_{\mathbf{T}_k} C_k{}^T Q(\tau) C_k d\tau$$

$$(\mathbf{1}_k{}^* d)(t) \triangleq d, \forall t \in \mathbf{T}_k$$

$$(\mathbf{R}_k \mathbf{q}_k)(t) \triangleq R(t)q(t), \forall t \in \mathbf{T}_k$$

$$\langle \mathbf{q}, \bar{\mathbf{q}}\rangle_k \triangleq \int_{\mathbf{T}_k} \langle q(t), \bar{q}(t)\rangle\, dt$$

and where $\mathbf{z}_k$ is the restriction of $\mathbf{z}$ to $\mathbf{T}_k$.

The optimising function $\mathbf{q}_k$ is therefore

$$\mathbf{q}_k{}^0 = -(\mathbf{R}_k + \mathbf{1}_k{}^* B_k{}^T P_{k+1} B_k \mathbf{1}_k)^{-1}\{\mathbf{1}_k{}^* B_k{}^T(b_{k+1} + P_{k+1}A_k x_k) - \mathbf{z}_k\} \triangleq \mathbf{e}_k + \mathbf{F}_k \tag{4.9}$$

Use of (4.9) in (4.8) enables $V_k^0(x_k)$ to be written as

$$V_k^0(x_k) = a_k + \langle b_k, x_k \rangle + \tfrac{1}{2}\langle x_k, P_k x_k \rangle \tag{4.10}$$

where

$$a_k = a_{k+1} - \tfrac{1}{2}\langle (z_k - \boldsymbol{1}_k{}^* B_k{}^T b_{k+1}), \boldsymbol{e}_k \rangle_k \tag{4.11}$$

$$b_k = A_k{}^T b_{k+1} + A_k{}^T P_{k+1} B_k \boldsymbol{1}_k \boldsymbol{e}_k \tag{4.12}$$

$$P_k = Q_k + A_k{}^T P_{k+1} A_k + A_k{}^T P_{k+1} B_k \boldsymbol{1}_k \boldsymbol{F}_k. \tag{4.13}$$

Obviously

$$V_{N+1}^0(x_{N+1}) = 0$$

so $V_{N+1}^0(x_{N+1})$ can be expanded as

$$V_{N+1}^0(x_{N+1}) = a_{N+1} + \langle b_{N+1}, x_{N+1} \rangle + \tfrac{1}{2}\langle x_{N+1}, P_{N+1} x_{N+1} \rangle \tag{4.14}$$

with

$$a_{N+1} = 0 \tag{4.15}$$

$$b_{N+1} = 0 \tag{4.16}$$

$$P_{N+1} = 0 \tag{4.17}$$

Equations (4.14), (4.6) and (4.10) justify the use of quadratic expansions for $V_{k+1}^0(x_{k+1})$.

The procedure for finding the function $\boldsymbol{q}$ which minimises $V^z(\boldsymbol{q})$ for the model $\boldsymbol{M}_D$ for a sequence of z's is therefore as follows:

(i) Solve (4.13) in reverse-time for the sequence $\{P_k\}$, using the initial condition (at the final time) of (4.17). This involves computing $\{\boldsymbol{F}_k\}$ using (4.9).

(ii) For each z, solve (4.12) in reverse time for the sequence $\{b_k\}$, using the initial condition of (4.16). This involves finding $\{\boldsymbol{e}_k\}$ using (4.9).

Then the minimising $\boldsymbol{q}$ can be generated using the feedback control law of (4.9) in the usual way.

A useful point to note in connection with the evaluation of $\boldsymbol{e}_k$ and $\boldsymbol{F}_k$ of (4.9) is that $(\boldsymbol{R}_k + \boldsymbol{1}_k{}^* B_k{}^T P_{k+1} B_k \boldsymbol{1}_k)^{-1} \boldsymbol{h}$ can be found quite easily for any $\boldsymbol{h}$ since

$$(\boldsymbol{R}_k + \boldsymbol{1}_k{}^* B_k{}^T P_{k+1} B_k \boldsymbol{1}_k)^{-1} \boldsymbol{h} = \boldsymbol{R}_k{}^{-1} \boldsymbol{h} - \boldsymbol{R}_k{}^{-1} \boldsymbol{1}_k{}^* B_k{}^T (I + P_{k+1} B_k \tilde{R}_k B_k{}^T)^{-1} \times P_{k+1} B_k \boldsymbol{1}_k \boldsymbol{R}_k{}^{-1} \boldsymbol{h}$$

where $$(\boldsymbol{R}_k^{-1}\boldsymbol{h})(t) = R(t)^{-1}h(t), t \in \mathbf{T}_k$$
$$\tilde{\boldsymbol{R}}_k = \mathbf{1}_k\boldsymbol{R}^{-1}\mathbf{1}_k^* = \int_{\mathbf{T}_k} R(t)^{-1}dt$$

and where $(I + P_{k+1}B_k\tilde{R}_kB_k^T)$ is relatively easy to invert because it is just an $\bar{n} \times \bar{n}$ matrix.

5. Optimisation using Models for more General Problems

For an optimal control problem which is more general than that of Section 3, in that a nonlinear system with a non-quadratic performance index is considered, say, the model used for optimisation purposes would probably be a linearisation of a nonlinear model of the system. A local quadratic expansion of the performance index would also be involved. The model and the second-order term in the quadratic expansion would therefore usually be iteration-dependent. Hence the approximation $\boldsymbol{K}$ (induced by the linear model and the quadratic expansion) of the actual second-derivative operator $\boldsymbol{G}$ associated with the system would be iteration dependent, as would $\boldsymbol{G}$. It would be possible to attempt to use the conjugate-gradient ($\boldsymbol{K}^{-1}$) algorithm with an iteration-dependent $\boldsymbol{K}$ (in much the same way as the usual conjugate-gradient algorithm can be used for minimising non-quadratic functions) but it would be unlikely to have the desirable convergence properties which occur for quadratics. The purpose of this section is to outline an algorithm for the more general case.

Suppose that at iteration j a control function $\boldsymbol{u}_j$ has been obtained and that it gives rise to an estimate $\boldsymbol{K}_j$ of the second-derivative operator (at $\boldsymbol{u}_j$) associated with the system, which is denoted $\boldsymbol{G}_j$. A number of control function changes $\Delta\boldsymbol{u}_i$, spanning a subspace $\Delta\mathbf{U}_j$, would have been made earlier and the same number of resulting gradient function changes $\Delta\boldsymbol{g}_i$ (for the system), spanning a subspace $\Delta\mathbf{G}_i$, would have been observed. By using this information it is easy to establish a linear map $\bar{\boldsymbol{G}}_j : \mathbf{U} \to \mathbf{U}$ which gives the predicted change in the gradient as $\Delta\boldsymbol{g} = \bar{\boldsymbol{G}}_j\Delta u \in \Delta\mathbf{G}_j$ for any $\Delta\boldsymbol{u} \in \Delta\mathbf{G}_j$ and as $\bar{\boldsymbol{G}}_j\Delta\boldsymbol{u} = \mathbf{0}$ if $\Delta\boldsymbol{u} \in (\Delta\mathbf{U}_j)^{\perp}$.

Expansion of V about $\boldsymbol{u}_j$ gives, to second-order,

$$V(\boldsymbol{u}_j + \boldsymbol{\delta u}) \cong V(\boldsymbol{u}_j) + \langle \boldsymbol{g}(\boldsymbol{u}_j), \boldsymbol{\delta u}\rangle + \tfrac{1}{2}\langle \boldsymbol{\delta u}, G_j\boldsymbol{\delta u}\rangle.$$

The gradient $\boldsymbol{g}(\boldsymbol{u}_j)$ can be split uniquely as

$$\boldsymbol{g}(\boldsymbol{u}_j) = \boldsymbol{g}_1 + \boldsymbol{g}_2$$

with $\boldsymbol{g}_1 \in \Delta\mathbf{G}_j$ and $\boldsymbol{g}_2 \in (\Delta\mathbf{G}_j)^{\perp}$.

The control function change from $\boldsymbol{u}_j$ which, based on information deduced from the system, should reduce $\boldsymbol{g}_1$ to zero is just $\boldsymbol{\delta\bar{u}} = -\bar{\boldsymbol{G}}_j^{+}\boldsymbol{g}_1$, where $^+$ denotes pseudoinverse. Here $\bar{G}_j{}^+ g_1$ is relatively easy to evaluate, as a result of the simple structure of $\bar{G}_j$. All the information deduced from past iterations has now been used. A predicted expansion of V about $V(u_j + \delta\bar{u})$ is, to second-order

$$\begin{aligned} V(u_j + \delta\bar{u} + \delta u) &\cong V(u_j + \delta\bar{u}) + \langle g_2, \delta u\rangle + \tfrac{1}{2}\langle \delta u, G_j \delta u\rangle \\ &\cong \bar{V}(u_j + \delta\bar{u} + \delta u) \\ &\triangleq V(u_j + \delta\bar{u}) + \langle g_2, \delta u\rangle + \tfrac{1}{2}\langle \delta u, K_j \delta u\rangle \quad (5.1) \end{aligned}$$

where the second-order information associated with the model has been used to give the approximation K_j to G_j. The function $\bar{V}(u_j + \delta\bar{u} + \delta u)$ could be minimised with respect to δu but it would not be desirable for the resulting control function change δu to destroy the predicted optimality of $u_j + \delta\bar{u}$ on the variety $\mathbf{u}_j + \Delta \mathbf{U}_j$. Hence it would seem to be desirable to minimise $\bar{V}$ of (5.1) with respect to δu subject to the linear equality constraint

$$\bar{G}_j \delta u = 0. \quad (5.2)$$

This just involves an equality-constrained optimisation of a quadratic performance index for the model, which, as a result of the simple structure of $\bar{G}_j$, is relatively easy to carry out. Denote the resulting control function change δu by $\delta\tilde{u}$.

The total control function change from u_j predicted using $\bar{G}_j$ and K_j is therefore

$$\delta u_p = \delta\bar{u} + \delta\tilde{u}.$$

The search direction s_j at iteration j can therefore be set equal to δu_p and a control function change Δu_j along it can then be made to give a new control function u_{j+1} with reduced performance index value.

This control function change will cause a gradient function change Δg_j. This information can next be added to $\bar{G}_j$ to give $\bar{G}_{j+1}$ with rank one greater than that of $\bar{G}_j$. The procedure can then be repeated.

The essence of an algorithm which uses second-derivative information deduced from the results of past iterations for the system and from a linear model has been described above. Various tests and provisions for restarting would be needed to give guaranteed convergence.

6. Conclusions

Several ways of using given simple models to alter (and, hopefully, improve) the conditioning of control function optimisation problems for systems have been discussed, for problems with no constraints (other than system dynamics). simple models of more complicated mathematical descriptions of systems will often be available for use in this way, since complicated models of systems often evolve from simpler models. In particular, the approach might be useful for using optimisations of coarsely-discretised finite-difference approximations to a distrubted system to speed the otpimisation of a more finely discretised (and therefore more accurate) approximation.

The approach can be extended to cases with linear equality constraints. If the constraint is $\boldsymbol{Du} = \boldsymbol{d}$, the natural way to use an approximation $\boldsymbol{K}$ to the second-derivative operator would be to minimise $V(\boldsymbol{K}^{-\frac{1}{2}}\tilde{\boldsymbol{u}})$ subject to the constraint $\boldsymbol{DK}^{-\frac{1}{2}}\tilde{\boldsymbol{u}} = \boldsymbol{d}$ using the projected gradient version of the usual conjugate-gradient algorithm. The resulting algorithm is then equivalent to the conjugate-gradient ($\bar{\boldsymbol{K}}$) algorithm for

$$\bar{\boldsymbol{K}} = \boldsymbol{K}^{-1} - \boldsymbol{K}^{-1}\boldsymbol{D}^*(\boldsymbol{DK}^{-1}\boldsymbol{D}^*)^{-1}\boldsymbol{DK}^{-1}$$

if the initial control function $\boldsymbol{u}_0$ satisfies the constraints. Functions $\bar{\boldsymbol{K}}\boldsymbol{z}$ can then be found as

$$\underset{\boldsymbol{q}}{\arg\min} \quad - \langle \boldsymbol{z}, \boldsymbol{q}\rangle + \tfrac{1}{2}\langle \boldsymbol{q}, \boldsymbol{Kq}\rangle$$

subject to the equality constraint

$$\boldsymbol{Dq} = \boldsymbol{0}$$

i.e. can be found by performing an equality constrained optimisation for the model.

The approach can also be used to enable an approximate solution of the Riccati equation for the actual system to help speed the determination of the exact optimal control for the system, as it yields an approximation to $(\boldsymbol{R} + \boldsymbol{W}^*\boldsymbol{QW})^{-1}$.

Computations performed by D. G. Chapman of Imperial College for several examples show that the use of simple models at least halves the computing time needed to obtain an excellent approximation to the optimal control. This indicates that control function optimisation of systems using simple models has promise.

References

1. Allwright, J. C., Duality and iterative control function optimisation for differentially-described dynamical systems. This Volume pp. 193–210.
2. Hestenes, M. R., The conjugate-gradient method for solving linear systems. *In* "Proc. Symposium on Applied Math., Vol. VI, Num. Anal." *Am. Math. Soc.* pp. 83–102 (1956).
3. Allwright, J. C., "Contraction-mapping and Conjugate-gradient Algorithms for Control Optimisation." Department of Computing and Control Publication 72/12, Imperial College, June (1972).

4. Horwitz, L. B. and Sarachik, P. E., Davidon's method in Hilbert space. *SIAM J. Appl. Math.* **6** (4) (1968), 676–695.
5. Lasdon, L. S., Conjugate direction methods for optimal control. *IEEE Trans. autom. Control* **AC-15** (1970), 267–268.
6. Myers, E., Properties of the conjugate-gradient and Davidon methods. *Journal of Optimisation Theory and Applications* **2** (4) (1968), 209–219.
7. Allwright, J. C., Generalised conjugate-gradient algorithm for control optimisation using models. Presented at the IEEE Decision and Control Conference, New Orleans, December 1972.
8. Freeman, E. A., On the optimisation of linear, time-variant multivariable control systems using the contraction-mapping principle. *Journal of Optimisation Theory and Applications* **3** (6) (1969), 416–443.
9. Allwright, J. C., Contraction-mapping algorithm with guaranteed convergence. Paper 34.3 of 5th IFAC World Congress, June 1972.
10. Freeman, E. A., Optimisation of contraction-mapping algorithm for calculating optimal controls. *Proc. I.E.E.* **119** (9) (1972), 1365–1371.
11. Freeman, E. A. and Todd, A., Suboptimal control for multivariable systems based on orthogonal decomposition technique. *Int. J. Control* **13** (4) (1971), 737–761.

Discussion

LEIGH (*British Steel Corporation*) Could you clarify the difference between the system and the reduced, or small model?

ALLWRIGHT. As far as the paper is concerned, the system is an equation $\dot{x} = Ax + Bu$ with an observation equation $y = Cx$. This would normally be a mathematical description of an engineering process. The model is a simpler (generally lower order) set of equations which approximates the above system equations. Here simplicity of equations would be measured, essentially, by the amount of computational effort needed to optimise them.

HEALEY (*Cardiff*). Why use weighting on the output vector y rather than on the state x?

ALLWRIGHT. In general, costing the output y corresponds to costing the state with the state cost matrix Q positive semi-definite. For the type of model considered in the paper, it is natural to cost the output, rather than the whole state of the system (which would give Q positive-definite). This arises because we want to use the model to help optimise the system and, as a result of the model being of lower-order than the system, the only common feature of the system and model is that both operate on the same input space to generate responses in the same output space (with, we hope, approximately equal transfer functions). The approach used in the paper could be extended easily to the case of full state cost for the system by letting the output y of the system be the full state for the system and by using a model of the form

$$y = C_0\bar{x} + D_0 u$$

for

$$\dot{\bar{x}} = \bar{A}\bar{x} + \bar{B}u$$

with the dynamic part ($\dot{\bar{x}} = \bar{A}\bar{x} + \bar{B}u$) of the model simpler than the dynamics of the system (so that the model is easier to optimise then the system).

SÁNCHEZ (*University of California*). Could you please summarise the purpose of using the simple model.

ALLWRIGHT. If the usual conjugate-gradient algorithm is used to optimise the complicated system equations, the convergence rate might be poor if the second derivative operator is badly conditioned. The point of the paper is that if a simple model of the system is available, it can be used in a generalized conjugate-gradient algorithm to (hopefully) improve the conditioning. Whether the resulting algorithm is actually more efficient depends on whether the potential increase in convergence rate (per iteration) is sufficiently large to offset the extra work needed at each iteration (to optimise the model). If the model is simple (in the snese that it is easy to optimise) and still gives a good approximation to the system, then this is likely to happen

DAVIES (*University College of Wales*). Is this work in the same spirit as that of Mitra?† Both you and he considered a large system and projected into a lower dimensional subspace, in order to facilitate the problem solution.

ALLWRIGHT. I believe that Dr. Mitra's work is concerned with *finding* a type of simple model. If this is true, it is probably not closely related because the aim here is to show how a *given* model can be used to try to improve the conditioning of a control function optimisation problem.

HEALEY. Is there any prospect of finding feedback controllers using your approach?

ALLWRIGHT. I'm working on it. The best approach seems to be to use the Riccati solution for the model to give an estimate of the system's Riccati solution, which can then be improved iteratively.

†References

1. Mitra, D., Theory of reduced systems. Paper given at Warwick "Exposition of Current Control Theory", July 1969.
2. Mitra, D., On the Reduction of Complexity of Linear Dynamical Models. A.E.E.W., reports R 520, 535; (1967).
3. Mitra, D., Analytical results on the use of reduced models in the control of linear dynamical systems. *Proc. I.E.E.* **116** (8) (1969), 1439–1444.
4. Mitra, D., *W* matrix and the geometry of model equivalence and reduction. *Proc. I.E.E.* **116** (6) (1969), 1101–1106.

13. Duality and Iterative Control Function Optimization for Differentially-described Dynamical Systems

J. C. ALLWRIGHT

Department of Computing and Control
Imperial College of Science and Technology, London, England

1. Introduction

In this paper, an alternative (dual) optimisation problem and a new lower-bound for the minimal value of the performance index are studied for control function optimisation (actually minimisation) problems.

Alternative optimisation problems are of interest when the solution of the original optimisation problem can be found easily from them. This arises because an alternative optimisation problem might have features which render it more attractive from the computational point of view than the direct numerical solution of the original optimisation problem.

The usefulness of lower-bounds for the minimal value of the performance index occurs in the following way. Many iterative algorithms for control optimisation generate a sequence $\{u_j\}$ of approximations to the unknown optimal control u^0. A control u_j might well be considered to be an adequate approximation to the optimal control if $V[u_j] - V[u^0] < \varepsilon$ for some pre-specified $\varepsilon > 0$, where V is the performance function to be minimised. Iterative optimisation could cease when u_j satisfies this condition, since it will then be an adequate approximation to u^0. Now $V[u^0]$ is not usually known beforehand, so this stopping condition is not useful immediately. However if a lower-bound $\check{V}[u_j]$ for the minimal value of the performance index, $V[u^0]$, can be found using little computational effort, then a simple and potentially useful sufficient condition for u_j to be an adequate approximation to u^0 is for $V[u_j] - \check{V}[u_j] < \varepsilon$. Thus lower-bounds can be used to give stopping conditions for iterative control function optimisation.

Pearson [1] was probably the first to consider the use of dual problems in optimal control studies. A summary of general duality results is given in the book by Luenberger [2].

In Section 2 an approach for finding lower-bounds and alternative optimisation problems is introduced which is based on "partial covers" of the (equality) constraints. The lower-bound result is shown to be better than the usual lower-bound result for quadratic programming problems with linear equality constraints, which is relevant because the optimal control problem considered is a special case. The approach can actually be applied to more general constrained optimisation problems [3].

The optimal control problem of interest is formulated in Section 3, where a description of the system differential equation in terms of a Volterra integral equation is mentioned.

An alternative optimisation problem is established in Section 4, using the aforementioned integral description of the system differential equation. A potentially desirable feature is that control function optimisation can be carried out using it without finding the solution of differential equations for a sequence of non-optimal control functions, as is necessary for most conventional iterative algorithms. The only integrations needed are quadratures.

In Section 5, using the Volterra integral equation, a lower-bound for the minimal value of the performance index with advantages over previous lower-bounds is developed. This bound and previous lower-bounds require the solution of an adjoint differential equation. If an exact solution of the adjoint equation is obtained, then the new lower-bound is usually better (i.e. more positive) than the earlier bounds and is at worst the same. However, if only an approximate solution of the adjoint equation is obtained, the formulae for the earlier bounds might not give numbers which are actually lower-bounds. The lower-bound of Section 5 overcomes this difficulty for an important class of problems. Further, it can actually provide information concerning whether a fine enough discretisation of the differential equations has been employed.

2. Partial Covers for Constraints and Lower-bounds

Let **H** and **G** be real Hilbert spaces. Consider the problem of minimising the function

$$\mathbf{J} : \mathbf{H} \to \mathbf{R}$$

subject to the constraint

$$\mathbf{C}[\boldsymbol{h}] = \boldsymbol{0} \tag{2.1}$$

where $\mathbf{C} : \mathbf{H} \to \mathbf{G}$.

For any constraint $\tilde{\mathbf{C}}[\boldsymbol{h}] = \boldsymbol{0}$, denote the minimal value of **J** (if it exists) on the feasible set $\mathbf{F}_{\tilde{\mathbf{C}}} \triangleq \{\boldsymbol{h} \in \mathbf{H} : \tilde{\mathbf{C}}[\boldsymbol{h}] = \boldsymbol{0}\}$ by $\hat{\mathbf{J}}_{\tilde{\mathbf{C}}}$ and denote a representative element of the equivalence class of minimising elements of **H** by $\hat{\boldsymbol{h}}_{\tilde{\mathbf{C}}}$.

DEFINITION 2.1. *A partial cover for the constraint* $\mathbf{C}$ *is defined to be a map* $\mathbf{P}:\mathbf{G}\rightarrow\tilde{\mathbf{G}}$ *for some real Hilbert space* $\tilde{\mathbf{G}}$ *such that* $\mathbf{F_C}\subset\mathbf{F_{PC}}$ *where* $\mathbf{PC}$ *denotes the composite map* $\mathbf{P}[\mathbf{C}[\cdot]]:\mathbf{H}\rightarrow\tilde{\mathbf{G}}$. *If* $\mathbf{F_C}=\mathbf{F_{PC}}$, *then* $\mathbf{P}$ *is called a cover for* $\mathbf{C}$. *Note that* $\mathbf{P}$ *is a partial cover if* $\mathbf{P}[0]=0$ *and a cover if the null-space of* $\mathbf{P}$ *is trivial as well.*

Since $\mathbf{F_C}\subset\mathbf{F_{PC}}$, the minimum of $\mathbf{J}$ on $\mathbf{F_{PC}}$ is the minimum of $\mathbf{J}$ on a set containing $\mathbf{F_C}$ and possibly 'larger' than $\mathbf{F_C}$, so

LEMMA 2.1. *If* $\mathbf{P}$ *is a partial cover for* $\mathbf{C}$,

$$\hat{\mathbf{J}}_{\mathbf{PC}}\leqslant\hat{\mathbf{J}}_{\mathbf{C}}. \tag{2.2}$$

This gives a way for finding lower-bounds. One usually requires the best (i.e. most positive) lower-bound which can be calculated easily. A way of obtaining a sequence of better lower-bounds would be to generate a sequence $\{\mathbf{P}_j\}$ of partial covers such that

$$\mathbf{F}_{\mathbf{P}_{j+1}\mathbf{C}}\subset\mathbf{F}_{\mathbf{P}_j\mathbf{C}}$$

so that

$$\hat{\mathbf{J}}_{\mathbf{P}_{j+1}\mathbf{C}}\geqslant\hat{\mathbf{J}}_{\mathbf{P}_j\mathbf{C}}$$

but this would usually involve a considerable amount of computational effort. Another way of obtaining better lower-bounds will be discussed next.

Consider the (possibly) simplest non-trivial partial cover, that when $\mathbf{P}[\mathbf{C}[\boldsymbol{h}]]=\langle\boldsymbol{p},\mathbf{C}[\boldsymbol{h}]\rangle=\boldsymbol{p}^*\mathbf{C}[\boldsymbol{h}]$ for $\boldsymbol{p}\in\mathbf{G}$. This makes the minimisation of $\mathbf{J}$ on $\mathbf{F_{PC}}$ rather simple, since there is just a scalar constraint.

Since $\hat{\mathbf{J}}_{\boldsymbol{p}^*\mathbf{C}}\leqslant\hat{\mathbf{J}}_{\mathbf{C}}$, it is natural to wonder if

$$\max_{\boldsymbol{p}\in\mathbf{G}}\hat{\mathbf{J}}_{\boldsymbol{p}^*\mathbf{C}}=\hat{\mathbf{J}}_{\mathbf{C}}?$$

In view of (2.2), this is assured if there exists $\boldsymbol{p}\in\mathbf{G}$ such that $\hat{\mathbf{J}}_{\boldsymbol{p}^*\mathbf{C}}=\hat{\mathbf{J}}_{\mathbf{C}}$. In fact:

THEOREM 2.1. *If*

(i) $\mathbf{J}$ *is a convex, Fréchet-differentiable function.*

(ii) $\mathbf{C}[\boldsymbol{h}]=\boldsymbol{D}\boldsymbol{h}-\boldsymbol{d}$ *with* $\boldsymbol{D}$ *a linear operator and* $\boldsymbol{d}\in\mathbf{G}$.

(iii) $\mathbf{R}[\boldsymbol{D}^*]$ *is closed or* $\boldsymbol{g}[\hat{\boldsymbol{h}}_{\mathbf{C}}]\in\mathbf{R}[\boldsymbol{D}^*]$,

then

$$\max_{\boldsymbol{p}\in\mathbf{G}}\hat{\mathbf{J}}_{\boldsymbol{p}^*\mathbf{C}}=\hat{\mathbf{J}}_{\mathbf{C}}.$$

Here D^* denotes the adjoint of D, $\mathrm{R}[D^*]$ the range of D^* and $g[h]$ denotes the gradient of $\mathbf{J}$ with respect to h.

The following standard result is used in the proof of Theorem 2.1, which is given later.

LEMMA 2.2. *For the function* $\mathbf{J}$ *and the constraint* $\mathbf{C}$ *of Theorem* 2.1, $\hat{h}_{\mathrm{C}}$ *minimises* $\mathbf{J}$ *on* $\mathbf{F}_{\mathrm{C}}$ *if and only if*

(i) $D\hat{h}_{\mathrm{C}} = d$

(ii) $g[\hat{h}_{\mathrm{C}}] \in \mathrm{N}[D]^{\perp}$

where $\mathrm{N}[D]^{\perp}$ *denotes the orthogonal complement of* $\mathrm{N}[D]$, *the null-space of* D.

Proof of Theorem 2.1. Now, from [4],

$$\mathrm{N}[D]^{\perp} = \overline{\mathrm{R}[D^*]}$$

(the overbar here denotes closure), so the conditions of Theorem 2.1 reveal that

$$g[\hat{h}_{\mathrm{C}}] = D^* z_{\mathrm{C}}$$

for some $z_{\mathrm{C}} \in \mathbf{G}$. If we take $\alpha p = z_{\mathrm{C}}$ for some $\alpha \in \mathbf{R}$, then

$$\mathbf{J}[\hat{h}_{\mathrm{C}} + \delta h] \geqslant \mathbf{J}[\hat{h}_{\mathrm{C}}] + \langle g[\hat{h}_{\mathrm{C}}], \delta h\rangle = \mathbf{J}[\hat{h}_{\mathrm{C}}] + \alpha\langle D^* p, \delta h\rangle = \mathbf{J}[\hat{h}_{\mathrm{C}}], \quad \forall\, \delta h \in \mathrm{N}[p^* D].$$

Hence $\hat{h}_{\mathrm{C}}$ minimises $\mathbf{J}$ on $\mathbf{F}_{p^*\mathrm{C}}$ as well as on $\mathbf{F}_{\mathrm{C}}$, so

$$\hat{\mathbf{J}}_{p^*\mathrm{C}} = \hat{\mathbf{J}}_{\mathrm{C}}$$

as required.

The lower-bound $\hat{\mathbf{J}}_{p^*\mathrm{C}}$ will next be compared with the standard lower-bound result for the following *Quadratic Programming Problem.* Here

$$\mathbf{J}[h] = a + \langle b, h\rangle + \tfrac{1}{2}\langle h, Kh\rangle$$

with $a \in \mathbf{R}$, $K : \mathbf{H} \to \mathbf{H}$ self-adjoint and positive semi-definite and with $b \in \mathrm{R}[K]$ (so that the minimum of $\mathbf{J}$ on $\mathbf{H}$ is finite). The constraint is

$$Dh = d$$

where D is a linear operator with $\mathrm{N}[D^*] = \{0\}$ (corresponding to linearly independent constraints).

The usual lower-bound approach [5] gives

THEOREM 2.2. *For the above quadratic programming problem, if* $h \in \mathbf{H}$ *and* $q \in \mathbf{G}$ satisfy

$$D^*q = b + Kh \tag{2.3}$$

then

$$\tilde{\mathbf{J}}[h, q] \leqslant \hat{\mathbf{J}}_C$$

where

$$\tilde{\mathbf{J}}[h, q] \triangleq a - \tfrac{1}{2}\langle h, Kh\rangle + \langle d, q\rangle.$$

By taking the partial cover p equal to q, it can be shown easily that the lower-bounds $\hat{\mathbf{J}}_{p*C}$ and $\tilde{\mathbf{J}}$ are related by

THEOREM 2.3. *For the above quadratic programming problem, if*

$$D^*p \in \mathbf{R}[K]$$

$$\begin{aligned}
\hat{\mathbf{J}}_{p*C} &= a - \tfrac{1}{2}\langle b, K^+b\rangle + \tfrac{1}{2}\langle p, [d + DK^+b]\rangle^2 / \langle p, DK^+D^*p\rangle \\
&= \max_{\beta \in \mathbf{R}} \tilde{\mathbf{J}}[h, \beta p] \\
&\geqslant \tilde{\mathbf{J}}[h, p] \\
&= a - \tfrac{1}{2}\langle b, K^+b\rangle + \langle p, [d + DK^+b]\rangle - \tfrac{1}{2}\langle p, DK^+D^*p\rangle
\end{aligned}$$

where K^+ *denotes the pseudo-inverse of* K.

The condition $D^*p \in \mathbf{R}[K]$ in Theorem 2.3 is necessary, since $b \in \mathbf{R}[K]$, if (2.3) is to have a solution h when $p = q$.

Remark 2.1. It is clear from Theorem 2.3 that the lower-bound $\hat{\mathbf{J}}_{p*C}$ established using the partial cover results is better (i.e. potentially more positive) than the usual lower-bound $\tilde{\mathbf{J}}[h, p]$ of Theorem 2.2, even though it is not more expensive to compute than $\tilde{\mathbf{J}}$.

The results will be applied to an optimal control problem in Section 5. As a result of the special structure of the control problem, the associated operator K is very easy to pseudo-invert and a choice for p can be associated with a control function in a simple way which involves little computation. Lower-bound evaluation is therefore rather inexpensive computationally for the optimal control case.

3. Formulation of the Optimal Control Problem

The optimal control problem considered is that of choosing the control function u to minimise the quadratic performance index

$$V[x, u] = \tfrac{1}{2}\int_{\mathbf{T}} \{\langle [x(t) - x_d(t)], Q(t)\,[x(t) - x_d(t)]\rangle + \langle u(t), R(t)u(t)\rangle\}\, dt \tag{3.1}$$

for a linear dynamical system described by the linear (vector) differential equation.

$$\dot{x}(t) = A(t)x(t) + B(t)u(t) : x(t_s) = x_s. \tag{3.2}$$

Here $u(t) \in \mathbf{R}^m$; $x_s, x_d(t), x(t) \in \mathbf{R}^n$; $Q(t) \geqslant 0$ and $R(t) > 0$ are symmetric matrices and $\mathbf{T} = [t_s, t_f], t_f - t_s < \infty$. The functions Q, R, A, B, x_d are all assumed to be continuous on $\mathbf{T}$.

It is well-known that the optimal control function is unique and is given by

$$u^0(t) = -R(t)^{-1}B(t)^T\lambda^0(t) \tag{3.3}$$

where λ^0 is the function λ^0 of the pair (x^0, λ^0) which solves the following two-point boundary-value problem of optimal control:

$$\dot{x}^0(t) = A(t)x^0(t) - B(t)R(t)^{-1}B(t)^T\lambda^0(t) : x^0(t_s) = x_s \tag{3.4}$$

$$-\dot{\lambda}^0(t) = A(t)^T\lambda^0(t) + Q(t)[x^0(t) - x_d(t)] : \lambda^0(t_f) = 0. \tag{3.5}$$

The optimisation problem will be considered here, however, in a Hilbert space context, in which the control function is to be chosen from the Hilbert space $\mathbf{U}$ of equivalence classes $\boldsymbol{u}$ which have square-integrable representative elements (i.e. control functions) $u : \mathbf{T} \to \mathbf{R}^m$. Similarly, $\mathbf{X}$ will denote the Hilbert space for square-integrable functions $x : \mathbf{T} \to \mathbf{R}^n$. For convenience, $\boldsymbol{u}$ will often be referred to as a control function and u will often be taken to denote a representative element of $\boldsymbol{u}$, etc.

The performance index V of (3.1) will therefore be rewritten as

$$V[\boldsymbol{x}, \boldsymbol{u}] = \tfrac{1}{2}\langle[\boldsymbol{x} - \boldsymbol{x}_d], \boldsymbol{Q}[\boldsymbol{x} - \boldsymbol{x}_d]\rangle + \tfrac{1}{2}\langle \boldsymbol{u}, \boldsymbol{R}\boldsymbol{u}\rangle \tag{3.6}$$

where

$$\boldsymbol{Q} : \mathbf{X} \to \mathbf{X} : [\boldsymbol{Q}\boldsymbol{x}](t) = Q(t)x(t) \tag{3.7}$$

$$\boldsymbol{R} : \mathbf{U} \to \mathbf{U} : [\boldsymbol{R}\boldsymbol{u}](t) = R(t)u(t), \qquad \langle \boldsymbol{x}, \boldsymbol{y}\rangle \triangleq \int_{\mathbf{T}} \langle x(t), y(t)\rangle \, dt. \tag{3.8}$$

As a result of the way $Q(t)$ and $R(t)$ are defined, the operators $\boldsymbol{Q}$ and $\boldsymbol{R}$ are self-adjoint with $\boldsymbol{Q} \geqslant 0$ and $\boldsymbol{R} > 0$. They are easy to pseudo-invert and invert, respectively, because it follows from (3.7) and (3.8) that $\boldsymbol{Q}^+$ and $\boldsymbol{R}^{-1}$ are defined by

$$[\boldsymbol{Q}^+\boldsymbol{x}](t) = Q(t)^+x(t), \qquad [\boldsymbol{R}^{-1}\boldsymbol{u}](t) = R(t)^{-1}u(t)$$

so that only operations on finite dimensional (and often diagonal) matrices are involved.

The differential system description of (3.2) can be replaced by integral descriptions in at least two ways.

The usual procedure is to write the solution $x(t)$ of (3.2) as

$$x(t) = x_0(t) + \int_{t_s}^{t} W(t, \tau)u(\tau)\, d\tau \tag{3.9}$$

where $x_0(t) \triangleq \phi(t, t_s)x_s$ and where the convolution kernel $W(t, \tau)$ is equal to $\phi(t, \tau)B(\tau)$. Here $\phi(t, \tau)$ is the state transition matrix for the system of (3.2), so this form of integral description has the disadvantage that it is necessary to find ϕ beforehand. The solution function x of (3.9) can also be written as

$$\boldsymbol{x} = \boldsymbol{x}_0 + \boldsymbol{W}\boldsymbol{u} \tag{3.10}$$

where

$$[\boldsymbol{x}_0](t) = x_0(t), \quad \boldsymbol{W} : \boldsymbol{U} \to \boldsymbol{X} : [\boldsymbol{W}\boldsymbol{u}](t) = \int_{t_s}^{t} W(t, \tau)u(\tau)\, d\tau. \tag{3.11}$$

An integral description can also be obtained by writing

$$x(t) = x(t_s) + \int_{t_s}^{t} \dot{x}(\tau)\, d\tau = x_s + \int_{t_s}^{t} [A(\tau)x(\tau) + B(\tau)u(\tau)]\, d\tau. \tag{3.12}$$

This may be rewritten as

$$\boldsymbol{L}\boldsymbol{x} + \boldsymbol{B}\boldsymbol{u} = \boldsymbol{I}x_s \tag{3.13}$$

where

$$\boldsymbol{L} : \mathbf{X} \to \mathbf{X} : [\boldsymbol{L}\boldsymbol{x}](t) = x(t) - \int_{t_s}^{t} A(\tau)x(\tau)\, d\tau \tag{3.14}$$

$$\boldsymbol{B} : \mathbf{U} \to \mathbf{X} : [\boldsymbol{B}\boldsymbol{u}](t) = -\int_{t_s}^{t} B(\tau)u(\tau)\, d\tau \tag{3.15}$$

$$\boldsymbol{I} : \mathbf{R}^n \to \mathbf{X} : [\boldsymbol{I}x_s](t) = x_s. \tag{3.16}$$

Equation (3.13) is a Volterra integral equation. Note that it is not necessary to find ϕ in this case.

Since $\boldsymbol{L}$ is a Volterra operator, its range is $\mathbf{X}$ and it is invertible on $\mathbf{X}$. It is interesting to note from (3.10) and (3.13) that

$$\boldsymbol{x}_0 = \boldsymbol{L}^{-1}\boldsymbol{I}x_s \qquad \boldsymbol{W} = -\boldsymbol{L}^{-1}\boldsymbol{B}.$$

The integrated version of the system differential equation given by (3.13) will be used as the system description in the rest of the paper.

4. An Alternative Optimisation Problem

An alternative optimisation problem is developed here which has potential advantages from the computational point of view over the direct minimisation of V of (3.1).

The optimisation problem of Section 3 may be viewed as the problem of

$$\min_{x \in \mathbf{X}, u \in \mathbf{U}} V[x, u] \tag{4.1}$$

subject to the linear equality constraint

$$Lx + Bu = Ix_s. \tag{4.2}$$

Let $\mathbf{H}$ be the product space $\mathbf{X} \times \mathbf{U}$, which is a real Hilbert space under the inner product induced by those on $\mathbf{X}$ and $\mathbf{U}$. Denote an element of $\mathbf{H}$ by the vector of functions

$$h = \begin{bmatrix} x \\ u \end{bmatrix} \tag{4.3}$$

and let $K : \mathbf{H} \to \mathbf{H}$ and $D : \mathbf{H} \to \mathbf{X}$ be the matrices of operators

$$K = \begin{bmatrix} Q & 0 \\ 0 & R \end{bmatrix}, \qquad D = [L \quad B]. \tag{4.4}$$

Then $V[x, u]$ of (3.6) can be written as

$$\mathbf{J}[h] = a + \langle b, h \rangle + \tfrac{1}{2}\langle h, Kh \rangle \tag{4.5}$$

with $b \in \mathbf{R}[K]$ and $K \geqslant 0$. The optimisation problem is therefore that of

$$\min_{h \in \mathbf{H}} \mathbf{J}[h]$$

subject to the linear equality constraint

$$Dh = Ix.$$

Since $\mathbf{R}[D^*] \subset \mathbf{N}[D]^{\perp}$, it follows from Lemma 2.2 that sufficient conditions for optimality at h^0 are

$$Dh^0 = Ix_s, \qquad b + Kh^0 = D^* z^0 \tag{4.6}$$

for some $z^0 \in \mathbf{X}$.

Use of (4.3) and (4.4) in (4.6) gives

$$Qx^0 = L^* z^0 + Qx_d \qquad Ru^0 = B^* z^0 \tag{4.7}$$

and so

$$u^0 = R^{-1} B^* z^0. \tag{4.8}$$

The constraint therefore becomes

$$\boldsymbol{L}\boldsymbol{x}^0 + \boldsymbol{B}\boldsymbol{R}^{-1}\boldsymbol{B}^*\boldsymbol{z}^0 = \boldsymbol{I}x_s. \tag{4.9}$$

The results of (4.7–4.9) can be related to the two-point boundary-value problem of optimal control of (3.4–3.5), as explained next.

One way to avoid the explicit two-point boundary-value problem would be to write

$$\lambda^0(t_f) = \lambda^0(t) + \int_t^{t_f} \dot{\lambda}^0(\tau)\, d\tau$$

from which

$$\lambda^0(t) = -\int_t^{t_f} \dot{\lambda}^0(\tau)\, d\tau. \tag{4.10}$$

Then (3.5) can be written (with a slight abuse of notation) as

$$\boldsymbol{L}^*[-\dot{\lambda}^0] = \boldsymbol{Q}[\boldsymbol{x}^0 - \boldsymbol{x}_d] \tag{4.11}$$

where $\boldsymbol{L}^*$ is given by

$$\boldsymbol{L}^* : \mathbf{X} \to \mathbf{X} : [\boldsymbol{L}^*x](t) = x(t) - A(t)^T \int_t^{t_f} x(\tau)\, d\tau \tag{4.12}$$

and (3.3) can be written as

$$\boldsymbol{u}^0 = \boldsymbol{R}^{-1}\boldsymbol{B}^*[-\dot{\lambda}^0] \tag{4.13}$$

where

$$\boldsymbol{B}^* : \mathbf{X} \to \mathbf{U} : [\boldsymbol{B}^*x](t) = -B(t)^\tau \int_t^{t_f} x(\tau)\, d\tau. \tag{4.14}$$

Integrating both sides of (3.4) then gives

$$\boldsymbol{L}\boldsymbol{x}^0 + \boldsymbol{B}\boldsymbol{R}^{-1}\boldsymbol{B}^*[-\dot{\lambda}^0] = \boldsymbol{I}x_s. \tag{4.15}$$

Hence, use of (4.10) has enabled the two-point boundary-value problem of (3.4–3.5) to be converted to the coupled pair of integral equations of (4.11–4.15) with no explicit two-point boundary-value difficulty. There are several similar ways [6] which can be used to convert the canonical equations of the Minimum Principle into equations which do not explicitly involve two-point boundary-value problems, but the procedure outlined above seems to be most suitable for the problem considered here.

It is rather interesting, but not really surprising, that (4.11–15–13) are the same as (4.7–9–8) when $\boldsymbol{z}^0$ is identified with $-\dot{\lambda}^0$. The known existence of a solution for the two-point boundary-value problem of optimal control ensures the existence of a solution pair (x^0, z^0) for (4.7–4.9).

An obvious way to attempt to solve (4.7–4.9) for x^0 and z^0 is to

$$\min_{x,z} ||Qx - L^*z - Qx_d||^2 + ||Lx + BR^{-1}B^*z - Ix_s||^2$$

but a more elegant procedure is developed next.

Equation (4.7) can be solved for x^0 to give

$$x^0 = Q^+[L^*z^0 + Qx_d] + x_1 \tag{4.16}$$

for some $x_1 \in \mathrm{N}[Q]$, if $L^*z^0 \in \mathrm{R}[Q]$, where Q^+ denotes the pseudo-inverse of Q. Use of (4.16) in (4.9) then gives

$$[LQ^+L^* + BR^{-1}B^*]z^0 = Ix_s - LQ^+Qx_d - Lx_1. \tag{4.17}$$

If x_1, the part of x^0 belonging to $\mathrm{N}[Q]$, were known, it would be possible to attempt to solve this for z^0. Usually x_1 will not be known unless $Q > 0$, when $x_1 = 0$ since then $\mathrm{N}[Q] = \{0\}$. The case $Q > 0$ (i.e. $Q(t) > 0$, $\forall t \in \mathbf{T}$) will therefore be considered in the rest of this section.

Now L is invertible on $\mathbf{X}$ so, as $Q, R > 0$

$$LQ^{-1}L^* + BR^{-1}B^* > 0.$$

Since $LQ^{-1}L^* + BR^{-1}B^*$ is self-adjoint and positive-definite, the solution z^0 of (4.7) can be found by solving a simple optimisation problem, as stated in

THEOREM 4.1. *If $Q > 0$, the optimal control function u^0 and the associated solution x^0 of the system differential equation are given by*

$$u^0 = R^{-1}B^*z^0, \qquad x^0 = Q^{-1}L^*z^0 + x_d \tag{4.18}$$

where z^0 either maximises

$$\mathrm{I}[z] = \langle [\mathrm{I}x_s - Lx_d], z\rangle - \tfrac{1}{2}\langle z, [LQ^{-1}L^* + BR^{-1}B^*]z\rangle \tag{4.19}$$

or minimises $-\mathrm{I}$.

Theorem 4.1 contains the alternative optimisation problem of [7], It could actually have been obtained from the theory underlying Theorem 2.2 but it is believed that the above derivation gives more insight. Various computational considerations will be mentioned next.

Remark 4.1. (Solution of the alternative problem).

The conjugate-gradient algorithm can be used to give an efficient procedure for solving the alternative optimisation problem and hence for finding the optimal control. Starting with a given initial estimate $z_0 \in \mathbf{X}$ of z^0, it generates a sequence $\{z_j\} \to z^0$. In view of (4.18), if

$$\boldsymbol{u}_j \triangleq \boldsymbol{R}^{-1}\boldsymbol{B}^* z_j, \qquad \boldsymbol{x}_j \triangleq \boldsymbol{Q}^{-1}\boldsymbol{L}^* z_j + \boldsymbol{x}_d$$

then the sequence $\{\boldsymbol{u}_j\}$ induced by the conjugate-gradient algorithm converges to the optimal control $\boldsymbol{u}^0$. Similarly, $\{\boldsymbol{x}_j\} \to \boldsymbol{x}^0$.

Two interesting points associated with the use of the conjugate-gradient algorithm are mentioned next.

The conjugate-gradient algorithm requires that the gradient of $\mathbf{I}$ with respect to z be computed at each iteration. The gradient is given by

$$\begin{aligned} \boldsymbol{g}[z_j] &= \boldsymbol{I}\boldsymbol{x}_s - \boldsymbol{L}\boldsymbol{x}_d - [\boldsymbol{L}\boldsymbol{Q}^{-1}\boldsymbol{L}^* + \boldsymbol{B}\boldsymbol{R}^{-1}\boldsymbol{B}^*]z_j \\ &= \boldsymbol{I}\boldsymbol{x}_s - \boldsymbol{L}\boldsymbol{x}_j - \boldsymbol{B}\boldsymbol{u}_j. \end{aligned}$$

This is just the error in satisfying the constraint of (4.2) and is easy to evaluate.

Since the conjugate-gradient algorithm optimises on an expanding sequence of varieties, the gradients $\boldsymbol{g}[z_0], \boldsymbol{g}[z_1], \ldots, \boldsymbol{g}[z_j], \ldots$ are orthogonal so that each additional iteration increases by one the dimension of the subspace on which the constraint is satisfied. Therefore if $z_0 = \boldsymbol{0}$ (which is a natural choice)

$$V[\boldsymbol{x}^0, \boldsymbol{u}^0] \geqslant \ldots \geqslant V[\boldsymbol{x}_{j+1}, \boldsymbol{u}_{j+1}] \geqslant V[\boldsymbol{x}_j, \boldsymbol{u}_j] \geqslant \ldots \geqslant V[\boldsymbol{x}_0, \boldsymbol{u}_0].$$

Remark 4.2. (Potential computational advantages of control function optimisation by solving the alternative problem).

All the operations involved in solving the alternative problem of Theorem 4.1 using the conjugate-gradient algorithm either involve no integration or only integrations of the type needed to evaluate

$$\int_{t_s}^{t} s(\tau)\, d\tau$$

as a function of t for given s. Such integrations seem to be more robust numerically than the solution of differential equations such as $\dot{x}(t) = f(x(t), u(t); t)$ using standard integration algorithms. Also such integrations can be evaluated recursively (in the sense that

$$\int_{t_s}^{t+\delta} s(\tau)\, d\tau = \int_{t_s}^{t} s(\tau)\, d\tau + \int_{t}^{t+\delta} s(\tau)\, d\tau)$$

unlike those needed for evaluating convolution integrals

$$\int_{t_s}^{t} W(t, \tau)u(\tau)\,d\tau$$

as a function of t.

Most conventional iterative algorithms for finding the optimal control require (ideally) the exact solution of the system and adjoint differential equations for the sequences of non-optimal control functions generated by the algorithms. This seems to be undesirable since there is usually no long-term interest in such solutions for non-optimal control functions. As explained above, the alternative optimisation problem does not require such solutions.

An additional point is that if z were to be optimised on a subspace spanned by functions which can be integrated analytically, say by piecewise polynomials, and if A, B, Q^{-1}, R^{-1} and x_d were to have elements which are also piecewise polynomials of t, then all the integrations involved could be carried out analytically.

Thus finding the optimal control using the alternative optimisation problem has potential advantages from the computational point of view over the use of the usual iterative algorithms for the direct minimisation of V. A detailed numerical analysis of control function optimisation using the alternative problem will be needed to establish whether the above potential advantages are in fact real advantages computationally.

Remark 4.3. (Comparison with ε-techniques).

Balakrishnan [8] has considered in detail the conversion of the solution of optimal control problems with their dynamic constraints into unconstrained optimisation problems by adding a penalty function term to the performance index to cost the error in satisfying the constraints. A complete cost-functional of the following type is constructed.

$$J[\varepsilon, x, u] = \frac{1}{2\varepsilon}\int_{\mathbf{T}} \|\dot{x}(t) - f[x(t), u(t); t]\|^2\,dt + V[x, u]$$

when the minimisation of $V[x, u]$ for the system

$$\dot{x}(t) = f[x(t), u(t); t] : x(t_s) = x_s$$

is being considered. Then J is minimised with respect to x and u for $\varepsilon > 0$. Denote the minimising pair by $(x_\varepsilon, u_\varepsilon)$. If $\{\varepsilon_j\}$ is a monotonically decreasing sequence of positive ε which converges to zero, then, under fairly general conditions,

$$\{x_{\varepsilon_j}\} \to x^0, \{u_{\varepsilon_j}\} \to u^0.$$

There are two superficial disadvantages associated with the nature of the penalty function term:

(i) If x is a numerically defined function, differentiation might be needed to obtain $\dot{x}$ from x—which would be undesirable numerically.

(ii) The initial condition $x(t_s) = x_s$ has to be incorporated in some way. The obvious way to overcome both objections is to constrain x to belong to a linear variety consisting of a translation of a finite-dimensional subspace spanned by functions with known derivatives such that every element of the variety has value x_s at time t_s. Then x could be chosen optimally from the variety by optimal choice of the components of the basis functions.

Following the use of the Volterra integral equation description of the system equations in [7], Frick [9] considered constructing a different penalty-function technique with composite cost functional defined by

$$J[\varepsilon, x, u] = \frac{1}{2\varepsilon}\int_{T}\left\|x(t) - x_s - \int_{t_s}^{t} f[x(\tau), u(\tau); \tau]\, d\tau\right\|^2 dt + V[x, u] \quad (4.20)$$

and obtained essentially the same convergence results as Balakrishnan. In this case the initial condition is incorporated automatically and no differentiation of x is required.

A general disadvantage of such penalty function methods is that optimisation is needed for a sequence (or at least several) values of ε. It will be shown next that for the function J of (4.20), each such optimisation is not essentially easier that the single optimisation needed to find the optimal control when the result of Theorem 4.1 is used.

The problem can be viewed as that of minimising, for $\boldsymbol{K} > 0$,

$$V[\boldsymbol{h}] = a + \langle \boldsymbol{b}, \boldsymbol{h}\rangle + \tfrac{1}{2}\langle \boldsymbol{h}, \boldsymbol{K}\boldsymbol{h}\rangle$$

subject to the linear equality constraint

$$\boldsymbol{D}\boldsymbol{h} = \boldsymbol{d}.$$

The ε-technique would involve minimising, for each $\varepsilon > 0$,

$$J[\varepsilon, \boldsymbol{h}] = \frac{1}{2\varepsilon}\|\boldsymbol{D}\boldsymbol{h} - \boldsymbol{d}\|^2 + V[\boldsymbol{h}].$$

Trivial manipulations reveal that the $\boldsymbol{h}$ which minimises $J[\varepsilon, \boldsymbol{h}]$ is

$$\boldsymbol{h}_\varepsilon = \boldsymbol{K}^{-1}\boldsymbol{D}^* z_\varepsilon - \boldsymbol{K}^{-1}\boldsymbol{b}$$

where z_ε satisfies

$$[\varepsilon \boldsymbol{I} + \boldsymbol{D}\boldsymbol{K}^{-1}\boldsymbol{D}^*]\, z_\varepsilon = \boldsymbol{d} + \boldsymbol{D}\boldsymbol{K}^{-1}\boldsymbol{b}.$$

The solution of the constrained optimisation problem is, however,

$$\boldsymbol{h}^0 = \boldsymbol{K}^{-1}\boldsymbol{D}^*\boldsymbol{z}^0 - \boldsymbol{K}^{-1}\boldsymbol{b}$$

where

$$\boldsymbol{D}\boldsymbol{K}^{-1}\boldsymbol{D}^*\boldsymbol{z}^0 = \boldsymbol{d} + \boldsymbol{D}\boldsymbol{K}^{-1}\boldsymbol{b}.$$

The result of Theorem 4.1 is equivalent to this.

Comparison of the formulae for z_ε and z^0 shows that it is not essentially easier to find $\boldsymbol{h}_\varepsilon$, for any given ε, than to find the solution $\boldsymbol{h}^0$ of the constrained optimisation problem.

Hence, for the control optimisation problem considered here with $\boldsymbol{Q} > 0$, the ε-technique does not seem to have advantages over the alternative optimisation problem of Theorem 4.1.

Remark 4.4. (Application to more general problems).

The result of Theorem 4.1 could be used for the case $\boldsymbol{Q} \geqslant \boldsymbol{0}$ by perturbing $\boldsymbol{Q}$ to $\tilde{\boldsymbol{Q}} > 0$ and then using the result to carry out the subsidiary optimisations which arise when the technique of [10] is used to find the optimal control for the original problem by solving a sequence of optimisation problems for the approximation to the original problem obtained by replacing $\boldsymbol{Q}$ by $\tilde{\boldsymbol{Q}}$. The procedure could also be used for nonlinear systems, etc., by using it to solve the sequence of linearised sub-problems which often arise in algorithms for solving nonlinear problems.

In this section an alternative optimisation problem has been discussed which is of some theoretical interest and has potential computational advantages, especially for problems so large that the Riccati technique is not feasible. The alternative problem obtained in Section 2 using partial cover analysis was not used here because it gives rise to a non-quadratic optimisation problem. The results of the partial cover analysis are used in the next section, however, where a new lower-bound is considered.

5. A Sharp Lower-Bound for the Minimal Value of the Performance Index

Let x_u be the solution of the system differential equation of (3.2) for a given control function u and let λ_u be the solution of the associated adjoint differential equation

$$-\dot{\lambda}_u(t) = A(t)^T\lambda_u(t) + Q(t)[x_u(t) - x_d(t)] : \lambda_u(t_f) = 0.$$

The gradient $g[u]$ of the performance index with respect to the control function u is then given by

$$g[u](t) = R(t)u(t) + B(t)^T\lambda_u(t).$$

Pearson's lower-bound for $V[x^0, u^0]$ is [1]

$$w[u] = \langle \lambda_u(t_s), x_s \rangle - \tfrac{1}{2}\int_{\mathbf{T}} \{\langle [x_u(t) + x_d(t)], Q(t)[x_u(t) - x_d(t)] \rangle + \langle \lambda_u(t), B(t)R(t)^{-1}B(t)^T \lambda_u(t) \rangle\}\, dt.$$

It can be shown [11] that a lower-bound is also given by

$$\tilde{V}[u] = V[x_u, u] - \tfrac{1}{2}\int_{\mathbf{T}} \langle g[u](t), R(t)^{-1} g[u](t) \rangle\, dt.$$

The main result of this section is

THEOREM 5.1. *A sharp lower-bound for the minimal value of the performance index is*

$$\hat{V}[u] = \tfrac{1}{2}\langle [\boldsymbol{I}x_s - \boldsymbol{L}x_d], \lambda_u \rangle^2 / \langle \lambda_u, [\boldsymbol{L}\boldsymbol{Q}^+\boldsymbol{L}^* + \boldsymbol{B}\boldsymbol{R}^{-1}\boldsymbol{B}^*]\lambda_u \rangle.$$

Further:

$$\hat{V}[u] \geqslant w[u] = \tilde{V}[u], \qquad \{\hat{V}[u_j]\} \to V[x^0, u^0] \quad \textit{if} \quad \{u_j\} \to u^0.$$

Proof. Use of the formulation of the optimal control problem of (4.1) and (4.2) with Theorem 2.3 gives the lower-bounds

$$\hat{\mathbf{J}} = \tfrac{1}{2}\langle \boldsymbol{p}, [\boldsymbol{I}x_s - \boldsymbol{L}\boldsymbol{Q}^+\boldsymbol{Q}x_d] \rangle^2 / \langle \boldsymbol{p}, [\boldsymbol{L}\boldsymbol{Q}^+\boldsymbol{L}^* + \boldsymbol{B}\boldsymbol{R}^{-1}\boldsymbol{B}^*]\boldsymbol{p} \rangle \tag{5.1}$$

and

$$\tilde{\mathbf{J}} = \langle \boldsymbol{p}, [\boldsymbol{I}x_s - \boldsymbol{L}\boldsymbol{Q}^+\boldsymbol{Q}x_d] \rangle - \tfrac{1}{2}\langle \boldsymbol{p}, [\boldsymbol{L}\boldsymbol{Q}^+\boldsymbol{L}^* + \boldsymbol{B}\boldsymbol{R}^{-1}\boldsymbol{B}^*]\boldsymbol{p} \rangle \tag{5.2}$$

with

$$\hat{\mathbf{J}} \geqslant \tilde{\mathbf{J}}$$

if $\boldsymbol{L}^*\boldsymbol{p} \in \mathbf{R}[\boldsymbol{Q}]$.

In fact, Eqn (2.3) gives, for this case,

$$\boldsymbol{L}^*\boldsymbol{p} = \boldsymbol{Q}[\boldsymbol{x} - \boldsymbol{x}_d].$$

This suggests a simple way of associating lower-bounds with any given control function u: set $\boldsymbol{p} = \boldsymbol{p}_u$ in $\hat{\mathbf{J}}$ of (5.1) and $\tilde{\mathbf{J}}$ of (5.2) where $\boldsymbol{p}_u$ satisfies

$$\boldsymbol{L}^*\boldsymbol{p}_u = \boldsymbol{Q}[\boldsymbol{x}_u - \boldsymbol{x}_d] \in \mathbf{R}[\boldsymbol{Q}]. \tag{5.3}$$

It can be shown that $\boldsymbol{p}_u = -\lambda_u$ satisfies (5.3). Use of this choice for $\boldsymbol{p}$ in $\hat{\mathbf{J}}$ of (5.1) then gives the lower-bound $\hat{V}[u]$ of Theorem 5.1. It is a simple matter to show that $w[u] = \tilde{V}[u]$ and that $\tilde{V}[u]$ is equal to $\tilde{\mathbf{J}}$ of (5.2) when $\boldsymbol{p}$ in (5.2) is replaced by $-\lambda_u$. Hence

$$V[x^0, u^0] \geqslant \hat{V}[u] \geqslant w[u] = \tilde{V}[u]. \tag{5.4}$$

When $u = u^0$, $g[u] = 0$ and so $\tilde{V}[u] = V[x^0, u^0]$. Setting $u = u^0$ in (5.4) therefore reveals that $\hat{V}$ is a sharp lower-bound. Also, if $\{u_j\} \to u^0$, $\{g[u_j]\} \to 0$ and $\tilde{V}[u_j] \to V[x^0, u^0]$. Therefore, from (5.4), $\{\hat{V}[u_j]\} \to V[x^0, u^0]$. This completes the proof.

Remark 5.1. (Superiority of $\hat{V}$).

The results of Theorem 5.1 show that the new bound $\hat{V}$ is a better lower-bound than the previous lower-bounds $w[u]$ and $\tilde{V}[u]$.

Remark 5.2. (Effects of errors in the solution of the adjoint equation).

In conventional iterative algorithms, the solution λ_u of the adjoint equation is usually needed for each u, for finding the gradient. Thus the *extra* work needed to compute any of the lower-bounds is not large. In practice, however, only approximations $\bar{x}_u$ and $\dot{\bar{\lambda}}_u$ to x_u and λ_u will be available, since x_u and λ_u are the solutions of differential equations. It can be shown that the above formulae for $w[u]$ and $\hat{V}[u]$ do not necessarily give numbers which are lower-bounds when such approximations are used in them. However, it follows from the proof of Theorem 5.1 that $\hat{V}[u]$ gives a lower-bound value so long as $\boldsymbol{L}^*[-\dot{\bar{\lambda}}_u] \in \mathbf{R}[Q]$. In the important case when $Q(t) > 0$, $\forall\, t \in \mathbf{T}$, this is true for any function $\dot{\bar{\lambda}}_u$, so $\hat{V}$ then gives a value which is a lower-bound in spite of (possible) errors in $\dot{\bar{\lambda}}_u$. This is a valuable feature of $\hat{V}$.

Remark 5.3. (Some computational points).

The first point is that if $\bar{\lambda}_u$ is absolutely continuous and $\bar{\lambda}_u(t_f) = 0$, then $[\boldsymbol{L}^*\dot{\bar{\lambda}}_u](t) = \dot{\bar{\lambda}}_u(t) + A(t)^T\bar{\lambda}_u(t)$ and $[\boldsymbol{B}^*\dot{\bar{\lambda}}_u](t) = B(t)^T\bar{\lambda}_u(t)$. These expressions could be used in the formula for $\hat{V}$ to avoid the integrations associated with $\boldsymbol{L}$ and $\boldsymbol{B}$. The second point is that the only integrations needed to evaluate $\hat{V}$ for given $\dot{\bar{\lambda}}_u$ are quadratures. In fact, if A, B, Q, R^{-1} and x_d are piecewise polynomials of t and if a piecewise polynomial approximation $\dot{\bar{\lambda}}$ is obtained, then all the integrations involved in the evaluation of $\hat{V}$ can be carried out analytically. The only numerical errors which might affect the lower-bound property of $\hat{V}$ are then round-off errors associated with manipulations of the coefficients of the polynomials.

Remark 5.4. (Is a finer discretisation of time required?).

The lower-bound $\hat{V}$ of Theorem 5.1 can be used to provide evidence concerning whether a fine enough discretisation of time has been used in numerical calculations, when $Q(t) > 0$, $\forall\, t \in \mathbf{T}$. Suppose u is the control function which satisfies the discretised optimality conditions and that $\dot{\bar{\lambda}}_u$ is a piecewise polynomial approximation to λ_u obtained from the discretised equations. Then $\hat{V}$ evaluated for $\dot{\bar{\lambda}}_u$ using analytical integrations would be a lower-bound for the *undiscretised* minimal value of the performance index (recall Remark 5.2). If the difference between this value of $\hat{V}$ and the calculated value of V were large, then it would be strong evidence that a finer discretisation is required.

Lower-bound results related to those mentioned above can also be obtained by using the convolution integral equation of (3.9) as the constraint, instead of the Volterra integral equation of (4.2). However, the insensitivity with respect to errors in the solution of the adjoint differential equation does not then occur.

6. Conclusions

A technique for finding lower-bounds for constrained minimisation problems has been presented which gives a better bound than the usual lower-bound result of quadratic programming. This result has been applied to an optimal control problem to give a new lower-bound with advantages over previous bounds. The lower-bound was obtained using a system description in terms of a Volterra integral equation equivalent to the usual differential equation description. Use of the Volterra integral equation gives rise, in a natural way, to an alternative (dual) optimisation problem, from the solution of which the required optimal control function can be found easily. Solution of the alternative problem has potential computational advantages over the direct solution of the original (primal) optimal control problem. The results can be extended easily to similar optimal control problems with terminal equality constraints.

References

1. Pearson, J. D., Reciprocity and duality in control programming problems. *J. Math. Anal. Appl.* **10** (1965), 388–408.
2. Luenberger, D. G., "Optimization by Vector Space Methods". Wiley, New York (1969.)
3. Allwright, J. C., "Lower-bounds for Constrained Minimisation Problems". Department of Computing and Control Publication, Forthcoming, Imperial College (1973).
4. Taylor, A. E., "Introduction to Functional Analysis". Wiley, New York (1958).
5. Dennis, J. B., "Mathematical Programming and Electrical Networks". Technology Press Research Monograph (1959).
6. Allwright, J. C., "Control Function Optimisation Using Quadrature Integrations and Pontryagin's Minimum Principle". Control and Automation Section Report 70/15, Imperial College (1970).
7. Allwright, J. C., Simultaneous solution of system dynamic equations and control function optimisation using the conjugate-gradient algorithm. *Electronics Lett.* **6** (11) (1970), 322–323.

8. Balakrishnan, A. V., On a new computing technique in optimal control. *SIAM J. Control* **6** (2) (1968), 149–173.
9. Frick, P. A. Iterative optimum control function determination without directly solving the system dynamical equations. Paper 34.2, Preprints of 5th IFAC World Congress, Paris (1972).
10. Allwright, J. C., Improving the conditioning of optimal control problems using simple models. This Volume, pp. 173–191.
11. Allwright, J. C., Stopping conditions for iterative control function optimisation. *Electronics Lett.* **6** (17) (1970), 545–547.

14. Conditions for Compactness of the Time-phase Space in Control Problems*

D. A. SÁNCHEZ

Department of Mathematics
University of California, Los Angeles, California, U.S.A.

In Lagrange or Mayer problems of control theory, where one is discussing the existence of a minimum for the cost functional, it is usually assumed that the time-phase space

$$A = \{(t, x)\} \subset R^1 \times R^n$$

is compact. Furthermore, if $U(t, x) \subset R^m$ is the control space, then it is assumed that the set

$$M = \{(t, x, u) \mid (t, x) \in A, \qquad u \in U(t, x)\}$$

is also compact.

If A is not compact but only closed then conditions must be given which insure that the trajectories $\{x_k(t)\}$ of an admissible minimizing sequence $\{x_k(t), u_k(t)\}$ lie in a compact subset $A_0 \subset A$, and then that the corresponding set

$$M_0 = \{(t, x, u) \mid (t, x) \in A_0, \qquad u \in U(t, x)\}$$

is compact. In the most common case, which we discuss here, A is contained in a slab $\{(t, x) \mid a \leqslant t \leqslant b, x \in R^n\}$ and it is first assumed that

(a) for every number $N \geqslant 0$ the set $M_N = \{(t, x, u) \mid (t, x, u) \in M, |x| \leqslant N\}$ is compact;

(b) there exists a compact subset $P \subset A$ such that every admissible trajectory $x(t)$ has at least one point $(t_0, x(t_0)) \in P$.

For the most frequent case, A a slab and $U(t, x) = \{u \mid |u| \leqslant k\}$, condition (a)

**Editor's Note.* This is an abstract of Professor Sánchez's paper. The full text is to appear in the *Boletín de la Sociedad Matemática Mexicana* (in Spanish). Permission was not obtained to publish the paper in these proceedings in any language.

is satisfied, and condition (b) is satisfied by requiring that all trajectories originate from a fixed point or fixed compact initial set.

For the Mayer problem of control a further "finite-escape time" or Liapunov type condition is imposed on the governing differential equations to insure that every admissible trajectory (i.e. solution) remains in a fixed, compact set. For the Lagrange problem with cost functional $I[x, u]$ either similar conditions are imposed, or one insures that given a set K of admissible pairs $\{(x_\alpha, u_\alpha)\}$ for which $I[x_\alpha, u_\alpha] \leqslant L$, then $|x_\alpha| \leqslant N$ for some constant N, e.g. if K is a minimizing sequence such that $\lim I[x_\alpha, u_\alpha] = \iota = \inf I[x, u]$ then one could let $L = \iota + 1$. Examples of some of these types of conditions may be found in the following papers: L. Cesari [1], [2], A. F. Filippov [4] and E. J. McShane [6].

In the early literature on the calculus of variations the case where the time-phase space is not compact was also discussed; in particular by L. Tonelli [7; Vol. II, pp. 307 ff.], E. J. McShane [5, pp. 302–304], and S. Cinquini [3, pp. 184–190]. When A was contained in a strip ($n = 1$) and hypothesis (b) above was satisfied, Tonelli gave conditions on the cost functional $I[x]$ which insured that $I[x] \to \infty$ as $x_{\max} \to \infty$ or $x_{\max} - x_{\min} \to \infty$ where

$$x_{\max} = \max_{a<t<b} |x(t)| \qquad \text{and} \qquad x_{\min} = \min_{a\leqslant t\leqslant b} |x(t)|.$$

Clearly, in view of (b), then any set of trajectories for which $I[x] \leqslant L$ must be uniformly bounded. Cinquini modified and extended the results of Tonelli to the case where $I[x]$ depended on derivatives of higher order and it is his results we wish to show are applicable to the Lagrange-type problems of control—the old wine of "il campo illimitato" into the new bottle of "noncompact time-phase space in optimal control".

References

1. Cesari, L., Existence theorems for optimal solutions in Pontryagin and Lagrange problems. *SIAM J. Control* **3** (1966), 475–498.
2. Cesari, L., Existence theorems for weak and usual optimal solutions in Lagrange problems with unilateral constraints, I. *Trans. Amer. Math. Soc.* **134** (1966), 369–412.
3. Cinquini, S., Sopra l'esistenza della soluzione nei problemi di calcolo delle variazioni di ordine *n*. *Ann. Scuola Norm. Super. Pisa* **5** (2) (1936), 169–190.
4. Filippov, A. F., On certain questions in the theory of optimal control. *SIAM J. Control* **1** (1962), 76–84.
5. McShane, E. J., Existence theorems for ordinary problems of the calculus of variations. *Ann. Scuola Norm. Super. Pisa* **3** (2) (1934), 181–211, 287–315.
6. McShane, E. J., Relaxed controls and variational problems, *SIAM J. Control* **5** (1967), 438–485.
7. Tonelli, L., Fondamenti di Calcolo delle Variazioni, (2 Vols.) Zanichelli, Bologna, Vol. 1, 1921, Vol. 2, 1923.

15. Two Recent Results in Nonlinear Filtering Theory

J. M. C. Clark

Department of Computing and Control,
Imperial College of Science and Technology, London, England

Introduction

This paper is in the nature of a supplement to the several recently published books and papers, for instance [1], [2] and [3], that survey the field of nonlinear filtering. We report on two very recent results that we feel deserve a wider audience than they are likely to obtain in the near future.

The first is a new proof by Fujisaki, Kallianpur and Kunita [4] of a filtering representation theorem. This is the latest in a long line of derivations that goes back to a basic paper [5] by Stratonovich in 1960, in which he posed the problem of determining the conditional probability density of a "signal" process (x_t) given an "observation" process (z_t) under the assumption that (x_t, z_t) formed a Markov diffusion process with prescribed coefficients. His solution took the form of a stochastic partial differential equation for the density of x_t. This, or its dual description as a system of ordinary differential equations for conditional moments of x_t, is a natural form for the solution since it describes, in principle at least, how the density or its moments change as new observations arise, and it is this feature that distinguishes the filtering problem from problems of estimation from observations over a fixed period, for which the density may be more appropriately described as a fixed functional of the observation process. The problem was also considered by Kushner [6] who rederived, with some minor corrections, Stratonovich's equation in the more familiar form of Ito stochastic differential equations. A number of papers followed with some important extensions, and most of these used in various ways the concept of a "generalised Bayes formula", first proposed by Bucy [7], that expressed the density as a functional on the observations. Perhaps the most mathematically complete statement of the formula has been given by Kallianpur and Striebel [8]. This approach has many applications in estimation theory and good accounts of it are given in [1] and [3]. Filtering equations can be obtained from such a formula by expansion according to Ito's transformation rule, but rigorous justification of this procedure seems to require a number of rather artificial moment conditions. A second approach is to show first that the conditional moments or density evolve according to Ito stochastic differential equations, and then to identify the coefficients in these equations directly. Though this idea has been known for some time, see for instance the formal discussion in [9], it was first used systematically by Frost and Kailath [10], [11] and [12] as part of their "innovations" approach to estimation theory. With it they derived in a natural fashion the equations of linear filtering and smoothing, but lack of an appropriate stochastic integral representation lemma prevented them from extending their results (in any generality) to the nonlinear filtering problem. In [4], however, Fujisaki, Kallianpur and Kunita have supplied the missing lemma, though not in quite the form conjectured by Frost and Kailath, and have presented a proof of the nonlinear filtering result that requires only the mildest of moment conditions. In the next section we give a brief account of their theorem and its proof. The proof, which, apart from the lemma referred to, requires little more than the composition of conditional expectations, emphasises the martingale nature of the result and seems to us to be the mathematically "correct" solution for the problem.

The second result, though not quite as fundamental as the first, is nevertheless interesting from a practical point of view. A disappointing feature of the filtering equations is that in general an infinite number of them are required to describe a filter; the equation for $E^t x_t$ involves $E^t[x_t h_t]$, the equation for this moment involves higher order moments and so on. There are, of course, examples of optimal filters that are finite-dimensional, the best known being the Kalman–Bucy filter for Gaussian Markov processes [1] and filter of Wonham [13] for finite-state Markov chains. Lo and Willsky [14] have shown that with suitable mathematical formulation the filtering problem for a large class of processes involving a rotation coordinate can also be resolved with a finite-dimensional filter. This has applications in problems of frequency and phase demodulation, frequency stability and gyroscopic analysis. We describe their results in the last section. Their papers cover many aspects of estimation for such processes, both in continuous and discrete time, but we concern ourselves only with the results on filtering.

The Filtering Theorem of Fujisaki, Kallianpur and Kunita

First we recall some definitions. The processes we consider will be assumed to be defined on some underlying probability space $(\Omega, \mathbf{B}, P)$. Suppose $(\mathbf{F}_t)$ is an increasing right-continuous family of σ-fields in $\mathbf{B}$. A stochastic process $(m_t(\omega))$, $t \geqslant 0$ is called a martingale on $(\mathbf{F}_t)$ if

$$E[m_t \mid \mathbf{F}_s] = m_s \quad \text{a.s.} \quad \text{for all} \quad s < t.$$

A process (x_t) is said to be adapted to $(\mathbf{F}_t)$ if, for each t, x_t is $\mathbf{F}_t$-measurable Though it is not standard terminology, we will say a vector Brownian motion (w_t) (that is, the Gaussian process with $Ew_t^i = 0$. $E[w_t^i w_s^j] = \min(t, s)\,\delta_{ij}$) is a Brownian motion on $(\mathbf{F}_t)$ if it is adapted to it and the σ-field generated by $(w_u - w_t)$ for fixed t and all $u > t$ is independent of $\mathbf{F}_t$. We assume the reader is familiar with the Ito stochastic integral and the Ito expansion rule for products of integrals etc. In the following analysis we frequently use the composition rules of conditional expectations. For instance if x is a random variable, then $\mathbf{F}_s \subset \mathbf{F}_t$ implies $E[x \mid \mathbf{F}_s] = E[E[x \mid \mathbf{F}_t] \mid \mathbf{F}_s]$, or more generally if (h_t) is integrable in (t, ω),

$$E\left[\int_s^t h_u\,du \mid \mathbf{F}_s\right] = E\left[\int_s^t E[h_u \mid \mathbf{F}_u]\,du \mid \mathbf{F}_s\right].$$

We remark that a (t, ω)-measurable modification of the process $E[h_u \mid \mathbf{F}_u]$ always exists.

Suppose a scalar signal process (y_t) and a vector observation process (z_t) are described on $0 \leqslant t \leqslant T$ by

$$y_t = y_0 + \int_0^t a_s\,ds + \int_0^t b_s\,dw_s$$

$$z_t = \int_0^t h_s\,ds + v_t$$

where the scalar process (a_t) and the vector processes (b_t) and (h_t) are (t, ω)-measurable and adapted to an increasing family of σ-fields $(\mathbf{F}_t)$ representing all past events, where (v_t) is a vector Brownian motion on $(\mathbf{F}_t)$ with $\dot{v}_t$ representing additive white noise, $(w\text{-})$ is a vector of not necessarily independent Brownian motions on $(\mathbf{F}_t)$, and where (w_t) and (v_t) are correlated by

$$E[(w_t - w_s)(v_t - v_s)'] = C(t - s), \qquad t > s.$$

Let $\mathbf{Z}_t$ be the σ-field generated by $(z_s, 0 \leqslant s \leqslant t)$; that is, the σ-field of observed events. Let $\hat{h}_t$ denote $E[h_t \mid \mathbf{Z}_t]$ and generally let $E^t[.]$ denote the expectation $E[. \mid \mathbf{Z}_t]$.

Theorem

If $Ey_0^2 < \infty$ and the processes (a_t), (b_t) and (h_t) are square-integrable on $[0, T] \times \Omega$; that is

$$\int_0^T E[a_t'\,a_t]\,dt < \infty$$

etc., then the process $E^t\,y_t = E[y_t \mid \mathbf{Z}_t]$ satisfies

$$d(E^t\,y_t) = E^t\,a_t\,dt + E^t[y_t(h_t - \hat{h}_t)' + b_t'C]\,d\bar{v}_t$$

where $(\bar{v}_t)$ is defined by

$$\bar{v}_t = z_t - \int_0^t \hat{h}_s \, ds$$

and is a vector Brownian motion on $(\mathbf{Z}_t)$.

This is a slightly specialised version of the main filtering theorem of Fujisaki, Kallianpur and Kunita. Its form and proof emphasise the fact that it is basically a statement about martingales rather than Markov processes. The following corollary, which is close to the original statement of Stratonovich, is a rephasing in terms of Markov processes, but its proof follows immediately from the theorem.

Corollary

If (x_t) and (z_t) form a joint vector Markov process

$$dx_t = a(x_t, z_t)\, dt + B(x_t, z_t)\, dw_t$$
$$dz_t = h(x_t, z_t)\, dt + dv_t, \qquad z_0 = 0,$$

where a, B and h are uniformly Lipschitz continuous in (x, z), where (w_t) and (v_t) are independent vector Brownian motions and where $Ex_0^2 < \infty$, then for any twice-continuously differentiable function f satisfying $Ef(x_t)^2 < \infty$,

$$d(E^t f(x_t)) = E^t[Af(x_t)]\, dt + E^t[f(x_t)(h_f - \hat{h}_f)'](dz_t - \hat{h}_t\, dt)$$

where $h_t = h(x_t, z_t)$, $\hat{h}_t = E[h(x_t, z_t) \mid \mathbf{Z}_t]$, and A is the differential operator

$$Af = a'f_x + \tfrac{1}{2} \operatorname{trace}(f_{xx} BB'),$$

f_x the vector first derivatives for x and f_{xx} the matrix of second derivatives.

To prove this, expand $f(x_t)$ by Ito's expansion rule

$$df(x_t) = Af\, dt + f_x' B\, dw_t$$

and then substitute $f(x_t)$ for y_t in the theorem.

The proof of the theorem relies on a representation lemma for martingales, but this in turn depends on the following lemma on the "innovations" process. To simplify the notation we assume from now on all processes (z_t), (v_t) etc., are scalar, but the vector case proceeds in a similar fashion.

Innovations lemma

The process

$$\bar{v}_t = z_t - \int_0^t \hat{h}_s \, ds$$

is a Brownian motion on $(\mathbf{Z}_t)$.

This in some ways rather surprising result ($\bar{v}_t$ is *not* a Brownian motion on $(\mathbf{F}_t)$) was proved, in essence, by Frost [10], but see also the proofs in [15] and [4]. The proof basically consists of showing $(\bar{v}_t)$ has the appropriate joint characteristic functions for a Brownian motion. The process $(\bar{v}_t)$ has been called by Kailath the "innovations" process since formally

$$dz_t = E[dz_t \mid \mathbf{Z}_t] + d\bar{v}_t$$

and $(\bar{v}_t)$ can be thought of as the sum of the unpredicted parts of the increments of (z_t). In their innovations method Frost and Kailath were able to show that if (h_t) was Gaussian, the σ-fields generated by $(\bar{v}_t)$ were the same as those generated by (z_t), and $(\bar{v}_t)$ could be regarded as an equivalent observation process. They could then appeal to a well-known result, basically due to Ito, that the square integrable martingales on the σ-fields generated by a Brownian motion could be represented as stochastic integrals, to obtain a representation lemma for martingales on the σ-fields of observations. However it is not known in general whether one-to-one correspondence between (z_t) and $(\bar{v}_t)$ holds when (h_t) is non-Gaussian. The following lemma, which is perhaps the key result in [4], does not require such an assumption.

Representation lemma

If (m_t) is a square-integrable ($Em_t^2 < \infty$) martingale on $(\mathbf{Z}_t)$ then it has the representation

$$m_t = m_0 + \int_0^t p_s \, d\bar{v}_s$$

where (p_t) is adapted to $(\mathbf{Z}_t)$ and

$$\int_0^T Ep_t^2 \, dt < \infty.$$

We cannot give the full proof of this as it is rather too delicate to give in condensed form, but it relies on a theorem of Girsanov [16] to make an absolutely continuous transformation of the probability measure P that makes (z_t) a Brownian motion. The result of Ito then expresses martingales in the transformed measure as stochastic integrals of (z_t) and it turns out that martingales in the original measure have the form above.

Proof of the theorem

The first step is to show that the process

$$m_t = E^t y_t - \int_0^t E^s a_s \, ds$$

is a square-integrable martingale on $(\mathbf{Z}_t)$. It is clearly square-integrable and by the composition rules of conditional expectations, we have, for $t > s$,

$$E^s[m_t - m_s] = E^s\left[E^t y_t - E^s y_s - \int_s^t E^u a_u \, du\right]$$

$$= E^s\left[y_t - y_s - \int_s^t a_u \, du\right]$$

$$= E^s\left[\int_s^t b_u \, dw_u\right]$$

$$= 0.$$

The last term is zero since the stochastic integral is a martingale on $(\mathbf{F}_t)$. The second step is to apply the representation lemma to (m_t). Let

$$m_t = m_0 + \int_0^t p_s \, d\bar{v}_s$$

m_T can be thought of as the projection of

$$y_T - \int_0^T E^s a_s \, ds$$

on $\mathbf{Z}_t$ and as such is determined by the relation

$$E[m_T n] = E\left[\left(y_T - \int_0^T E^s a_s \, ds\right) n\right]$$

where n ranges over some suitably dense set of $\mathbf{Z}_t$-measurable random variables in $L^2(\Omega)$ with $E[n] = 0$. Let n_t denote the martingale $E[n \mid \mathbf{Z}_t]$ and let

$$\int_0^t q_s \, d\bar{v}_s$$

denote its stochastic integral representation with (q_t) adapted to $(\mathbf{Z}_t)$. The variables n for which the corresponding (q_t) are bounded can easily be shown to be such a dense set. From the expansions

$$y_t = y_0 + \int_0^t a_s \, ds + \int_0^t b_s \, dw_s$$

$$n_t = \int_0^t (h_s - \hat{h}_s) q_s \, ds + \int_0^t q_s \, dv_s,$$

the Ito expansion formula for products of integrals, and the composition rules for conditional expectations, we have

$$E[m_T n] = E\left[\left(y_T - \int_0^T E^t a_t\, dt\right) n_T\right]$$

$$= E\left[\int_0^T (n_t(a_t - E^t a_t) + y_t(h_t - \hat{h}_t)\, q_t + b_t\, Cq_t)\, dt\right]$$

$$= E\left[\int_0^T E^t[y_t(h_t - \hat{h}_t) + b_t\, C]q_t\, dt\right].$$

But also

$$E[m_T n] = E\left[\int_0^T p_t\, q_t\, dt\right].$$

Since the set of bounded $q_t(\omega)$ adapted to $(\mathbf{Z}_t)$ is dense in the set of functions in $L^2([0, T] \times \Omega)$ adapted to $(\mathbf{Z}_t)$, the identity of the last two expressions determines (p_t) uniquely and the proof is complete.

The Finite-dimensional Filters of Lo and Willsky

The following example of phase demodulation serves to illustrate the class of filtering problems considered by Lo and Willsky. The problem is to track a modulating phase process $x_1(t)$ (or $x_1(t)$ modulo (2π)) from noise-corrupted measurements of a signal $A \cos(\omega_0 t + x_1(t))$. For the sake of concreteness we suppose the process x_1 can be modelled as a second-order diffusion process such that the phase-rate $x_2 = \dot{x}_1$ is a Brownian motion. In frequency modulation problems the process $x_2(t)$ would be the process of interest. In one commonly used model of the measurements the noise enters additively and removal of the carrier frequency ω_0 produces an equivalent two-dimensional observation process

$$\begin{aligned} dy_1(t) &= \cos(x_1(t))\, dt + dv_1(t) \\ dy_2(t) &= \sin(x_1(t))\, dt + dv_2(t) \end{aligned} \tag{1}$$

in which the Brownian motions v_1, v_2, or more precisely $\dot{v}_1, \dot{v}_2$, represent "additive white noise". The observation process is then nonlinear in x_1 and the resulting optimal nonlinear filter is infinite-dimensional; a remarkable study has been made of this particular filter in [17]. However, an alternative model for the observation process is of the form

$$\begin{aligned} y_1(t) &= \cos(x_1(t) + v_t) \\ y_2(t) &= \sin(x_2(t) + v_t) \end{aligned} \tag{2}$$

where (v_t) represents a Brownian motion drift, and it is this sort of model that is considered by Lo and Willsky.

The natural space for the process is the unit circle S^1 rather than R^2. A point on S^1 can be represented in various ways, such as an angle on $[-\pi, \pi]$, but also as a complex variable of unit length. In this representation the process (2) is

$$Z = y_1(t) + iy_2(t) = e^{iz_t} \tag{3}$$

where

$$z_t = x_1(t) + v_t. \tag{4}$$

Since also $Z_t = e^{ix_1(t)} e^{iv_t}$ the noise can be said to be entering "multiplicatively" into observations.

It is clear at this stage that if observations were available for (z_t) rather than (Z_t), the normal Kalman–Bucy filter would provide an algorithm for the conditional density of $x_1(t)$. But the relation (3) between z_t and Z_t is many-to-one and it is not immediately apparent that this is the case.

However, let Z^t and z^t denote the sample paths of (Z_t) and (z_t) over the period $[0, t]$ (assume that $Z_0 = 1$ and $z_0 = 0$). Then there does exist a "one-to-one onto" mapping $Z^t = J(z^t)$ between the spaces of sample paths of Z^t and z^t and its inverse J^{-1} has the following stochastic integral representation. Ito's rule applied to (3) gives

$$dZ_t = Z_t(idz_t - \tfrac{1}{2}dt).$$

Since $Z_t^{-1} = \bar{Z}_t$, the complex conjugate of Z_t,

$$dz_t = \operatorname{Im}(\bar{Z}_t\, dZ_t) = y_1(t)\, dy_2(t) - y_2(t)\, dy_1(t). \tag{5}$$

Moreover the σ-fields generated by z^t and Z^t are the same and the process (z_t) can be regarded as an equivalent observation process. Kalman–Bucy filtering theory then shows the conditional probability density of the vector process

$$x_t = \begin{pmatrix} x_1(t) \\ x_2(t) \end{pmatrix}$$

is normal where the conditional mean $\hat{x}_t$ and matrix covariance satisfies the usual linear filtering equations, which, after substitution by (5), become

$$d\hat{x} = A\hat{x} + KC'(y_1\, dy_2 - y_2\, dy_1)$$
$$K = AK + KA' - KC'CK + Q$$

where

$$A = \begin{bmatrix} 0 & 1 \\ 0 & 0 \end{bmatrix}, \qquad C = [1, 0], \qquad Q = \begin{bmatrix} 0 & 0 \\ 0 & 1 \end{bmatrix}$$

and where K_0 and $\hat{x}_0$ are assumed to be known *a priori*. In practical terms the data would first be transformed by a nonlinear processor specified by (5) and then passed through a linear Kalman–Bucy filter. The conditional density of $x_1(t)$, $x_2(t)$ and hence of the phase $\theta_t = x_1(t) \bmod(2\pi)$ on $[-\pi, \pi]$ is thus determined. A natural point estimate for $x_2(t)$ is the least-squares estimate $x_2(t)$, but the choice is not so clear for θ_t for which there is no immediately comparable error criterion. However, its density, which is the "folded" normal density on $[-\pi, \pi]$,

$$p_\theta = \frac{1}{(2\pi k_{11})^{\frac{1}{2}}} \sum_{n=-\infty}^{\infty} \exp\left[-\frac{1}{2k_{11}}(\theta + 2n\pi - \hat{x}_1)\right]$$

has a single mode at $\hat{x}_1 \bmod(2\pi)$. Moreover, for any natural error $f(\theta)$ that is symmetric, increasing on $[0, \pi]$ and decreasing on $[-\pi, 0]$, such as $(1 - \cos\theta)$, the mode $\hat{x}_1 \bmod(2\pi)$ is the best estimate in the sense that, for all other estimates $\tilde{\theta}(Z^t)$

$$E[f(\theta - \hat{x}_1 \bmod(2\pi) \mid Z^t] \leqslant E[f(\theta_t - \tilde{\theta}) \mid Z^t]$$

and so is a natural estimate for θ_t.

A similar sort of analysis can be made of any filtering problem in which the signal process and observation process evolve on a state space

$$R^n \times (S^1)^m$$

consisting of a direct product of copies of the real line and the unit circle; that is, Euclidean space times an m-dimensional "torus", and in which the probabilistic behaviour can be regarded as being induced by a linear diffusion process in the manner described. Various examples, such as joint AM-FM demodulation problem and the frequency of an oscillation, are discussed in detail in [14].

Acknowledgement

Slide illustrations of the evolution of the conditional probability density in the phase demodulation problem were taken from [17].

References

1. Bucy, R. S. and Joseph, P. D., "Filtering for Stochastic Processes with Applications to Guidance". John Wiley, New York (1968).
2. Jazwinski, A. H., "Stochastic Processes and Filtering Theory". Academic Press, New York (1971).
3. Wong,E., "Stochastic Processes in Information and Dynamical Systems". McGraw-Hill, New York (1971).

4. Fujisaki, M., Kallianpur, G. and Kunita, H., Stochastic differential equations for the nonlinear filtering problem. *Osaka J. Math.* **9** (1) (1972), 19–40.
5. Stratonovich, R. L., Conditional Markov process theory. *Theory of Prob. Appl.* (*USSR*) **5** (1960), 156–178.
6. Kushner, H. J., On differential equations satisfied by conditional probability densities of Markov processes. *SIAM J. Control* **2** (1964), 106–119.
7. Bucy, R. S., Nonlinear filtering. *I.E.E.E. Trans. autom. Control* **AC-10** (1965), 198.
8. Kallianpur, G. and Striebel, C., Estimation of stochastic systems: Arbitrary system process with additive white noise observation errors. *Ann. Math. Statist.* **39** (1968), 785–801.
9. Clark, J. M. C., "The Representation of Nonlinear Stochastic Systems with Applications to Filtering". Ph.D. Thesis, Imperial College, University of London (1966).
10. Frost, P. A., "Nonlinear Estimation in Continuous Time Systems". Ph.D. Dissertation, Stanford University, California, 1968.
11. Kailath, T., An innovations approach to least-squares estimation, I: Linear filtering in additive white noise. *I.E.E.E. Trans. autom. Control* **AC-13** (1968), 646–654.
12. Frost, P. A. and Kailath, T., An innovations approach to least-squares estimation, III: Nonlinear estimation in white Gaussian noise. *I.E.E.E. Trans. autom. Control* **AC-16** (1971), 217–226.
13. Wonham, W. M., Some applications of stochastic differential equations to optimal nonlinear filtering. *SIAM J. Control* **2** (1965), 347–369.
14. Lo, J. T. and Willsky, A. S., "Estimation for Rotational Processes With One Degree of Freedom". Tech. Report No. 635, Division of Engineering and Applied Physics, Harvard University, July, 1972.
15. Kailath, T., A general likelihood-ratio formula for random signals in Gaussian noise. *I.E.E.E. Trans. Information Th.* **IT-15** (1969), 350–361.
16. Girsanov, I. V., On transforming a certain class of stochastic processes by absolutely continuous substitution of measures. *Theory of Prob. and its Appl.* **5** (3) (1960), 285–301.
17. Bucy, R. S., Hecht, C. and Senne, K. D., "An Engineer's Guide to Building Non-linear Filters". TR-72-0004, Frank J. Seiler, Res. Lab. USAF Acad., Colorado, May, 1972.

16. On Time-dependent Linear Stochastic Control Systems

T. SUBBA RAO and H. TONG
Department of Mathematics,
University of Manchester Institute of Science and Technology
Manchester, England

1. Introduction

The assumption of time-invariance is made in most studies of linear stochastic systems (see e.g. [1,2,3,6]) since such an assumption usually simplifies the statistical analysis. For example, the classical cross-spectral analysis is a convenient tool for the identification of an open-loop system [1,4,6]. However, in many practical applications, the assumption of time-invariance of the system may be open to question. As an illustration consider the following time dependent systems. [Here $\{W(t)\}$ denotes the output process and $\{X(t)\}$ the input process.]

Example 1

$$W(t) = c(t) \int_0^\infty d(u) X(t-u)\, du \tag{1}$$

where $c(t)$ is a real-valued deterministic function of t.

Example 2

$$W(t) = \begin{cases} X(t-1) & \text{for } \quad 0 \leqslant t \leqslant \Delta \\ X(t-2) & \text{for } \quad \Delta < t < \infty \end{cases} \tag{2}$$

where Δ is a constant.

Example 3

$$W(t+1) = X(t) + \tfrac{1}{2}\cos(t/100) X(t-1). \tag{3}$$

It has been shown [7,11] that under suitable conditions pertinent to the system the classical cross-spectral analysis may be extended to cover the following time-dependent systems:

$$Y(t) = \int_{-\infty}^{\infty} d_t(u)X(t-u)du + e(t) \tag{4}$$

where $\{X(t)\}$ is the input *(non-stationary) oscillatory stochastic process* and $\{e(t)\}$ is the noise stochastic process of a similar type uncorrelated with $\{X(t)\}$. Under suitable conditions on $d_t(u)$ it can be shown that the process $\{Y(t)\}$ is oscillatory [7,11]. Oscillatory processes as considered by Priestley [5,8] possess "*evolutionary spectral*" analysis similar to the classical spectral analysis of second-order stationary stochastic processes, which are, in fact, special cases of the former.

2. Identification of System Dynamics

We may view $\{d_t(u)\}$ of Eqn (4) as describing the system dynamics of a time-dependent system. Assuming that $d_t(u)$ is absolutely integrable for each t, we may term its Fourier transform (for each t), $D_t(\omega)$, the *time-dependent transfer function.* Assuming further that $D_t(\omega)$ is "slowly varying over t", it has been shown [7,11] that

$$f_{t,yx}(\omega) \approx D_t(\omega) f_{t,xx}(\omega) \tag{5}$$

where $f_{t,yx}$ is the evolutionary cross spectrum of the processes $Y(t)$ and $X(t)$ and $f_{t,xx}$ is the evolutionary autospectrum of the process $X(t)$. Thus, one obvious estimate of $D_t(\omega)$ is given by (in an obvious notation)

$$\hat{D}_t(\omega) = \hat{f}_{t,yx}(\omega)/\hat{f}_{t,xx}(\omega) \tag{6}$$

provided $\hat{f}_{t,xx}(\omega) \neq 0$.

Given a sample record of $\{X(t), Y(t): t \in [0, T_0]\}$ of finite length T_0, we may estimate $\hat{f}_{t,yx}(\omega)$ and $\hat{f}_{t,xx}(\omega)$ by the modified "window technique" in evolutionary spectral analysis [5,8]. Briefly, we choose a window $\{g(u)\}$ satisfying the usual conditions, with 'width' B_g ($\ll T_0$), and form

$$U_x(t, \omega) = \int_{t-T_0}^{t} g(u)X(t-u)e^{-i\omega(t-u)}du \tag{7}$$

and

$$U_y(t, \omega) = \int_{t-T_0}^{t} g(u)Y(t-u)e^{-i\omega(t-u)}du. \tag{8}$$

Then we choose a second window $\{w_{T'}(u)\}$ depending on the parameter T' and satisfying the usual conditions and form

$$\hat{f}_{t,yx}(\omega) = \int_{t-T_0}^{t} w_{T'}(u)\,[U_x(t-u,\omega)U_y^*(t-u,\omega)]\,du. \tag{9}$$

An estimate of $f_{t,xx}(\omega)$ may be similarly obtained.

Since $D_t(\omega)$ is usually complex-valued, we find it more convenient to work with the vector $\{\log_e G_t(\omega), \phi_t(\omega)\}$, where $G_t(\omega)$ and $\phi_t(\omega)$ are the polar co-ordinates of $D_t(\omega)$, for each t and ω, such that

$$D_t(\omega) = G_t(\omega)\exp\{-\hat{\phi}_t(\omega)\}. \tag{10}$$

We may similarly write $\hat{D}_t(\omega)$ as

$$\hat{D}_t(\omega) = \hat{G}_t(\omega)\exp\{-\hat{\phi}_t(\omega)\}. \tag{11}$$

It has been shown that $\log \hat{G}_t(\omega)$ and $\hat{\phi}_t(\omega)$ are approximately unbiased estimates of $\log G_t(\omega)$ and $\phi_t(\omega)$ respectively [9,11]. Further, the variance-covariance matrix, Σ, of the vector random variable $\{\log \hat{f}_{t,xx}(\omega), \log \hat{f}_{t,yy}(\omega), \log \hat{G}_t(\omega), \hat{\phi}_t(\omega)\}$ may be approximately given [11,12] by

$$\Sigma = K \begin{bmatrix} 1 & W^2(\omega) & 0 & 0 \\ W^2(\omega) & 1 & 0 & 0 \\ 0 & 0 & \dfrac{1}{W^2(\omega)} - 1 & 0 \\ 0 & 0 & 0 & \dfrac{1}{W^2(\omega)} - 1 \end{bmatrix} \tag{12}$$

where K is a constant depending on the widths of the windows chosen and $W^2(\omega)$ is the *square coherency spectrum* given by

$$W^2(\omega) = |f_{t,xy}(\omega)|^2 / \{f_{t,xx}(\omega) f_{t,yy}(\omega)\}$$

provided $f_{t,xx}(\omega) \neq 0$ and $f_{t,yy}(\omega) \neq 0$.

3. Tests of Hypotheses

In a practical situation, we are usually given only a sample record of the input process $\{X(t)\}$ and the observed output process $\{Y(t)\}$ of finite length T_0, say. We may identify the system dynamics at a selection of time points. (See [9,11] for a discussion on the selection of points.) It is now clear that a test of the "uniformity" of the set of estimates of the system dynamics at different time points is equivalent to a test of the time dependence of the system. One approach is via the multivariate statistical analysis, which yields at least two alternative tests:

($T1$) A two-factor multivariate analysis of variance (MANOVA) based on the approximate variance-covariance matrix $\boldsymbol{\Sigma}$ given by (12);

($T2$) A two-factor MANOVA with $\boldsymbol{\Sigma}$ being unknown [9,10].

In both tests, we consider the random vector $(\eta_1(t, \omega), \eta_2(t, \omega))'$, where

$$\eta_1(t, \omega) = \log \hat{G}_t(\omega)/\{1/W^2(\omega) - 1\}^{\frac{1}{2}} \tag{13}$$

and

$$\eta_2(t, \omega) = \hat{\phi}_t(\omega)/\{1/W^2(\omega) - 1\}^{\frac{1}{2}} \tag{14}$$

$(W^2(\omega) \neq 1)$.

For simplicity of notation, we denote $(\eta_1(t_i, \omega_j), \eta_2(t_i, \omega_j))'$ by $\boldsymbol{\eta}_{ij}$. We then set up the MANOVA model

$$\boldsymbol{\eta}_{ij} = \boldsymbol{\mu} + \boldsymbol{\tau}_i + \boldsymbol{\phi}_j + \boldsymbol{\gamma}_{ij} + \boldsymbol{\varepsilon}_{ij}, \quad \left\{\begin{array}{l} i = 1, 2, \ldots, I \\ \\ j = 1, 2, \ldots, J \end{array}\right\} \tag{15}$$

where $\sum \boldsymbol{\tau}_i = \sum \boldsymbol{\phi}_j = \sum_i \boldsymbol{\gamma}_{ij} = \sum_j \boldsymbol{\gamma}_{ij} = \mathbf{0}$, and $\boldsymbol{\varepsilon}_{ij} \sim N_2(\mathbf{0}, \boldsymbol{\Sigma})$.

We may interpret $\boldsymbol{\tau}_i$ as the 'time effect', $\boldsymbol{\phi}_j$ as the 'frequency effect' and $\boldsymbol{\gamma}_{ij}$ as the interaction between time and frequency. The testing of time invariance of $D_t(\omega)$ is equivalent to the testing of $\boldsymbol{\gamma}_{ij} = \mathbf{0}$ for all i and j and $\boldsymbol{\tau}_i = \mathbf{0}$ for all i. Simulation studies of Example 3 have been carried out on the ATLAS computer at Manchester and results have been reported elsewhere [8,9,11]. The computation is presented in the table opposite.

($T1$): We test the first hypothesis $h_1 : \boldsymbol{\gamma}_{ij} = \mathbf{0}$, for all i and j. For this, we refer $\text{tr}(\mathbf{E}/K)$ to the chi-square distribution with $2(I-1)(J-1)$ degrees of freedom. If the hypothesis is rejected we conclude that the transfer function is time dependent. Otherwise we proceed to test the hypothesis $h_2 : \boldsymbol{\tau}_i = \mathbf{0}$, for all i. For this we refer the $\text{tr}(\mathbf{H}_1/K)$ to the chi-square distribution with $2(I-1)$ degrees of freedom. If h_2 is rejected then we conclude that the transfer function is time dependent.

If neither h_1 nor h_2 is rejected then we may accept the hypothesis that the transfer function is time-invariant.

($T2$) The basic steps are similar to ($T1$) except for some modification to take account of the fact that $\mathbf{\Sigma}$ is assumed unknown [10].

Source of Variation	Matrix	Sum of squares and products
Between Times	$\mathbf{H}_1$	$J \sum_{i=1}^{I} (\boldsymbol{\eta}_{i.} - \boldsymbol{\eta}_{..})(\boldsymbol{\eta}_{i.} - \boldsymbol{\eta}_{..})'$
Between Frequencies	$\mathbf{H}_2$	$I \sum_{j=1}^{J} (\boldsymbol{\eta}_{.j} - \boldsymbol{\eta}_{..})(\boldsymbol{\eta}_{.j} - \boldsymbol{\eta}_{..})'$
Interaction + Residual	$\mathbf{E}$	$\sum_i \sum_j (\boldsymbol{\eta}_{ij} - \boldsymbol{\eta}_{i.} - \boldsymbol{\eta}_{.j} + \boldsymbol{\eta}_{..}) \times (\boldsymbol{\eta}_{ij} - \boldsymbol{\eta}_{i.} - \boldsymbol{\eta}_{.j} + \boldsymbol{\eta}_{..})'$

[$\boldsymbol{\eta}_{i.}$ denotes the average of $\boldsymbol{\eta}_{ij}$ over j with similar definition for the others.]

References

1. Akaike, H., Some problems in the application of cross-spectral method. *In* "Advanced Seminar on Spectral Analysis of Time Series" (ed. B. Harris). John Wiley, New York (1967).
2. Åström, K. J. and Bohlin, T., "Numerical Identification of Linear Dynamic Systems from Normal Operating Records". I.B.M. Nordic Laboratory Technical Paper TP/8 159.
3. Box, G. E. P. and Jenkins, G. M., Some statistical aspects of adaptive optimization and control. *J. Roy. Statist. Soc.* **B24** (1962), 297–343.
4. Jenkins, G. M., Cross-spectral analysis and the estimation of linear open loop transfer functions. *In* (ed. M. Rosenblatt) "Symposium on Time Series Analysis". John Wiley, New York (1963), 267–278.
5. Priestley, M. B., Evolutionary spectra and non-stationary processes. *J. Roy. Statist. Soc.* **B27** (1965), 204–237.
6. Priestley, M. B., Estimation of transfer function in closed loop stochastic systems. *Automatica* **5** (1969), 623–632.
7. Priestley, M. B. and Tong, H., On the analysis of bivariate non-stationary processes. To appear in *J. Roy. Statist. Soc.* **B**.
8. Priestley, M. B., Subba Rao, T. and Tong, H., "Spectral Analysis of Non-stationary Time Series". Monograph to be published.
9. Subba Rao, T. and Tong, H., A test for time-dependence of linear open loop systems, *J. Roy. Statist. Soc.* **B34** (2) (1972), 235–250.

10. Subba Rao, T. and Tong, H., On some tests for time dependence of a transfer function. *Biometrika* **60** (3) (1973).
11. Tong, H., "Some Problems in the Spectral Analysis of Bivariate Non-stationary Stochastic Processes". Ph.D. Thesis, University of Manchester (unpublished).
12. Tong, H., Evolutionary cross-spectral analysis of time-dependent linear open-loop systems. Submitted to *J. Roy. Statist. Soc.* **B** for publication.

Discussion

RUDZINSKI (*Central Electricity Generating Board*). Would Dr. Tong tell me if an alternative to his approach would be differencing input and output to decide whether one has a time-invariant system? If I have a stationary input then the auto-correlation will go to zero. If the input and output of the system are non-stationary then after a number of differences their respective auto-correlations will go to zero.

TONG. The non-stationary process which you mention is called an accumulated process which was first studied by A. M. Yaglom in Russia in 1955. We did not study that type of non-stationary time series. The non-stationary series there is really generated by a mechanism which is time independent. Suppose we take the simple case where an auto-regressive/moving-average process is the result after n differences. There are coefficients are really time invariant, whereas the type of model we considered has been generated by a mechanism the coefficients of which depend on time.

HAMMOND. (*Portsmouth Polytechnic*) You said that if a non-stationary input is fed into a time invariant system, then the output will have a representation with respect to the same family.

TONG. Approximately the same family. You have to modify it by the transfer function.

HAMMOND. But surely you have to consider the bandwidth of the family of functions.

TONG. Yes, the consideration is pertinent to the order of approximation.

HENERY (*University of Strathclyde*). Would you care to amplify the frequency-domain approach. Why do you not adopt the time-domain approach when looking for time-dependence?

TONG. If you do the whole analysis over the time domain, you probably have to look at the cross-correlation function or the cross co-variance function. In this case, you have a function of two variables. You will probably require a large number of observations and the beauty of spectral analysis is that it is a non-parametric approach. The spectral functions themselves give you the information.

DIPROSE. (*University of Bath*) Have you tested your method to see if the decision as to whether you operate stationary or not depends on the number of samples of time into which you divide your record?

TONG. The estimation would in fact depend on the length of the record. If you have a very short length of record then you cannot do very much estimation. After all, the estimation depends upon some kind of averaging over the neighbouring time points. If you have a very short time series then the whole thing is hopeless.

DIPROSE. Yes, indeed. What I was getting at was if you have a given length of record, does it make any difference to your decision whether you decide to divide it into two pieces, four pieces, or whatever?

TONG. There are restrictions on the number of points that you can take because if you look at the variance-covariance matrix then you will notice it depends on the spacing of the time points. There is a similar type of restriction on the frequency points. If the time points are very close then the two estimates will be highly correlated and the theory will break down. So, that would put a restriction on the minimum spacing you have. The next thing is the length of record you have; again that would determine the maximum number of spacings you can get.

BRYANT (*Imperial College, London*). Have you applied this method to any practical cases?

TONG. We have only applied it so far to simulated data. A student at Manchester is at present working on real data.†

CLARK (*Imperial College, London*). From where were the readings taken?

TONG. The readings have been taken from some United Nation handbooks.

† *Editors Note*. This work has now been completed.

17. A Decentralised Filter for Certain Time Lag Systems

M. G. SINGH

*Control Engineering Group, Dept. of Engineering,
University of Cambridge, England*

1. Introduction

The computational burden associated with the estimation and control for linear dynamical systems is proportional to the cube of the system order. This is the main reason why the practical application of optimal estimation and control theory has met with limited success. In most industrial and other applications, the system can only be modelled adequately by a high order state vector. In addition most practical systems are distributed parameter systems and it is necessary to include the effects of the associated time lags in the system description thus further increasing the dimensionality. The theory of hierarchical control has recently been developed [1, 2] to avoid this problem of dimensionality. This is done by splitting the overall system into subsystems of lower dimension, optimising the subsystems individually by their own controllers on the first level and accounting for the interactions by the operations of a higher level controller called the coordinator. This structure is shown in Fig. 1. Effectively, the computational burden is being shared between the controllers on the first level and the coordinator. The filtering problem of a system comprising many subsystems could also be viewed within a hierarchical structure where on the first level independent filters solve their individual filtering problems and the coordinator iterates with the first level successively providing better estimates of the interactions until an overall optimum is achieved. It should be noted that there is no computational saving in using optimal hierarchical filtering or control since the decreased computational burden at the first level is made up by extra computations during the inter-level iterations. In this paper a simple sub-optimal hierarchical strategy is developed for systems comprising lumped parameter subsystems separated by time delays. In this strategy, the computational load on the second level is minimal thus resulting in substantial computational savings. The time-lags do not increase the overall dimensionality; rather they help in the solution of the on-line filtering problem by providing extra time for computations. The savings in computation that occur are at the expense of some loss of optimality. However, for a number of cases of practical interest, particularly when the measurement noise spectral density is low, the filter is near optimal.

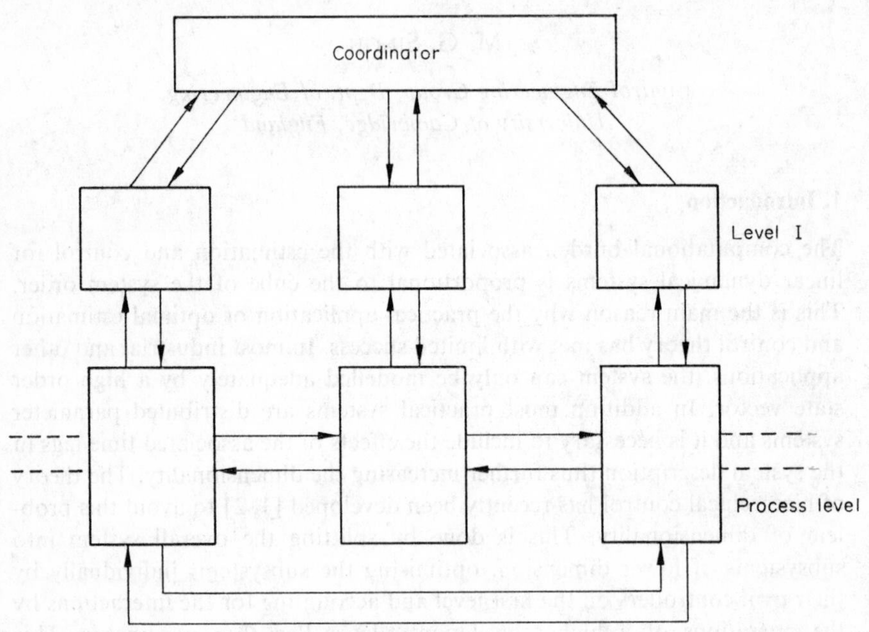

FIG. 1.

A similar filter was recently developed by Shah [3] for discrete linear dynamical systems. However, Shah's work is incomplete since a valid [4] theoretical justification for his results has never been published. This paper provides such a theoretical justification and it is from this that the proofs of Shah's theorems follow. A critique of his filtering results is given in Appendix I. In the main body of this paper, the time-lag filter is developed. Expressions are developed for its computational advantage and it is shown how it is possible to establish its suboptimality for a given system. The filter is illustrated by two numerical studies. The first of these is a hypothetical example and this is a time-lag version of the one used by Shah. For this example it is seen that the slope of the autocorrelation function of the subsystem interaction noise terms is closely related to the degree of suboptimality of the filter. The second study is a more practical example from the water pollution field.

2. Serial Time-lag Systems

The general class of systems to be considered is shown in Fig. 2. For the purposes of this development it will suffice to consider a system comprising two subsystems with time lags between them as shown in Fig. 3. The dynamical equations of the system could be written as:

$$\begin{aligned} \dot{x}_1(t) &= A_{11}(t)\,x_1(t) + A_{12}(t)\,x_2(t-\theta_2) + B_{11}(t)\,u_1(t) \\ y_1(t) &= M_{11}(t)\,x_1(t) + v_1(t) \\ \dot{x}_2(t) &= A_{21}(t)\,x_1(t-\theta_1) + A_{22}(t)\,x_2(t) + B_{22}(t)\,u_2(t) \\ y_2(t) &= M_{22}(t)\,x_2(t) + v_2(t) \end{aligned} \tag{1}$$

where the symbols have the following meaning:

x_1 is the $l_1 \times 1$ state vector of subsystem I

x_2 is the $l_2 \times 1$ state vector of subsystem II

$A_{11}, A_{12}, A_{21}, A_{22}$, are respectively $l_1 \times l_1, l_1 \times l_2, l_2 \times l_1, l_2 \times l_2$ matrices.

u_1, u_2 are white noise vectors of dimensions $m_1 \times 1$, $m_2 \times 1$ respectively.

B_{11}, B_{22} are $l_1 \times m_1$ and $l_2 \times m_2$ noise driving matrices for the system.

y_1, y_2 are the output vectors of the two subsystems and are $p_1 \times 1$ and $p_2 \times 1$ respectively.

v_1, v_2, are $p_1 \times 1$ and $p_2 \times 1$ vectors of measurement noise for the two subsystems.

M_{11}, M_{22}, are measurement noise driving matrices of the two subsystems and are of dimension $p_1 \times l_1$ and $p_2 \times l_2$ respectively.

The stochastic processes $[u_i(t); t \geqslant t_0, i = 1, 2]$ and $[v_i(t); \geqslant t_0, i = 1, 2]$ are zero mean Gaussian white noise of known covariance, i.e.

$$E[u_i(t)\, u_i^T(\tau)] = Q_i(t)\delta_i(t-\tau); \qquad (i = 1, 2)$$

and (2)

$$E[v_i(t) v_i^T(\tau)] = R_i(t)\delta_i(t-\tau); \qquad (i = 1, 2)$$

for all $t, \tau \geqslant t_0$, where E denotes the expected value and δ is the Dirac delta function. The $(m_i \times m_i)$ $(i = 1, 2)$ matrices Q_i are positive semidefinite for $t \geqslant t_0$ while the $(p_i \times p_i)(i = 1, 2)$ matrices R_i are continuous and positive definite for $t \geqslant t_0$. It is assumed that the stochastic processes v_i, u_i are independent of each other, i.e.

$$E[u_i(t)\, v_i^T(\tau)] = 0 \tag{3}$$

for all $\tau, t \geqslant t_0$.

The initial states $x_i(t_0)$ $(i = 1, 2)$ are zero mean Gaussian random $(l_i \times 1)$ vectors. They are independent of $u_i(t)\, t \geqslant t_0$ and of $v_i(t)\, t \geqslant t_0$ and their $(l_i \times l_i)$ covariance matrices

$$E[x_i(t_0)\, x_i^T(t_0)] = P(t_0) \text{ are positive semidefinite.}$$

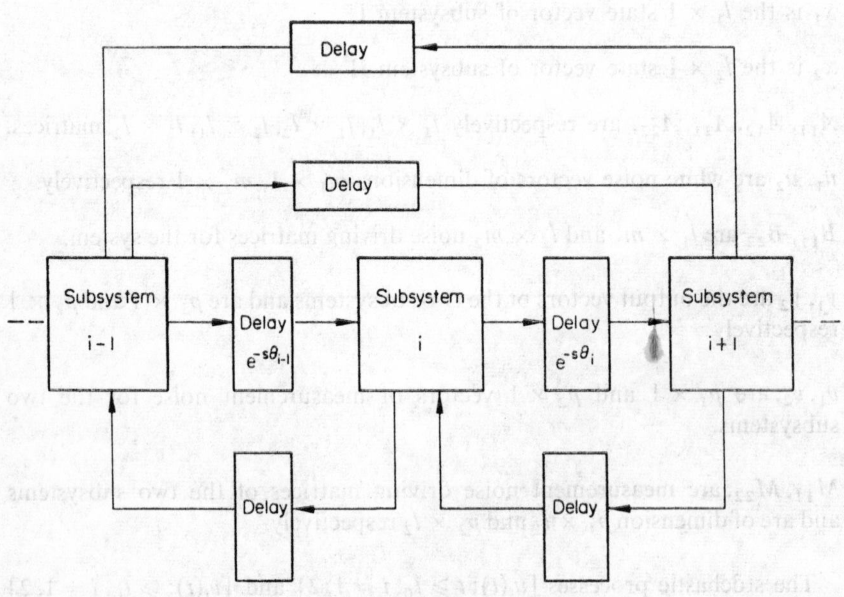

FIG. 2.

Following Fuller [5], the states of the pure delays θ_1, θ_2 could be defined as

$$w_1(\sigma_1, t) = y_1(t - \theta_1 + \sigma_1) \begin{cases} \theta_1 \leqslant t < \infty \\ 0 \leqslant \sigma_1 < \theta_1 \end{cases}$$

where w_1 is the state of θ_1, σ_1 being its distributed parameter. For $(0 \leqslant t < \theta_1)$,

$$w_1(\sigma_1, t) = \begin{cases} y_1(t - \theta_1 + \sigma_1); & \theta_1 - t \leqslant \sigma_1 < \theta_1 \\ \phi_1(t - \theta_1 + \sigma_1); & 0 \leqslant \sigma < \theta_1 - t \end{cases}$$

where ϕ_1 is an arbitrary function which defines the initial conditions of the delay θ_1.

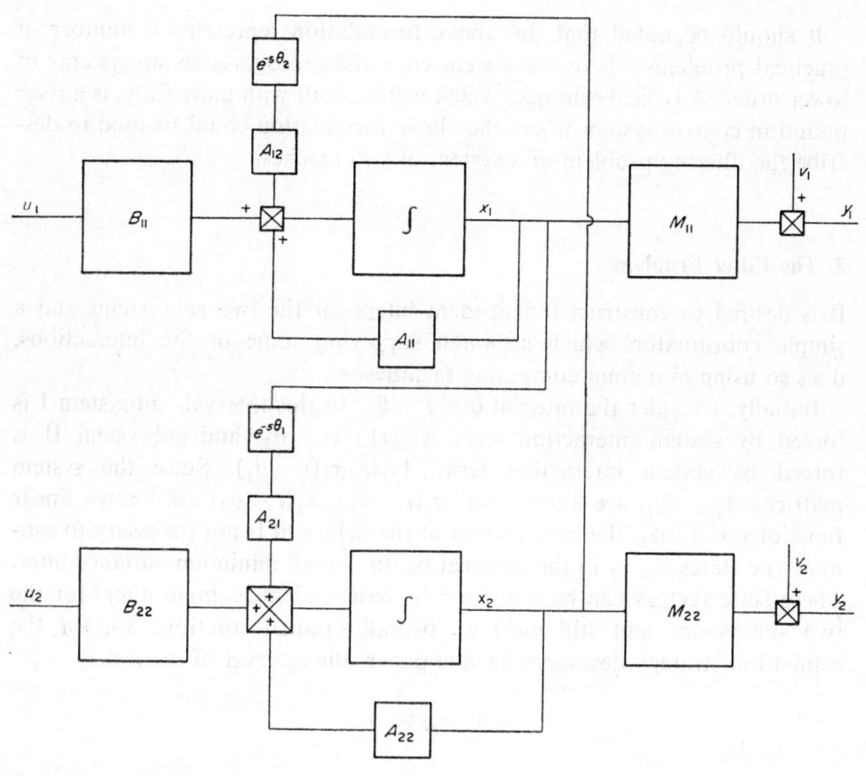

FIG. 3.

Similarly, the state w_2 of the delay θ_2 could be defined as

$$w_2(\sigma_2, t) = y_2(t - \theta_2 + \sigma_2); \qquad \begin{cases} \theta_2 \leqslant t < \infty \\ 0 \leqslant \sigma_2 < \theta_2 \end{cases}$$

and for $0 \leqslant t < \theta_2$

$$w_2(\sigma_2, t) = \begin{cases} y_2(t - \theta_2 + \sigma_2); & \theta_2 - t \leqslant \sigma_2 < \theta_2 \\ \phi_2(t - \theta_2 + \sigma_2); & 0 \leqslant \sigma_2 < \theta_2 - t. \end{cases}$$

The initial states ϕ_1, ϕ_2 of the delays θ_1, θ_2 can be assumed to be uncorrelated with the stochastic processes u_i, v_i, i.e.

$$E[\phi_i(t)\, u_i^T(\tau)] = 0; \qquad (i = 1, 2)$$

$$E[\phi_i(t)\, v_i^T(\tau)] = 0; \qquad (i = 1, 2).$$

It should be noted that the above formulation represents a number of practical problems where the system comprises recognisable subsystems of lower order. A typical example, which will be dealt with more fully, is a river pollution control system where the above formulation could be used to describe the filtering problem of a system of two reaches.

3. The Filter Problem

It is desired to construct independent filters for the two subsystems and a simple coordinator, which although supplying some of the interactions, does so using minimum computing facilities.

Initially, consider the interval $0 \leqslant t < \theta_2$. In this interval, subsystem I is forced by system interaction term $A_{12}(t)\, x_2(t - \theta_2)$ and subsystem II is forced by system interaction term $A_{21}(t)\, x_1(t - \theta_1)$. Since the system matrices A_{12}, A_{21}, are known and $x_1(t - \theta_1)$, $x_2(t - \theta_2)$ are known linear functions of ϕ_1, ϕ_2, the initial states of the delays, it is not necessary to estimate the states x_1, x_2 in this interval by an overall minimum variance filter. These state vectors can be estimated by independent Kalman filters for the two subsystems and still yield an overall optimal solution. So, for the estimation strategy developed in this paper, the interval of interest is

$$\theta_2 \leqslant t < \infty$$

In what follows, it will be assumed that t_0, the time when filtering starts, is equal to θ_2.

The estimation error is defined by the expression

$$\tilde{x}_i(t_1/t) = x_i(t_1) - \hat{x}_i(t_1/t) \qquad (i = 1, 2) \tag{4}$$

where $\hat{x}_i(t_1/t)$ is the estimate at time t_1.

Before tackling the filtering problem, it is further assumed that the overall system noise covariance matrix is block diagonal, i.e.

$$E[u_1(t)\, u_2{}^T(\tau)] = 0 \quad \text{for all} \quad t, \tau.$$

This is a realistic assumption for practical systems. Consider subsystem II. Since $x_1(t - \theta_1) = \hat{x}_1(t - \theta_1) + \tilde{x}_1(t - \theta_1)$ by definition of the estimation error, the subsystem equations could be written as

$$\dot{x}_2(t) = A_{21}(t)\hat{x}_1(t - \theta_1) + A_{21}\tilde{x}_1(t - \theta_1) + A_{22}(t)x_2(t) + B_{22}(t)\, u_2(t)$$

$$y_2(t) = M_{22}(t)\, x_2(t) + v_2(t)$$

or

$$\dot{x}_2(t) = A_{21}(t)\, \hat{x}_1(t - \theta_1) + A_{22}(t)\, x_2(t) + u_2{}^*(t) \tag{5}$$

where

$$u_2{}^*(t) = A_{21}(t)\tilde{x}_1(t - \theta_1) + B_{22}(t)\, u_2(t). \tag{6}$$

Similarly, subsystem I equations could be written as

$$\dot{x}_1(t) = A_{11}(t)x_1(t) + A_{12}(t)\, \hat{x}_2(t - \theta_2) + A_{12}(t)\, \tilde{x}_2(t - \theta_2) + B_{11}(t)\, u_1(t)$$

or

$$\dot{x}_1(t) = A_{11}(t)\, x_1(t) + A_{12}(t)\, \hat{x}_2(t - \theta_2) + u_1{}^*(t) \tag{7}$$

where

$$u_1{}^*(t) = A_{12}(t)\, \tilde{x}_2(t - \theta_2) + B_{11}(t)\, u_1(t) \tag{8}$$

Now, consider the following simple filter structure for this two subsystem system:

$$\dot{\hat{x}}_1(t) = A_{11}\, \hat{x}_1(t) + K_1[y_1(t) - M_{11}\, \hat{x}_1(t)] + A_{12}\, \hat{x}_2(t - \theta_2) \tag{9a}$$

$$\dot{\hat{x}}_2(t) = A_{22}(t)\, \hat{x}_2(t) + K_2[y_2(t) - M_{22}\, \hat{x}_2(t)] + A_{21}\, \hat{x}_1(t - \theta_1) \tag{9b}$$

where

$$K_1 = P_{11}M_{11}{}^T R_1{}^{-1} \tag{9c}$$

$$K_2 = P_{22}M_{22}{}^T R_2{}^{-1} \tag{9d}$$

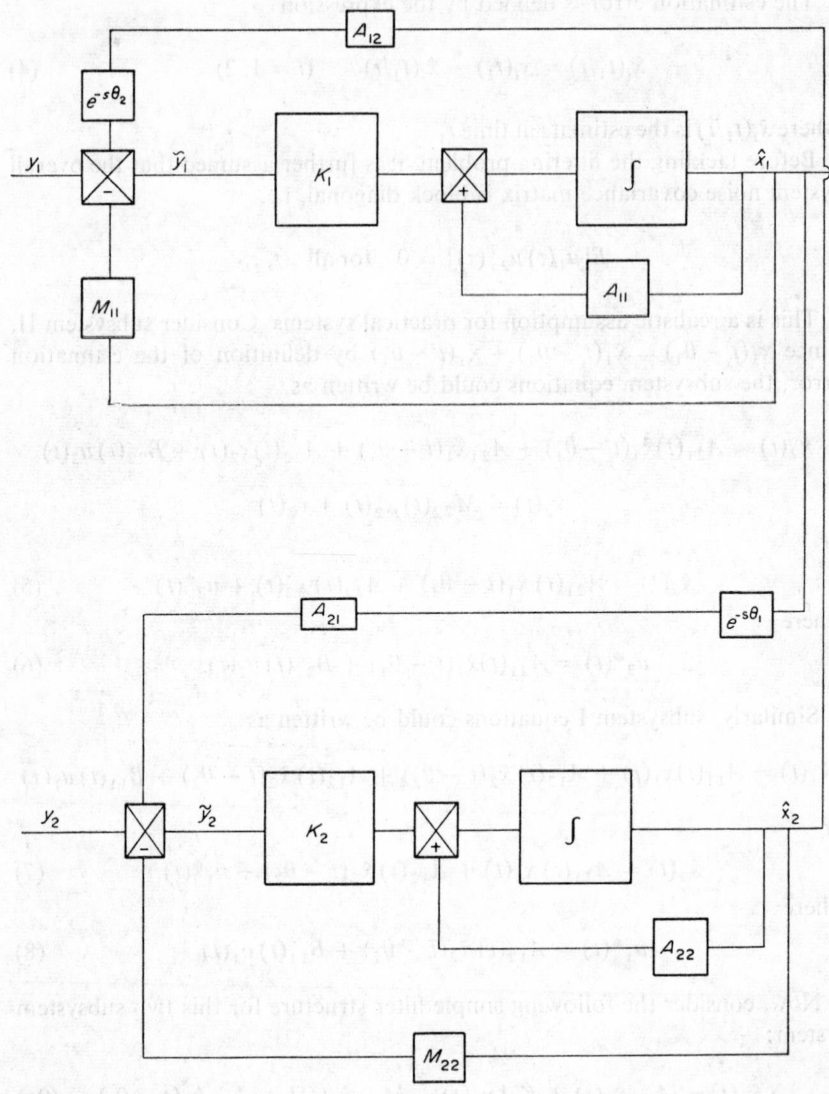

FIG. 4.

and P_{11}, P_{22}, are the solutions of the matrix Riccati equations:

$$\dot{P}_{11} = A_{11}P_{11} + P_{11}\,A_{11}{}^T - P_{11}M_{11}{}^T R_1{}^{-1}M_{11}P_{11} + B_{11}Q_1B_{11}{}^T \quad \text{(9e)}$$

$$\dot{P}_{22} = A_{22}P_{22} + P_{22}\,A_{22}{}^T - P_{22}M_{22}{}^T R_2{}^{-1}M_{22}P_{22} + B_{22}Q_2B_{22}{}^T \quad \text{(9f)}$$

Figure 4 shows the structure of the filter. It should be noted that such a filter does not explicitly account for the interactions which are expressed in the optimal filter by the inclusion of the off diagonal terms (like P_{12}) in the overall system covariance matrix P. It should be of considerable interest to see if it is nevertheless possible to get near optimal filtering. This leads to a simple proposition that was assumed to be correct by Shah for discrete linear dynamical systems.

3.1 *Proposition* 1

The filter described by Eqns (9) *is an optimal filter for the systems described by Eqns* (1) *if and only if* u_1^*, u_2^* *as given by Eqns* (6) *and* (8) *are white noise vectors.*

Proof. If u_1^*, u_2^* are white, then all the information about subsystem I interactions is contained in $A_{12}\,\hat{x}_2(t-\theta_2)$ and similarly for subsystem II, all the information about its interactions with subsystem I is contained in $A_{21}\,\hat{x}_1(t-\theta_1)$. Since the interactions are accounted for, and K_1, K_2 are calculated from optimal Riccati equations, the filter must be optimal.

To prove necessity, let it be assumed that u_1^*, u_2^* are not white. Then they contain interaction information which can be extracted from these coloured processes by modelling them as differential equations and augmenting the individual state vectors of the subsystems. The filter which ignores these interactions cannot be optimal.

The above proposition shows that it is necessary for u_1^*, u_2^* to be white for the simple filter structure described by Eqns (9) to provide optimal filtering. Clearly such a structure is very attractive computationally since the second level of filtering in this case does virtually no work.

In the following it will be shown that for a number of cases of practical interest, u_1^*, u_2^* are indeed good approximations to white noise. This was never established by Shah for discrete dynamical systems and also Shah assumed that "the interactions between the subsystems are weak" and implied that this meant that if P_{12}, the covariance of the interactions, is zero, the filter is optimal. This is obviously true. However, if $P_{12} = 0$ there is no hierarchical filtering problem since it is then possible to construct independent filters.

3.2 *The whiteness of* u_1^*, u_2^*:

Consider subsystem II. It is desired to see under what conditions if any,

$$E[u_2^*(t_2)\,u_2^*(t_1)^T] = \delta(t_2, t_1)\,Q_2^*$$

where Q_2^* is a matrix of known covariance and δ is the Dirac delta function. Consider when $t_2 = t_1 = t$. Then

$$E\{u_2^*(t)\, u_2^{*T}(t_1)\} = E\{A_{21}(t)\tilde{x}_1(t-\theta_1) + B_{22}(t)u_2(t)\}\,\{A_{21}(t)\tilde{x}_1(t-\theta_1)$$
$$+ B_{22}(t)u_2(t)\}^T = E\,\{A_{21}(t)\,\tilde{x}_1(t-\theta_1)\,\tilde{x}_1^T(t-\theta_1)A_{21}^T(t)\}$$
$$+ E\{A_{21}(t)\tilde{x}_1(t-\theta_1)\,u_2^T(t)\,B_{22}^T(t)\} + E\{B_{22}(t)\,u_2(t)\tilde{x}_1^T(t-\theta_1)A_{21}^T(t)\}$$
$$+ E\{B_{22}(t)\,u_2(t)u_2^T(t)B_{22}^T(t)\}.$$

The second and third terms on the right-hand side of the above expression vanish because θ_1 is positive and future noise is uncorrelated with past signal and estimation error. Therefore

$$E\{u_2^*(t)\, u_2^{*T}(t)\} = A_{21}(t)\, P_{11}(t-\theta) A_{21}^T(t) + B_{22}(t)\, Q_2(t)\, B_{22}^T(t) \quad (10)$$

Thus u_2^* would be approximately white if $E\{u_2^*(t_2)\, u_2^{*T}(t_1)\}$ $(t_2 \neq t_1)$ is much smaller than the expression in Eqn (10) or if the spectral density of u_2^* is approximately constant over a range larger than that for subsystem II.

Consider the case when $t_2 > t_1$. Then

$$E[u_2^*(t_2)\, u_2^{*T}(t_1)] = E[A_{21}(t_2)\,\tilde{x}_1(t_2-\theta_1)$$
$$+ B_{22}(t_2)\, u_2(t_2)]\,[A_{21}(t_1)\,\tilde{x}_1(t_1-\theta_1 + B_{22}(t_1)\, u_2(t_1)]^T$$
$$= E[A_{21}(t_2)\,\tilde{x}_1(t_2-\theta_1)\tilde{x}_1^T(t_1-\theta_1)\, A_{21}^T(t_1)]$$
$$+ E[A_{21}(t_2)\,\tilde{x}_1(t_2-\theta_1)\, u_2^T(t_1)\, B_{22}^T(t_1)]$$
$$+ E[B_{22}(t_2)\, u_2(t_2)\,\tilde{x}_1^T(t_1-\theta_1)\, A_{22}^T(t_1)]$$
$$+ E[B_{22}(t_2)u_2(t_2)\, u_2^T(t_1)\, B_{22}^T(t_1)] \quad (11)$$

The fourth term is zero from Eqn (2). The third term is zero since future noise is uncorrelated with past errors in the state estimates.

Now, depending upon the value of the delay, θ_1, two cases of interest arise:

(a) when $t_2 - \theta_1 > t_1$

(b) when $t_2 - \theta_1 \leqslant t_1$.

Case (a)

For this case:

$$E[u_2^*(t_2)\, u_2^{*T}(t_1)] = E[A_{21}(t_2)\, \tilde{x}_1(t_2 - \theta_1)\, \tilde{x}_1^T(t_1 - \theta_1)\, A_{21}^T(t_1)] + E[A_{21}(t_2)\, \tilde{x}_1(t_2 - \theta_1)\, u_2^T(t_1)\, B_{22}^T(t_1)] \tag{12}$$

and for *case (b)*

(i) when $t_2 - \theta_1 < t_1$, the second term in Eqn (11) is zero since future noise is uncorrelated with past signal.

(ii) when $t_2 - \theta_1 = t_1$, the second term is Eqn (11) is zero since present error is uncorrelated with present noise whereas the third term also vanishes since θ_1 is a constant so that the argument of u_2 is greater than the argument of $\tilde{x}_1$ and future noise is uncorrelated with past signal.

Thus for case (b)

$$E[u_2^*(t_2)\, u_2^{*T}(t_1)] = E[A_{21}(t_2)\, \tilde{x}_1(t_2 - \theta_1)\, \tilde{x}_1^T(t_1 - \theta_1)\, A_{21}^T(t_1)] \tag{13}$$

From the above analysis, it is clear that in order to evaluate $E[u_2^*(t_2)\, u_2^{*T}(t_1)]$ $(t_2 \neq t_1)$, it is necessary to relate $\tilde{x}_1(t_2 - \theta_1)$ with $\tilde{x}_1(t_1 - \theta_1)$ and with $u_2(t_1)$. In order to do this, consider the overall optimal filter

$$\dot{\hat{x}} = A\,\hat{x} + K[Y - M\hat{x}]. \tag{14}$$

The overall system equations could also be written as

$$\dot{x} = Ax + Bu \tag{15}$$

$$Y = Mx + v. \tag{16}$$

Subtracting Eqn (14) from (15)

$$(\dot{x} - \dot{\hat{x}}) = A\,(x - \hat{x}) + Bu - K[Y - M\hat{x}]$$

i.e.

$$\dot{\tilde{x}} = A\tilde{x} + Bu - K[Mx + v - M\hat{x}]$$

i.e.

$$\dot{\tilde{x}} = (A - KM)\,\tilde{x} + Bu - Kv. \tag{17}$$

Equation (17) can be integrated between $(t_1 - \theta_1)$ and $(t_2 - \theta_1)$ to give

$$\tilde{x}\,(t_2 - \theta_1) = \exp\,[(A - KM)(t_2 - \theta_1 - t_1 + \theta_1)]\,\tilde{x}_1(t_1 - \theta_1) + \int_{t_1-\theta_1}^{t_2-\theta_1} \Phi\,(t_2 - \theta_1, \tau)\,[\,-K(\tau)\,v(\tau) + B(\tau)u(\tau)]\,d\tau$$

Then

$$E[\tilde{x}(t_2-\theta_1)\,\tilde{x}^T(t_1-\theta_1)] = \Phi(t_2,t_1)\,E[\tilde{x}(t_1-\theta_1)\,\tilde{x}^T(t_1-\theta_1)]$$
$$= \Phi\,(t_2,t_1)\,P\,(t_1-\theta_1) \qquad (18)$$

where $\Phi\,(t_2, t_1)$ is the transition matrix associated with $(A - KM)$.

To relate $\tilde{x}_1(t_2-\theta_1)$ with $u_2(t_1)$, Eqn (17) can be integrated between $(t_2-\theta_1)$ and t_1 to give

$$\tilde{x}\,(t_2-\theta_1) = \exp\,[(A-KM)(t_2-\theta_1-t_1)]\,\tilde{x}(t_1)$$
$$+\int_{t_1}^{t_2-\theta_1}\Phi(t_2-\theta_1,\tau)[-K(\tau)\,v(\tau)+B(\tau)\,u(\tau)]\,d\tau.$$

Then

$$E[\tilde{x}(t_2-\theta_1)\,u^T(t_1)] = [\Phi(t_2-\theta_1,t_1)\,B(t_1)\,Q(t_1)]. \qquad (19)$$

Equation (18) could be written in terms of the component subsystems as

$$E\begin{Bmatrix}\tilde{x}_1(t_2-\theta_1)\,\tilde{x}_1{}^T(t_1-\theta_1) & \tilde{x}_1(t_2-\theta_1)\,\tilde{x}_2{}^T(t_1-\theta_1)\\ \tilde{x}_2(t_2-\theta_1)\,\tilde{x}_1{}^T(t_1-\theta_1) & \tilde{x}_2(t_2-\theta_1)\,\tilde{x}_2{}^T(t_1-\theta_1)\end{Bmatrix}$$

$$= E\begin{Bmatrix}\Phi_{11}(t_2,t_1) & \Phi_{12}(t_2,t_1)\\ \Phi_{21}(t_2,t_1) & \Phi_{22}(t_2,t_1)\end{Bmatrix}\begin{Bmatrix}P_{11}(t_1-\theta_1) & P_{12}(t_1-\theta_1)\\ P_{12}{}^T(t_1-\theta_1) & P_{22}(t_1-\theta_1)\end{Bmatrix}.$$

Then

$$E\,[\tilde{x}_1(t_2-\theta_1)\,\tilde{x}_1{}^T(t_1-\theta_1)] = \Phi_{11}(t_2,t_1)\,P_{11}(t_1-\theta_1) + \Phi_{12}(t_2,t_1)$$
$$P_{12}{}^T\,(t_1-\theta_1).$$

Similarly, in component form

$$E\begin{Bmatrix}\tilde{x}_1(t_2-\theta_1)u_1(t_1) & \tilde{x}_1(t_2-\theta_1)u_2(t_1)\\ \tilde{x}_2(t_2-\theta_1)u_1(t_1) & \tilde{x}_2(t_2-\theta_1)u_2(t_1)\end{Bmatrix}$$

$$= E\begin{Bmatrix}\Phi_{11}(t_2-\theta_1,t_1) & \Phi_{12}(t_2-\theta_1,t_1)\\ \Phi_{21}(t_2-\theta_1,t_1) & \Phi_{22}(t_2-\theta_1,t_1)\end{Bmatrix}\begin{Bmatrix}B_{11} & 0\\ 0 & B_{22}\end{Bmatrix}\begin{Bmatrix}Q_1 & 0\\ 0 & Q_2\end{Bmatrix}$$

from the assumed block diagonal form of the overall system noise covariance matrix. Then

$$E\,[\tilde{x}_1(t_2-\theta_1)u_2{}^T(t_1)] = \Phi_{12}(t_2-\theta_1,t_1)B_{22}Q_2$$

Thus from Eqn (12) for $t_2 - \theta_1 > t_1$,

$$E\,[u_2{}^*(t_2)u_2{}^{*T}(t_1)]$$

$$= A_{21}(t_2)[\Phi_{11}(t_2, t_1)\,P_{11}(t_1 - \theta_1) + \Phi_{12}(t_2, t_1)P_{12}{}^T(t_1 - \theta_1)]\,A_{21}{}^T(t_1)$$

$$+ A_{21}(t_2)\,\Phi_{12}(t_2 - \theta_1, t_1)B_{22}Q_2B_{22} \tag{20a}$$

and from Eqn (13) for $t_2 - \theta_1 \leqslant t_1$,

$$E\,[u_2{}^*(t_2)u_2{}^{*T}(t_1)] = A_{21}(t_2)\,[\Phi_{11}(t_2, t_1)P_{11}(t_1 - \theta_1)$$

$$+ \Phi_{12}P_{12}{}^T(t_1 - \theta_1)]\,A_{21}{}^T(t_1) \tag{20b}$$

Equations (10) and (20) describe the autocorrelation function of the subsystem noise forcing terms. The conditions under which the expression on the right-hand side of Eqn (20) is very much smaller than the expression on the right-hand side of Eqn (10) will now be examined.

It should be noted that the subsystem equations which produced $P_{12} > P_{11}$, P_{22} would not represent a good decomposition of the overall system and that it should always be possible to find a decomposition where $P_{11}, P_{22} > P_{12}$. In any practical situation in fact, P_{11}, P_{22} would be much larger than P_{12}. Consider when $t_2 - \theta_1 > t_1$. Then the elements of the $\Phi(t_2, t_1) = \exp\,(A - KM)(t_2 - t_1)$ matrix will be small if K is large. In that case if

$$A_{21}(t_1)P_{11}(t_1 - \theta_1)A_{21}{}^T(t_1) + B_{22}Q_{21}B_{22}{}^T \quad \text{is} \quad O(\delta),$$

then

$$A_{21}(t_2)[\Phi_{11}(t_2, t_1)P_{11}(t_1 - \theta_1) + \Phi_{12}(t_2, t_1)P_{12}(t_1 - \theta_1)]A_{21}{}^T(t_1)$$

$$+ A_{21}\Phi_{12}(t_2 - \theta_1, t_1)B_{22}Q_2B_{22}{}^T$$

would be $O(\delta^2)$. When $t_2 - \theta_1 \leqslant t_1$, then again, if K is large, and if

$$A_{21}P_{11}(t_1 - \theta_1)A_{21}{}^T$$

is $O(\delta)$, then

$$A_{21}[\Phi_{11}(t_2, t_1)P_{11}(t_1 - \theta_1) + \Phi_{12}(t_2, t_1)P_{12}]A_{21}{}^T$$

will be $O(\delta^2)$ so that $u_2{}^*$ is a good approximation to white noise.

Thus the interaction noise term is approximately white whenever K is large. One case of considerable practical significance is when the measurement noise spectral density is low since in that case, the overall steady state gain is quite large. This is best illustrated by the scalar case.

Consider the scalar system:

$$\dot{x} = ax + bu, \qquad y = mx + v.$$

In this case, the scalar Ricatti equation becomes

$$\dot{p} = 2ap - \frac{p^2 m^2}{r} + b^2 q.$$

In the steady state, $\dot{p} \to 0$ so that

$$p^2 - \frac{2ar}{m^2} p - \frac{b^2 r}{m^2} q = 0$$

or

$$p = \frac{ar}{m^2} \pm \frac{1}{m^2} \sqrt{a^2 r^2 + b^2 q m^2 r};$$

then

$$K = \frac{pm}{r} = \frac{a}{m} \pm \frac{1}{m} \sqrt{a^2 + b^2 m^2 q/r}.$$

From this equation it is clear that K is large when r is small, or when q is much larger than r.

In a practical case, for a given system, all that is really required is that the initial slope of the autocorrelation function of the interaction noise be much greater than the subsystem time constant. This will be illustrated in the numerical study. It should be noted that the covariance of the subsystem noise for the two subsystems from Eqn (10) is

$$Q_2{}^* = B_{22} Q_2 B_{22}{}^T + A_{21}(t) P_{11}(t - \theta_1) A_{21}{}^T(t)$$

$$Q_1{}^* = B_{11} Q_1 B_{11}{}^T + A_{12}(t) P_{22}(t - \theta_2) A_{12}{}^T(t)$$

and this is the value of the subsystem noise which is used in the filter of Eqn (9).

4. Computational Requirements

The main advantage of using the suboptimal filter is its computational simplicity. If the overall system matrix is of order l, the measurement vector of order p and the system noise vector is of order r, then it can be shown that using a fourth order Runge–Kutta procedure, the number of elementary multiplication operations required per iteration for the optimal and suboptimal filters are respectively

optimal:

$$20l^3 + 4p^3 + 5l^2p + 5lp^2 + 4lm^2 + 8l^2 + 12lp + 4lm + 4m^2 \qquad (21)$$

suboptimal:

$$\left[\sum_{i=1}^{2} 20l_i^3 + 4p_i^3 + 5l_ip_i^2 + 5l_i^2p_i + 5l_im_i^2 + 5m_i^2 + 8l_i^2 + 4l_im_i + 12l_ip_i\right]$$

$$+10l_1^2l_2 + 10l_1l_2^2 \qquad (22)$$

5. Examples

5.1 *A hypothetical example*

This example is a time lag version of one of the studies of Shah [3]. The overall system equations are

$$\dot{x}_1(t) = -0{\cdot}8\,x_1(t) + 0{\cdot}6\,x_2(t-\theta) + u_1$$

$$y_1 = x_1 + v_1$$

$$\dot{x}_2(t) = -0{\cdot}8x_2(t) + 0{\cdot}6x_1(t-\theta) + u_2$$

$$y_2 = x_2 + v_2$$

The optimal and suboptimal filters were simulated for this example on the ICL 4130 digital computer of Cambridge University. The delays in the overall filter were approximated by a second order Taylor series expansion

i.e. $$x_1(t-\theta) = x_1(t)\,e^{-s\theta} = x_1(t)/(1 + s\theta + s^2\theta^2/2)$$

so that the overall system becomes sixth order.

5.1.1 *Figure of merit.* To compare the performance of the optimal and suboptimal filters the square of the estimation error was taken as the criterion function i.e. at each iteration, the optimal and suboptimal estimates were compared with the states obtained from a simulation of the overall system. This is a more valid criterion than the trace of the covariance matrix as used by Shah for discrete dynamical systems since in that case, the subsystem covariance matrix does not represent the true covariance of the suboptimal estimates since it is obtained from a suboptimal Riccati equation.

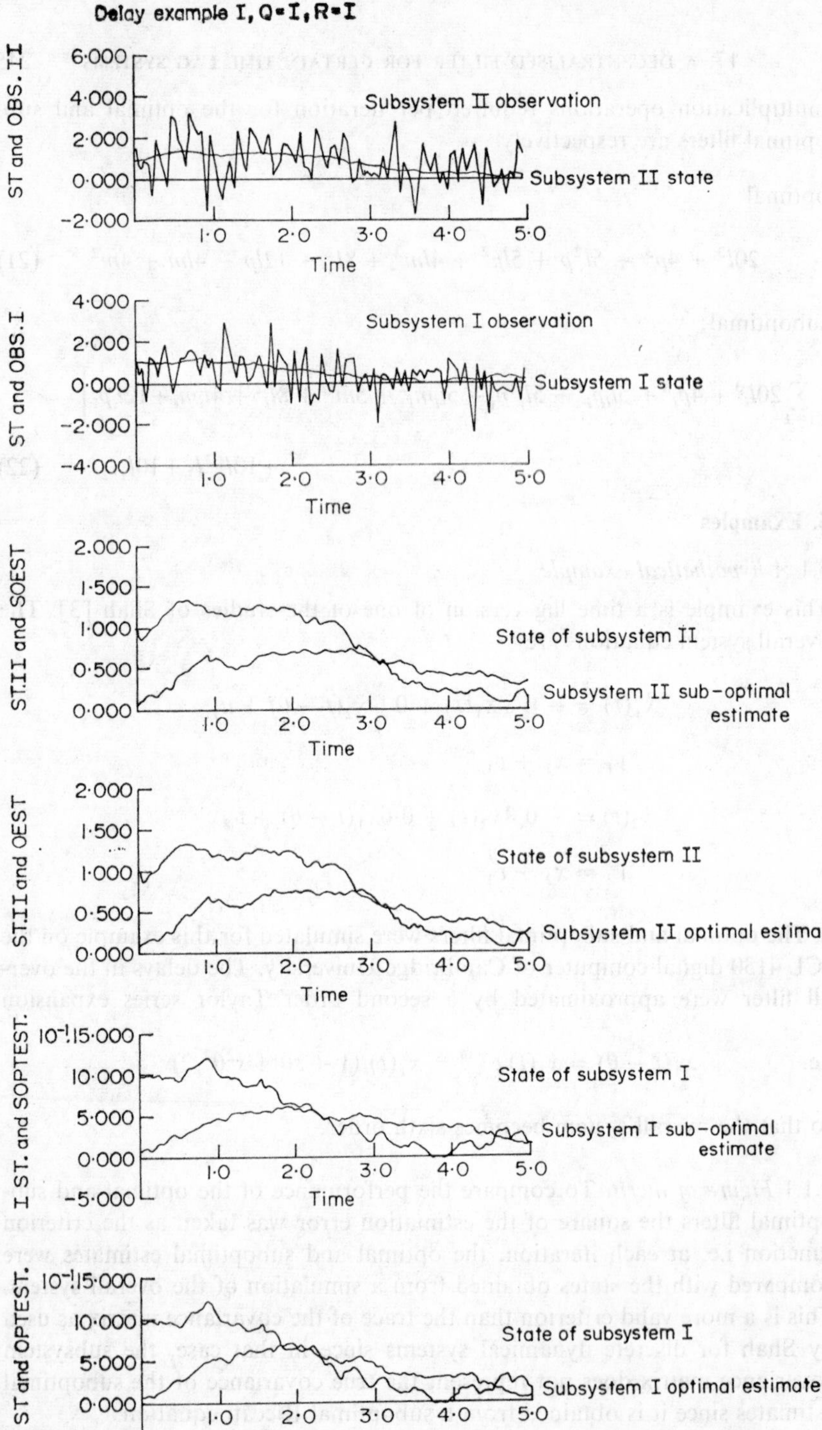

Delay example I, Q=I, R=I
ST and OBS. II
6·000
4·000
2·000
0·000
−2·000
Subsystem II observation
Subsystem II state
1·0
2·0
3·0
4·0
5·0
Time
ST and OBS. I
4·000
2·000
0·000
−2·000
−4·000
Subsystem I observation
Subsystem I state
1·0
2·0
3·0
4·0
5·0
Time
ST. II and SOEST
2·000
1·500
1·000
0·500
0·000
State of subsystem II
Subsystem II sub-optimal estimate
1·0
2·0
3·0
4·0
5·0
Time
ST. II and OEST
2·000
1·500
1·000
0·500
0·000
State of subsystem II
Subsystem II optimal estimate
1·0
2·0
3·0
4·0
5·0
Time
I ST. and SOPTEST.
10⁻¹ 15·000
10·000
5·000
0·000
−5·000
State of subsystem I
Subsystem I sub-optimal estimate
1·0
2·0
3·0
4·0
5·0
Time
I ST and OPTEST.
10⁻¹ 15·000
10·000
5·000
0·000
−5·000
State of subsystem I
Subsystem I optimal estimate
1·0
2·0
3·0
4·0
5·0
Time

FIG. 5.

TABLE 1

Q	R	Optimal average cost	Suboptimal average cost	Loss of optimality %
I	0·1I	5·360	5·362	0·04
I	0·5I	16·494	17·753	7·5
I	I	19·100	21·230	9
I	2I	27·508	30·643	10

5.1.2 *Results.* The optimal filter is optimal in a statistical sense. Hence it is necessary to compare the performances of the optimal and suboptimal filters for a number of sample functions of the noise and to take the average. Table I gives such an average for five different simulations for varying values of R, the measurement noise. All the costs are for a hundred iterations.

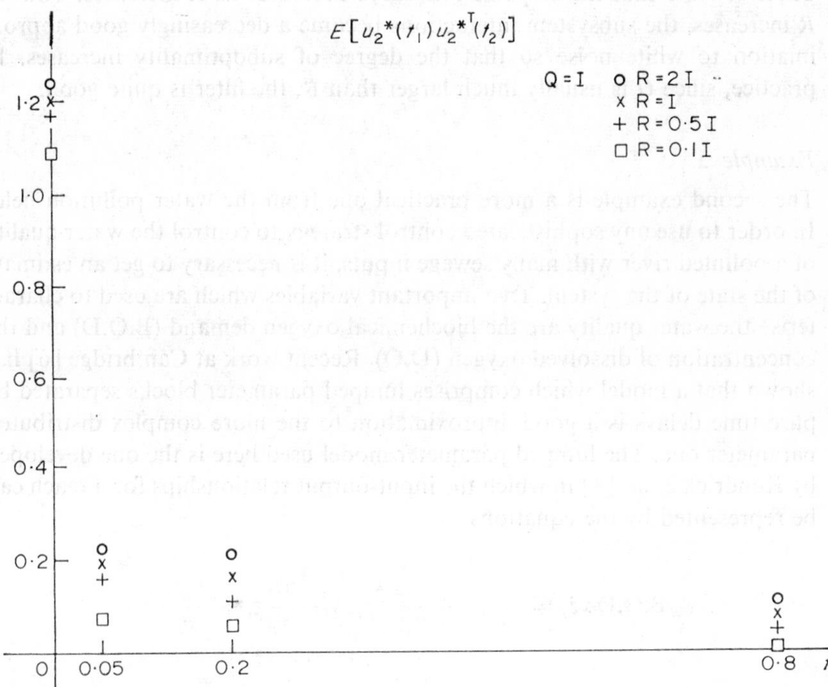

FIG. 6.

In each case the estimate starts from zero and the state from one. The initial value of the covariance matrix $P(0)$ is known exactly. Figure 5 shows the performance of the filters for $Q = I$, $R = I$ for one of the sample functions of the noise. It should be noted that the top two curves which show the observations are on a different scale and that the measurement noise levels are quite high. Even for this case of high measurement noise, the suboptimal filter behaves quite well.

From expressions (21) and (22), the computational requirements for the two filters are

Optimal = 5390 multiplications
Suboptimal = 156 multiplications
Savings = 35 *times*

In addition, Eqns (10) and (20) were programmed. Figure 6 shows the steady state values of $E[u_2{}^*(t_2)u_2{}^{*T}(t_1)]$ where $t_1 = 0$ and $0 \leqslant t_2 \leqslant 0{\cdot}8$. Since the time constant of subsystem II is 0.8, Fig. 6 shows that the suboptimal filters should be near optimal and this has already been confirmed by Table I. Note that the slope of this curve decreases as R increases. Thus as R increases, the subsystem interactions become a decreasingly good approximation to white noise so that the degree of suboptimality increases. In practice, since Q is usually much larger than R, the filter is quite good.

Example 2.

The second example is a more practical one from the water pollution field. In order to use any sophisticated control strategy to control the water quality of a polluted river with many sewage inputs, it is necessary to get an estimate of the state of the system. Two important variables which are used to characterise the water quality are the biochemical oxygen demand (B.O.D) and the concentration of dissolved oxygen (D.O). Recent work at Cambridge [6] has shown that a model which comprises lumped parameter blocks separated by pure time delays is a good approximation to the more complex distributed parameter one. The lumped parameter model used here is the one developed by Kendrick *et al.* [8] in which the input-output relationships for a reach can be represented by the equations

$$\text{B.O.D}: \dot{z}_i = -\alpha_i z_i + \frac{\psi_{i-1}}{v_{i-1}} z_{i-1} - \frac{\psi_i}{v_i} z_i + \frac{m_i s_i}{v_i}$$

$$\text{D.O.}\ \dot{q}_i = \beta_i(q^s - q_i) + \frac{\psi_{i-1}}{v_{i-1}} q_{i-1} - \frac{Q_i}{v_i} q_i - \alpha_i z_i - \frac{\eta_i}{v_i}$$

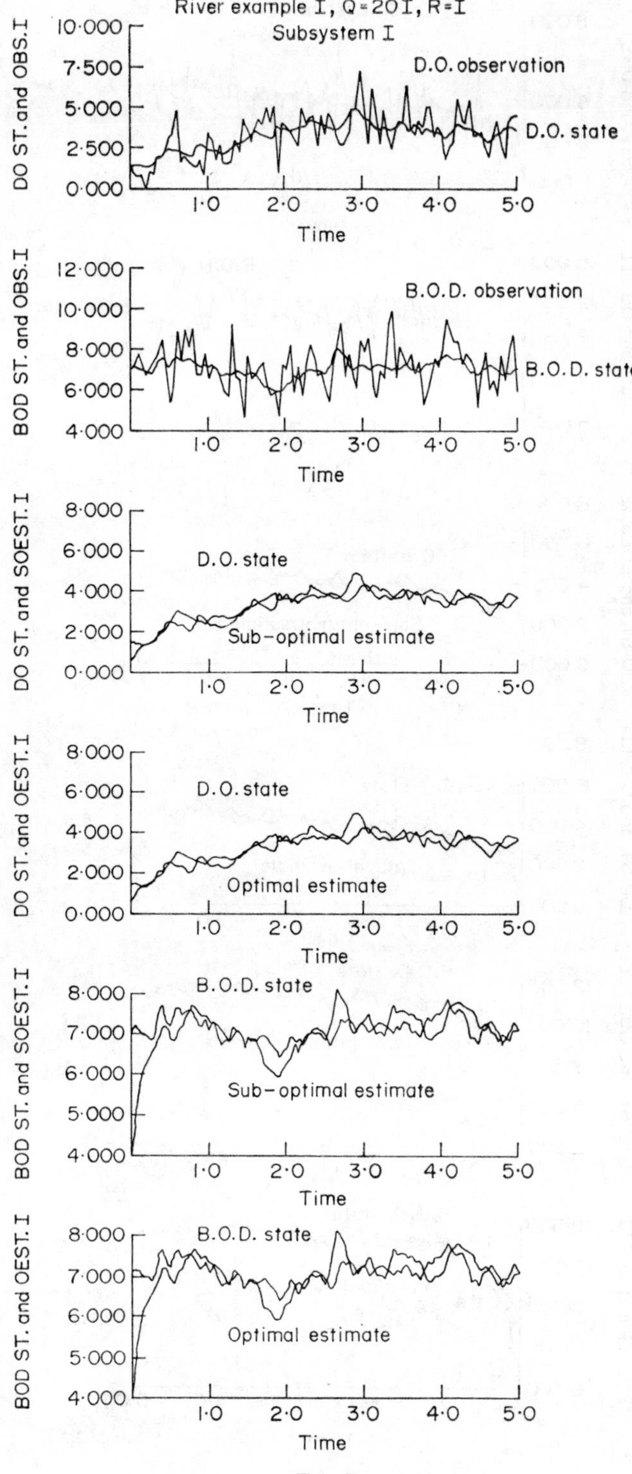

River example I, Q=20I, R=I
Subsystem I
DO ST. and OBS.I
10·000
7·500
5·000
2·500
0·000
D.O. observation
D.O. state
1·0
2·0
3·0
4·0
5·0
Time
BOD ST. and OBS.I
12·000
10·000
8·000
6·000
4·000
B.O.D. observation
B.O.D. state
1·0
2·0
3·0
4·0
5·0
Time
DO ST. and SOEST.I
8·000
6·000
4·000
2·000
0·000
D.O. state
Sub-optimal estimate
1·0
2·0
3·0
4·0
5·0
Time
DO ST. and OEST.I
8·000
6·000
4·000
2·000
0·000
D.O. state
Optimal estimate
1·0
2·0
3·0
4·0
5·0
Time
BOD ST. and SOEST.I
8·000
7·000
6·000
5·000
4·000
B.O.D. state
Sub-optimal estimate
1·0
2·0
3·0
4·0
5·0
Time
BOD ST. and OEST.I
8·000
7·000
6·000
5·000
4·000
B.O.D. state
Optimal estimate
1·0
2·0
3·0
4·0
5·0
Time

FIG. 7.

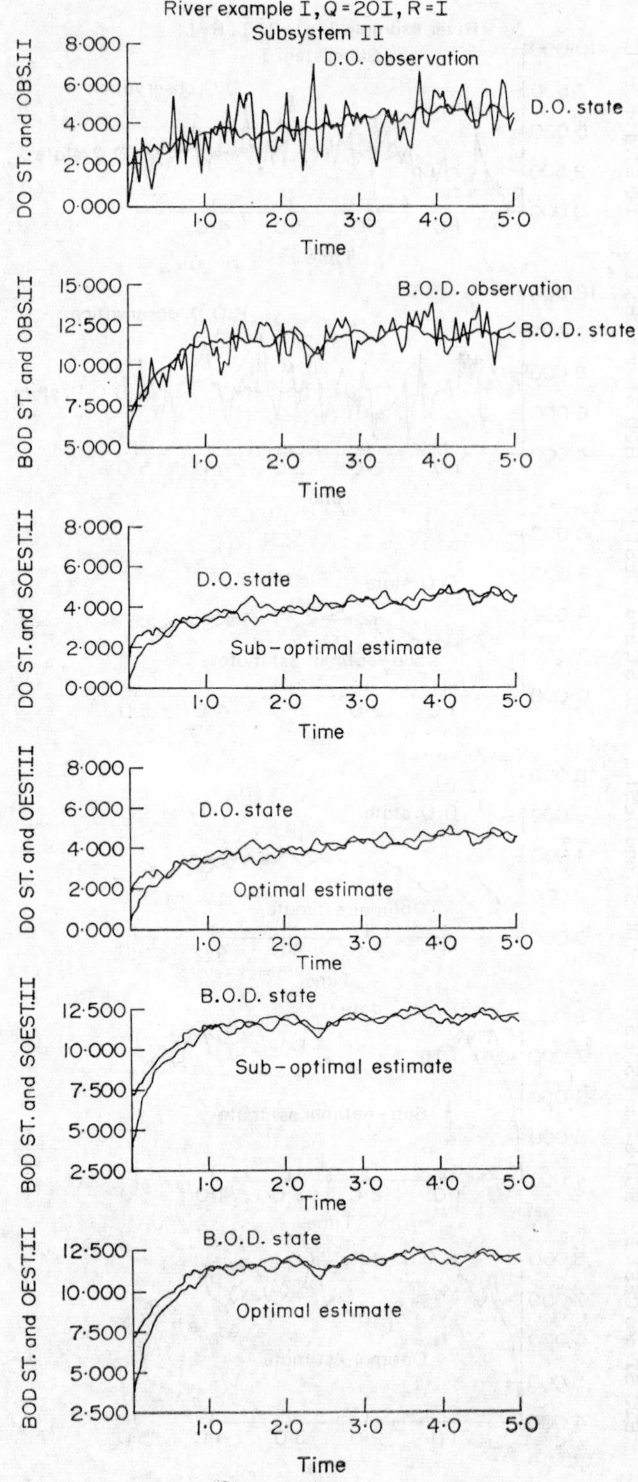

River example I, Q = 20I, R = I
Subsystem II
D.O. observation
D.O. state
DO ST. and OBS.II
8·000
6·000
4·000
2·000
0·000
1·0
2·0
3·0
4·0
5·0
Time
B.O.D. observation
B.O.D. state
BOD ST. and OBS.II
15·000
12·500
10·000
7·500
5·000
Time
DO ST. and SOEST.II
D.O. state
Sub-optimal estimate
Time
DO ST. and OEST.II
D.O. state
Optimal estimate
Time
BOD ST. and SOEST.II
B.O.D. state
Sub-optimal estimate
12·500
10·000
7·500
5·000
2·500
Time
BOD ST. and OEST.II
B.O.D. state
Optimal estimate
Time

FIG. 8.

where the symbols have the following meaning

$z_i(t)$ B.O.D. concentration at the ith reach (mg/litre)

$q_i(t)$ D.O. concentration at the ith reach (mg/litre)

$\alpha_i(t)$ B.O.D. decay rate in reach i (per hour)

$\beta_i(t)$ D.O. Recovery rate in reach i (per hour)

$\psi_i(t)$ Stream flow rate in reach i (litres/hour)

$v_i(t)$ Volume of water in reach i (M litres)

$m_i(t)$ Treated effluent flow rate into the river.

$s_i(t)$ B.O.D. concentration of effluent going into the river.

$\eta_i(t)$ D.O. requirements of the bottom sludge (MG/hr)

$q^s(t)$ D.O. concentration saturation level of the stream.

Using a second order approximation to the delays as in the last example, the system equations of the two reaches became of 8th order. The particular parameter values used were from a section of the River Cam where

$$\psi_i/v_i = 1, \quad \alpha_i = 0{\cdot}42, \quad \beta_i = 0{\cdot}98, \quad q_i^s = 10, \quad \eta/v = 2.$$

The overall system was further forced by input B.O.D. and D.O. levels of 10 mg/litre for each reach. The states B.O.D. at each reach were started from 7 and the states D.O. from 1·55.

Results:

Figure (7) shows the performance of the optimal and suboptimal filters for subsystem I for $Q = 20I$, $R = I$ and Fig. 8 for subsystem II. In each case, the estimates converge very quickly and the optimal and suboptimal filters are quite close.

A combined suboptimal filtering and control strategy for a more practical version of this problem, i.e. where the easily measureable variable D.O. is used for the measurements only, is described elsewhere [7].

6. Conclusions

A simple decentralised filter has been developed for serial time-lag systems with and without recycles. The filter is near optimal for a number of cases of practical interest and it has minimal computing requirements. For a given system, from a knowledge of the maximum values of the system and measurement noise levels, it is possible to assess whether in fact the filter will provide good estimates if implemented by using Eqns (10) and (20);

Acknowledgements

The author is indebted to his research supervisor Prof. J. F. Coales for encouragement and guidance and to Dr. P. C. Young, Dr. A. T. Fuller, Dr. G. White, and Mr. D. B. Reid for many helpful discussions.

References

1. Mesarovic, M. D., Macko, D. and Takahara, Y., "Theory of Hierarchical Multi-level Control". Academic Press, New York (1970).
2. Aoki, M., Some iterative techniques for decentralised dynamic systems control. "Proc. 5th IFAC Congress", Paris (1972).
3. Shah, M., "Suboptimal Filtering Theory for Interacting Control Systems". Ph.D. Thesis, University of Cambridge (1971).
4. Singh, M. G., Discussion on the paper of J. Atary and M. Shah *In* "Proc. 5th IFAC Congress". Paris (1972).
5. Fuller, A. T., Optimal non-linear control of systems with pure delays. *Int. J. Control* **8** (2) (1968), 145–168.
6. Beck, M. B., University of Cambridge Internal Note.
7. Singh, M. G. and Young, P. C., "Multi-level Control of Waste Water Treatment Facilities". Cambridge University Engineering Department Report No. CUED/B-CONTROL TR 32 (1972).
8. Kendrick, D. A., Rao, H. S. and Wells C. H., Optimal operation of a system of water waste treatment facilities' *In* "I.E.E.E. Symposium on Adaptive Processes" (1970).

Appendix I

The Filter of Shah:

M. M. Shah has recently made significant contributions to filtering techniques by developing the SPA filter for discrete dynamical systems. His filter is practical. He has justified it purely by numerical examples so that although he states that the system interaction terms are white noise sequences he never proves this. Nor does he examine the conditions under which this is a good approximation except by stating that he assumes that the subsystems are weakly correlated and from this assumption

$$E\,[\tilde{x}_1(t_2)\tilde{x}_1{}^T(t_1)] = 0$$

This in fact implies that $P_{12} = 0$ and if this is so, there is no hierarchical filtering problem since in that case independent filters can be constructed for the subsystems.

The analysis of this paper can be applied to continuous linear dynamical systems by setting $\theta_1 = \theta_2 = 0$.

The main criticism of Shah's computational results is that he uses as a measure of the performance of the filters, the trace of the covariance matrix. This is valid for the optimal filter but not for the suboptimal one since in that case, by solving the partial Riccati equation, the subsystem covariance matrices do not represent the true covariances of the suboptimal estimates.

18. Optimal Control of a Degenerate Markovian System

M. H. A. DAVIS

Department of Computing and Control,
Imperial College of Science and Technology, London, England

1. Introduction

This paper concerns a system described by stochastic differential equations of the form

$$dx_t = f(t, x_t, y_t)dt$$

$$dy_t = g(t, x_t, y_t, u_t)dt + \sigma(t, y_t)dw_t \tag{1.1}$$

where $t \in [0, 1]$, $\{w_t\}$ is a Brownian motion and the control function u_t is of the form $u(t, x_t, y_t)$. For a given control function, $\{x_t, y_t\}$ is a Markov process with degenerate (non-uniformly elliptic) differential operator. The objective is to choose u so as to minimize the cost:

$$J(u) = E\int_0^1 h(t, x_t, y_t, u_t)dt.$$

The main result of the paper is Theorem 3 which says that an optimal control policy exists for this problem, that this policy is in fact optimal in the class of all controls depending on the complete past of the process, and that it is obtained by minimizing a certain Hamiltonian function.

This result is based on a representation (Theorem 2) of the *value function* $\phi(t, x, y)$ which, roughly speaking, is the minimum cost starting at $x_t = x$, $y_t = y$. Theorem 2 relies on some fairly recent results in martingale theory (in particular on the Meyer decomposition of supermartingales); the reader is referred to [1]–[4] for details of the terminology and results used here.

A similar problem has been considered from the point of view of partial differential equations in a recent paper [11] by Rishel. Some comments on the relationship of [11] to this paper will be found in Section 5.

Since the function g in (1.1) is not assumed to be Lipschitz, (1.1) may not have a solution in the usual Ito sense. The solution is defined instead by the Girsanov measure transformation technique; this is outlined in the next section and has been previously used in several papers, including [6], [11] and [12].

2. Preliminaries

The various system functions are as follows. $f:[0,1] \times R^n \times R^m \to R^n$ is continuous and satisfies a uniform Lipschitz condition:

$$|f(t,x,y) - f(t,x',y')| \leqslant K(|x-x'| + |y-y'|). \tag{2.1}$$

$g:[0,1] \times R^n \times R^m \times U \to R^m$ (here U, the *control set*, is a compact subset of R^k) is measurable, continuous on U for each fixed (t,x,y) and there is a constant K' such that for all $(t,x,y,u) \in [0,1] \times R^n \times R^m \times U$,

$$|g(t,x,y,u)| \leqslant K'(1 + |x| + |y|). \tag{2.2}$$

The cost rate h is a positive real-valued function, bounded by a constant k. $\sigma(t,y)$ is an $m \times m$ matrix for each $(t,y) \in [0,1] \times R^m$ whose elements $\sigma^{ij}(t,y)$ are measurable and uniformly Lipschitz continuous in y. The inverse matrix $\sigma^{-1}(t,y)$ is assumed to exist and be bounded for $(t,y) \in [0,1] \times R^m$.

Let $(\Omega, \mathscr{A}, \mu)$ be a probability space carrying an m-dimensional separable Brownian motion process $\{w_t\}$. Let C be the space of continuous functions from $[0,1]$ to R^m, and $\{\mathscr{F}_t\}$ the increasing family of σ-fields of subsets of C generated by the evaluation maps. With the above conditions the following (Ito) equation has a unique solution (with a fixed initial value $y_0 \in R^m$).

$$dy_t = \sigma(t, y_t)dw_t. \tag{2.3}$$

Regarding C as the sample space of $\{y_t\}$, $\{y_t\}$ induces a measure P on $(C, \mathscr{F}_1)$. The probability space $(C, \mathscr{F}_1, P)$ will be taken as the "basic space" from now on. In view of (2.1) the equation

$$dx_t = f(t, x_t, y_t(\omega))dt, \qquad x(0) = x_0 \in R^n \tag{2.4}$$

has a unique solution for each $\omega \in \Omega$, defining a function $\chi_t: C \to R^n$ such that $\chi_t(y) = x_t$. (Henceforth y denotes an element of C). Evidently χ_t is $\mathscr{F}_t$-measurable for each t.

Let $\mathscr{M}$ denote the set of measurable functions $u:[0,1] \times R^n \times R^m \to U$

(the *Markov control policies*) and $\mathscr{U}$ (the *non-anticipative control policies*) the set of measurable functions $\tilde{u}:[0, 1] \times C \to U$ such that $\tilde{u}(t_1, .)$ is $\mathscr{F}_t$-measurable for each t. Notice that $\mathscr{M} \subset \mathscr{U}$ via the formula $\tilde{u}(t, y) = u(t, \chi_t(y), y_t)$. For $u \in \mathscr{U}$ define the following random variable:

$$\zeta_s^t(u) = \int_s^t (g_\tau, a_\tau^{-1}\, dy_\tau) - \tfrac{1}{2}\int_s^t (g_\tau, a_\tau^{-1}\, g_\tau)d\tau\dagger$$

where
$$g_\tau = g\big(\tau, \chi_\tau(y), y_\tau, u(\tau, \chi_\tau(y), y_\tau)\big)$$

and
$$a_\tau = a(\tau, y_\tau) = \sigma(\tau, y_\tau)\sigma'(\tau, y_\tau).$$

THEOREM 1. (Girsanov [10], Beneš [12])
Let the measure P_u on $(C, \mathscr{F}_1)$ be defined by

$$P_u A = \int_A \exp[\zeta_0{}^1(u)]dP \qquad A \in \mathscr{F}_1;$$

then

(i) *P_u is a probability measure.*

(ii) *P_u is mutually absolutely continuous with respect to P.*

(iii) *$\{B_t, t \in [0, 1]\}$ is a Brownian motion under P_u,*

where

$$dB = \sigma^{-1}(dy - g\, dt). \tag{2.5}$$

Thus P_u is the measure corresponding to the solution of (1.1), *in the sense that under P_u*

$$dx_t = f(t, x_t, y_t)dt$$

$$dy_t = g(t, x_t, y_t, u_t)dt + \sigma(t, y_t)dB_t \tag{2.6}$$

where $\{B_t\}$ is a Brownian motion. Denoting integration with respect to P_u by E_u the cost corresponding to $u \in \mathscr{U}$ is

$$J(u) = E_u \int_0^1 h(t, \chi_t(y), y_t)dt$$

LEMMA 1. *Suppose $u \in \mathscr{M}$. Then the process $(x_t. y_t)$ is Markov under P_u.*

Proof: Let $\mathscr{B}_t = \sigma(x_t, y_t)\ddagger$. It suffices to prove that for $t > s, \gamma$ a bounded

† $(.,.)$ is the inner product in R^m.
‡ The σ-field generated by the r.v.s. x_t, y_t.

Borel function,

$$E_u[\gamma(x_t, y_t) \mid \mathscr{F}_s] = E_u[\gamma(x_t, y_t) \mid \mathscr{B}_s] \quad \text{a.s.} \tag{2.7}$$

First, (x_t, y_t) is Markov under P, as a standard argument ([2] p.155) shows. Now denote $\rho_s^t = \exp[\zeta_s^t(u)]$ for $u \in \mathscr{M}$. $(\rho_0^t, \mathscr{F}_t)$ is a martingale ([10] Lemma 1). Since $dP_u/dP = \rho_0^1$,

$$\begin{aligned} E_u[\gamma(x_t, y_t) \mid \mathscr{F}_s] &= \frac{E[\rho_0^1 \gamma(x_t, y_t) \mid \mathscr{F}_s]}{E[\rho_0^1 \mid \mathscr{F}_s]} \\ &= E[\rho_s^1 \gamma(x_t, y_t) \mid \mathscr{F}_s] \quad \text{a.s.} \\ &= E[\rho_s^1 \gamma(x_t, y_t) \mid \mathscr{B}_s] \quad \text{a.s.} \end{aligned} \tag{2.8}$$

The last line follows from the Markov property of (x_t, y_t) under P, since $\rho_s^1 \gamma(x_t, y_t)$ is $\sigma(x_\tau, y_\tau, s \leqslant \tau \leqslant 1)$-measurable, and (2.7) is immediate from (2.8).

3. Structure of the Value Function

Let $u \in \mathscr{M}$. The expected remaining cost from time t on is

$$\begin{aligned} \psi_u(t) &= E_u\left[\int_t^1 h(\tau, x_\tau, y_\tau, u_\tau) d\tau \mid \mathscr{F}_t\right] \\ &= E\left[\rho_t^1 \int_t^1 h(\tau, x_\tau, y_\tau, u_\tau) d\tau \mid \mathscr{B}_t\right] \end{aligned}$$

by the same calculation as in Lemma 1. Notice that $\psi_u(t)$ does not depend on what control was used on $[0, t)$. Since $\psi_u(t)$ is $\mathscr{B}_t$-measurable there is a function $\phi^u : [0,1] \times R^n \times R^m \to R$ such that $\psi_u(t) = \phi^u(t, x_t, y_t)$, and $0 \leqslant \phi^u \leqslant k$ since $0 \leqslant J(u) \leqslant k$. Thus $\phi^u(t,.,.) \in L_\infty(R^{m+n})$ for each $t \in [0,1]$. Since L_∞ is a complete lattice the following infimum exists in L_∞:

$$\phi(t,.) = \bigwedge_{u \in \mathscr{M}} \phi^u(t,.).$$

Notice that

$$\phi(1, \xi, \eta) = 0$$

$$\phi(0, x_0, y_0) = J^* \triangleq \inf_{u \in \mathscr{M}} J(u). \tag{3.1}$$

This is the *value function* for the control problem: i.e. $\phi(t, \xi, \eta)$ is the minimum cost starting at time t with $x_t = \xi, y_t = \eta$. An argument similar to that of [6, Lemma 6.1] shows that $\phi(t, \xi, \eta)$ satisfies the following "principle of optimality" (where $\phi_t \triangleq \phi(t, x_t, y_t)$)

$$\phi_t \leqslant E_u\left[\int_t^{t+\delta} h(s, x_s, y_s, u_s)ds \mid \mathscr{B}_t\right] + E_u[\phi_{t+\delta} \mid \mathscr{B}_t] \quad \text{a.s.} \tag{3.2}$$

for each $t \in [0, 1]$ and $\delta \in [0, 1 - t]$.

THEOREM 2. *ϕ has the representation*

$$\begin{aligned}\phi(t, x_t, y_t) = J^* &+ \int_0^t \Lambda\phi(s, x_s, y_s)ds \\ &+ \int_0^t \nabla\phi(s, x_s, y_s)dy_s\end{aligned} \tag{3.2a}$$

a.s. under measure P, where $\Lambda\phi(\nabla\phi)$ is a measurable function from $[0, 1] \times R^n \times R^m \to R\ (R^m)$ *and*

$$\int_0^1 |\nabla\phi(t, x_t, y_t)|^2 dt < \infty \qquad \text{a.s.} \tag{3.3}$$

Proof: Let $\theta_t = \phi_t + kt$ and fix $u \in \mathscr{M}$. Since $h \leqslant k$ it follows from (3.2) that

$$\theta_t \leqslant E_u[\theta_{t+\delta} \mid \mathscr{F}_t] \quad \text{a.s.}$$

i.e. $(\theta_t, \mathscr{F}_t, P_u)$ is a submartingale. Since $\theta_t \leqslant k$, θ is in class (D), so from the Meyer decomposition of supermartingales,

$$\theta_t = M_t + A_t \tag{3.4}$$

where $\{M_t\}$ and $\{A_t\}$ are adapted to $\mathscr{F}_t$, $\{M_t\}$ is a P_u-martingale and $\{A_t\}$ is an integrable increasing process. (i.e. $E_u[A_1] < \infty$).

LEMMA 2. *For almost all $y \in C$, $\{A_t(y)\} \ll$ Lebesgue measure i.e. there exists an adapted process $\{\alpha_t\}$, $\alpha_t \geqslant 0$ a.s. such that*

$$A_t = \int_0^t \alpha_s ds.$$

Proof: The proof is that of [6, Lemma 4.2].

LEMMA 3. *$\{M_t\}$ has the representation*

$$M_t = M_0 + \int_0^t \beta_s dB_s$$

where $\{B_t\}$ is as in (2.4) and $\{\beta_t\}$ is adapted to $\mathscr{F}_t$ with

$$\int_0^t |\beta_s|^2 \, ds < \infty \qquad \text{a.s.}$$

Proof: This is an application of [6, Theorem 2.3], the basic fact being that all martingales on the σ-fields of a Brownian motion are stochastic integrals.

Combining the above facts and using (2.5) and (3.1) ϕ_t is seen to have the representation

$$\phi_t = J^* + \int_0^t m_s \, ds + \int_0^t q_s \, dy_s \tag{3.5}$$

(the latter integral being evaluated under measure P) where

$$m_s = \alpha_s - k - \beta_s \sigma_s^{-1} g_s$$
$$q_s = \beta_s \sigma_s^{-1}.$$

The representation (3.5) is easily seen to be unique since if (3.5) is also satisfied with m' and q' then

$$\int_0^t (m_s - m_s')ds = \int_0^t (q_s' - q_s)dy_s$$

so that

$$\int_0^t (m_s - m_s')ds$$

is a bounded variation continuous local martingale and hence necessarily zero.

It remains to show that m_t, q_t are just functions of (t, x_t, y_t)—i.e. are $\mathscr{B}_t$-measurable. In the following the terminology and notation are those of [3]; see also [4]. Recall that for any local L_2-martingale X, $\langle X, X\rangle_t$ is the unique predictable increasing process such that $X_t^2 - \langle X, X\rangle_t$ is a local martingale, and for $\xi \in L_2(\langle X, X\rangle)$ the stochastic integral

$$Y_t = \int_0^t \xi_s dX_s$$

is characterised by the property that it is the unique local martingale satisfying

$$\langle Y, Z\rangle_t = \int_0^t \xi_s \, d\langle X, Z\rangle_s$$

for each local martingale Z. If $\{w\text{-}\}$ is a Brownian motion then

$$\langle w^i, w^j \rangle_t = \delta_{ij}\, t.$$

From (2.3),

$$y_t = y_0 + \int_0^t \sigma_s \, dw_s.$$

Thus

$$\begin{aligned}\langle y^i, w^j \rangle_t &= \left\langle \sum_k \int_0^t \sigma^{ik}\, dw^k, w^j \right\rangle_t \\ &= \sum_k \int_0^t \sigma^{ik}\, d\langle w^k, w^j \rangle_s \\ &= \int_0^t \sigma^{ij} ds.\end{aligned}$$

Let

$$M_t = \int_0^t q_s dy_s.$$

Then

$$\begin{aligned}\langle M, y^i \rangle_t &= \left\langle \sum_j \int_0^t q^j\, dy^j, y^i \right\rangle_t \\ &= \left\langle \sum_j \int_0^t (q'\,\sigma)^j\, dw^j, y^i \right\rangle_t \\ &= \sum_j \int_0^t (q'\sigma)^j\, d\langle w^j, y^i \rangle_s \\ &= \sum_j \int_0^t (q'\sigma)^j \sigma^{ij}\, ds = \sum_j \int_0^t q^j a^{ij}\, ds.\end{aligned}$$

Thus, defining

$$\langle M, y \rangle_t = (\langle M, y^1 \rangle_t \ldots \langle M, y^m \rangle_t), \qquad \langle M, y \rangle_t = \int_0^t q'\, a\, ds$$

i.e.

$$q_t = a^{-1} \frac{d}{dt}(\langle M, y \rangle_t). \tag{3.6}$$

Applying the stochastic differential formula [3] to the product $\phi_t y_t^i$, where

ϕ_t is given by (3.5), gives

$$d(\phi_s y_s^i) = \phi_s dy_s^i + y_s^i d\phi_s + d\langle M, y^i\rangle_s.$$

Thus

$$\langle M, y^i\rangle_{t+\delta} - \langle M, y^i\rangle_t = \phi_{t+\delta} y_{t+\delta}^i - \phi_t y_t^i - \int_t^{t+\delta} \phi_s dy_s^i - \int_t^{t+\delta} y_s^i d\phi_s. \tag{3.7}$$

Each term on the right of (3.7) is measurable with respect to $\mathscr{B}_t^{t+\delta} \triangleq \sigma\{x_s, y_s, t \leqslant s \leqslant t+\delta\}$. Referring to (3.6) one sees that q_t is $\mathscr{B}_t^{t+\delta}$-measurable for each $\delta > 0$, and is therefore measurable with respect to $\bigcap_{\delta>0} \mathscr{B}_t^{t+\delta} = \mathscr{B}_t$. There therefore exists a function $\nabla\phi$ such that

$$q_t(y) = \nabla\phi(t, x_t, y_t)$$

and (3.3) follows from Lemma 3.

The existence of $\Lambda\phi$ such that

$$m_t(y) = \Lambda\phi(t, x_t, y_t)$$

now follows by a similar argument using the fact that

$$\int_s^t m_\tau d\tau = \phi_t - \phi_s - \int_s^t \nabla\phi_\tau\, dy_\tau.$$

This completes the proof.

4. Applications

Using the representation of Theorem 2 one can easily derive a "Hamilton–Jacobi" condition for optimality (Lemma 4) and then show (Theorem 3) that an optimal Markov control exists and that this is in fact the optimal non-anticipative control.

LEMMA 4. *For almost all* $(t, \xi, \eta, u) \in [0, 1] \times R^{n+m} \times U$,

$$\Lambda\phi(t, \xi, \eta) + \nabla\phi(t, \xi, \eta)\, g(t, \xi, \eta, u) + h(t, \xi, \eta, u) \geqslant 0. \tag{4.1}$$

$u^0 \in \mathscr{M}$ *is optimal if and only if equality holds a.e. in* (4.1) *with* $u = u^0(t, \xi, \eta)$.

Proof: Let $u \in \mathscr{M}$. Combining (3.2a) and (2.5) it is seen that ϕ has the representation

$$\phi_t = J^* + \int_0^t (\Lambda\phi + \nabla\phi . g)dt + \int_0^t \nabla\phi . \sigma dB \tag{4.2}$$

where B is Brownian under P_u. Now the last term is a martingale (this follows from the uniqueness of the decomposition (3.4)) so, using (3.2),

$$E_u\left[\int_t^{t+\delta} (\Lambda\phi + \nabla\phi . g + h)ds \mid \mathscr{F}_t\right] \geqslant 0 \quad \text{a.s.} \tag{4.3}$$

It is easily seen that the left side of (4.3) converges to $\Lambda\phi_t + \nabla\phi_t . g_t + h_t$ weakly in L_1 for almost all t and this gives (4.1) after noticing that the value of $u(t, x_t, y_t)$ could be anywhere in U.

One shows without difficulty that, if u^0 is optimal, then equality holds in (3.2) and hence in (4.1). Conversely, suppose equality holds in (4.1) for $u = u^0$. Evaluating (4.2) at $t = 1$ gives

$$0 = J^* + \int_0^1 (\Lambda\phi + \nabla\phi . g)dt + \int_0^1 \nabla\phi . \sigma\, dB$$

$$= J^* - \int_0^1 h dt + \int_0^1 \nabla\phi . \sigma dB. \qquad \text{(from (4.1))}$$

Thus

$$E_{u^0}\int_0^1 h(t, x_t, y_t, u_t^0)dt = J^* = \inf_{u\in\mathscr{M}} J(u)$$

so that u^0 is optimal in $\mathscr{M}$.

LEMMA 5. $$\inf_{u\in\mathscr{U}} J(u) = \inf_{u\in\mathscr{M}} J(u)$$

Proof: The same calculation as above, using an arbitrary $u \in \mathscr{U}$, will show that $J^* \leqslant J(u)$. This gives the result in view of the definition of J^*.

This shows, what is intuitively clear, that there is no advantage in using non-Markovian control policies.

THEOREM 3. *There exists a control policy $u^0 \in \mathscr{M}$ which is optimal in $\mathscr{U}$.*

Proof: For $p \in R^m$, let

$$\mathscr{H}'(t, \xi, \eta, u, p) = p . g(t, \xi, \eta, u) + h(t, \xi, \eta, u)$$

and

$$\mathscr{H}(t, \xi, \eta, p) = \inf_{u\in U} \mathscr{H}'(t, \xi, \eta, u, p). \tag{4.4}$$

Let S be a countable dense subset of U. $\mathscr{H}'$ is continuous in u for fixed (t, ξ, η, p), so that for any $b \in R$,

$$\{(t, \xi, \eta, p) : \mathscr{H}(t, \xi, \eta, p) < b\} = \bigcup_{u\in S} \{(t, \xi, \eta, p) : \mathscr{H}'(t, \xi, \eta, u, p) < b\}$$

so that $\mathscr{H}$ is measurable. The infimum in (4.4) is attained since U is compact. An implicit function lemma of Beneš [5] shows that there is a measurable function u^* such that

$$\mathscr{H}'(t,\xi,\eta,u^*(t,\xi,\eta,p),p) = \mathscr{H}(t,\xi,\eta,p).$$

Now define

$$u^0(t,\xi,\eta) = u^*(t,\xi,\eta,\nabla\phi(t,\xi,\eta)).$$

Evidently $u^0 \in \mathscr{M}$. In fact, u^0 is optimal. To show this, define

$$\alpha(t,\xi,\eta) = \Lambda\phi(t,\xi,\eta) + \mathscr{H}(t,\xi,\eta,\nabla\phi(t,\xi,\eta)).$$

From (4.1), $\alpha \geqslant 0$ a.e. Calculations similar to those in the proof of Lemma 4 show that for any $u \in \mathscr{U}$,

$$J^* \leqslant E_u \int_0^1 h(t,x_t,y_t,u_t)dt + E_u \int_0^1 \alpha_s ds \tag{4.5}$$

while for u^0,

$$J^* = E_u \int_0^1 h(t,x_t,y_t,u_t^0)dt + Eu^0 \int_0^1 \alpha_s ds. \tag{4.6}$$

Now $\alpha = 0$ a.e.: the proof is exactly that of the corresponding theorem in [7]; consequently, (4.5) and (4.6) say that u^0 is optimal in $\mathscr{U}$. This completes the proof.

Remark: If we regard the system (1.1) as a non-Markovian, completely observable system (via the formula $x_t = \chi_t(y)$), then the result of [7] is applicable and one concludes immediately that there exists an optimal $u^0 \in \mathscr{U}$. However, there is no direct way of showing that $u^0 \in \mathscr{M}$.

5. The Hamilton–Jacobi Equation

If $\phi \in C^{1,2}([0,1] \times R^{n+m})$ (the space of real-valued functions on $[0,1]$ R^{n+m} with continuous first (first and second) derivatives in t (x and y)) then an application of the stochastic differential formula, using (2.6) with $u \in \mathscr{M}$, gives:

$$\phi(t,x_t,y_t) = J^* + \int_0^t \left(\frac{\partial\phi}{\partial t} \phi_x . f + \phi_y . g + \tfrac{1}{2}\phi_{yy} . a\right) dt + \int_0^t \phi_y . \sigma\, dB$$

where

$$\phi_x = \frac{\partial\phi}{\partial x},$$

etc., and

$$\phi_{yy}.a = \sum_{i,j} \phi_{y_i y_j}.a^{ij}.$$

Then calculations similar to those of Lemma 4 lead to the Hamilton–Jacobi equation

$$\frac{\partial \phi}{\partial t} + \phi_x . f(t, \xi, \eta) + \tfrac{1}{2}\phi_{yy}.a(t, \eta) + \min_{u \in U} \{\phi_y . g(t, \xi, \eta, u) + h(t, \xi, \eta, u)\}$$

$$= 0 \qquad (t, \xi, \eta) \in [0, 1] \times R^{n+m} \qquad (5.1)$$

$$\phi(1, \xi, \eta) = 0.$$

The question arises whether this equation has a solution, and if so whether the solution is in fact the value function for the control problem. The differential generator of the Markov process (x_t, y_t) under P_u (i.e. satisfying (2.6)) is degenerate (i.e. non-uniformly elliptic; see [8]) so that a solution of (5.1) cannot be expected to exist in $C^{1,2}$; however Fleming shows in [9] that there is a solution in $W_2^{1,2}$ (i.e. the derivatives of ϕ appearing in (5.1) are interpreted in the distribution sense and are functions in L_2 such that (5.1) is satisfied almost everywhere).

In [11], Rishel shows that this weak solution is the value function for the control problem, but only if the distributions of the process (x_t, y_t) under measure P_u (the solution of (2.3), (2.4)) have densities $p(t, x, y)$ which are in $L_{2+\alpha}([0, 1] \times R^n \times R^m)$ for some $\alpha > 0$. So in that case the functions $\Lambda\phi, \nabla\phi$ are given by

$$\Lambda\phi = \frac{\partial \phi}{\partial t} + \phi_x . f + \tfrac{1}{2}\phi_{yy}.a$$

$$\nabla\phi = \phi_y.$$

But it is not known in general whether the moment condition holds or not. The theory would be complete if one could show that the function value ϕ defined by (3.1) is always in $W_2^{1,2}$, but this may not be true without extra conditions.

References

1. Meyer, P. A., "Probability and Potentials". Blaisdell, Waltham, Mass. (1966).
2. Wong, E., "Stochastic Processes in Information and Dynamical Systems". McGraw-Hill, New York (1971).
3. Kunita, H. and Watanabe, S., On square integrable martingales. *Nagoya Math. J.* **30** (1967), 209–245.
4. Wong, E., Representation of martingales, quadratic variation and applications. *SIAM J. Control* **9** (1971), 621–633.

5. Beneš, V. E., Existence of optimal strategies based on specified information, for a class of stochastic decision problems. *SIAM J. Control* **8** (1970), 179–188.
6. Davis, M. H. A. and Varaiya, P. P., Dynamic programming conditions for partially observable stochastic systems. *SIAM J. Control* **11** No. 2 (1973).
7. Davis, M. H. A., "On the Existence of Optimal Policies in Stochastic Control". *SIAM J. Control* **11** No. 4 (1973).
8. Fleming, W. H., Optimal continuous-parameter stochastic control. *SIAM Review* **11** (1969), 470–509.
9. Fleming, W. H., Duality and a priori estimates in Markovian optimization problems. *J. Math. Anal. Appl.* **16** (1966), 254–279.
10. Girsanov, I. V., On transforming a certain class of stochastic processes by absolutely continuous substitution of measures, *Theory of Probability and Applications* **5** (1960), 285–301.
11. Rishel, R. W., Weak solutions of a partial differential equation of dynamic programming. *SIAM J. Control* **9** (1971), 519–528.
12. Beneš, V. E., Existence of optimal stochastic control laws, *SIAM J. Control* **9** (1971), 446–475.

19. On How to Shake a Piece of String to a Standstill

P. C. PARKS

Control Theory Centre,
University of Warwick, Coventry, England

Alice laughed. "There's no use trying," she said, "one can't believe impossible things." "I daresay you haven't had much practice," said the Queen. "When I was your age, I always did it for half-an-hour a day. Why, sometimes I've believed as many as six impossible things before breakfast."

"Alice through the Looking Glass",
Lewis Carroll, Chapter 4, (1872).

1. Introduction

In the last five years or so many of the ideas developed for the control of systems described by ordinary differential equations in state space—particularly $\dot{\mathbf{x}} = \mathbf{Ax} + \mathbf{Bu}$, $\mathbf{y} = \mathbf{Cx} + \mathbf{Du}$—have been extended to more complex situations and especially to the control of systems described by partial differential equations—"distributed parameter" systems in the language of control engineers.

It is not the purpose of this paper to survey these developments—the more important books and papers are listed [1–5]—but rather we shall concentrate on one or two ideas that arise, illustrated by a particular problem which gives the paper its title. These ideas are particularly inspired by the work of D. L. Russell [6] and A. G. Butkovskii [7] and they rely heavily on theorems of R. E. A. C. Paley and N. Wiener [8].

2. The Problem

We shall consider lateral planar vibrations of a string under tension which obeys the one-dimensional wave equation written in non-dimensional form as

$$\frac{\partial^2 y}{\partial t^2}(x, t) = \frac{\partial^2 y}{\partial x^2}(x, t) \tag{1}$$

where $y\ (x, t)$ is the lateral displacement of the string. We shall suppose the string extends from $x = 0$ to $x = 1$, that the end $x = 0$ is fixed and that control is applied at the end $x = 1$ so that the boundary conditions associated with equation (1) are

$$\left.\begin{aligned} y\,(0, t) &= 0 \\ y\,(1, t) &= u(t) \end{aligned}\right\} \text{ for all } t. \tag{2}$$

We are thus considering a "boundary control" problem.

From a general initial state at $t = 0$ when the displacement and velocity at each point of the string are known we are asked to find a control $u\ (t)$ which will bring the string to rest in time T, to be as small as possible.

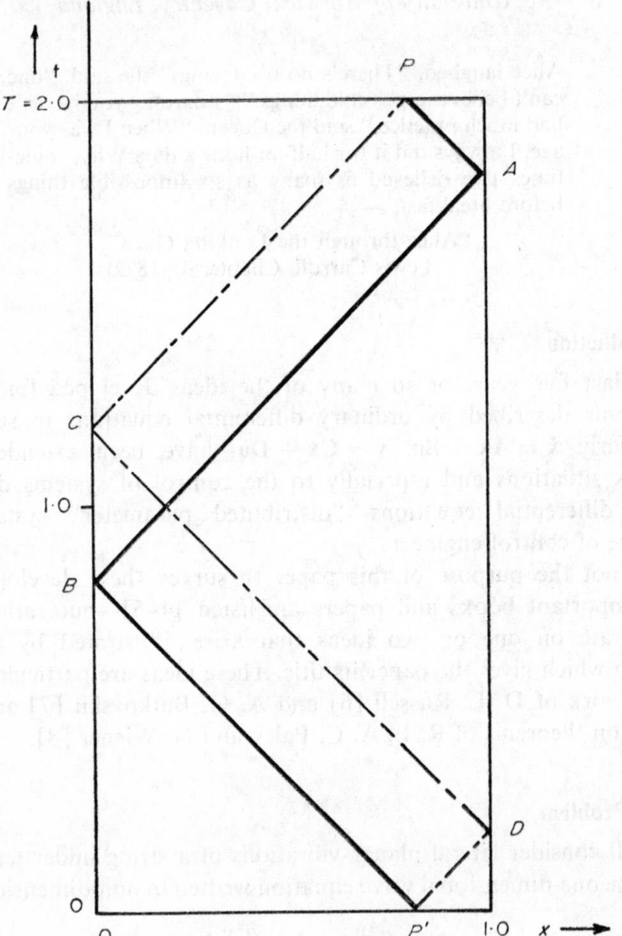

FIG. 1. Characteristics Diagram for $T = 2$.

3. First Solution—Using the Method of Characteristics

Consider the plane of x and t shown in Fig. 1. In this plane we know, from standard wave equation theory, that $\partial y/\partial t + \partial y/\partial x$ is constant along straight lines with a slope of -1 and that $\partial y/\partial t - \partial y/\partial x$ is constant along straight lines with a slope of $+1$. These lines are called "characteristics", and enable us to solve the problem very neatly when the time scale runs from 0 to 2 as shown in Fig. 1.

The two characteristics from the general point P on the line $t = 2$ run to, and are reflected at, the lines $x = 0$ and $x = 1$ to arrive at P' on the $t = 0$ line as shown.

Now we suppose tentatively that our objective is achieved at time $t = T = 2$, that is

$$y(x, 2) = 0, \qquad \frac{\partial y}{\partial t}(x, 2) = 0$$

for all x. Along the characteristic PC we have

$$\frac{\partial y}{\partial t} - \frac{\partial y}{\partial x} = \text{constant} = 0$$

since these derivatives are both zero at P. Similarly along PA

$$\frac{\partial y}{\partial t} + \frac{\partial y}{\partial x} = 0.$$

At C

$$\frac{\partial y}{\partial t} = 0 \quad \text{and} \quad \frac{\partial y}{\partial x} = 0$$

also, so that

$$\frac{\partial y}{\partial t} + \frac{\partial y}{\partial x} = 0$$

along CD.

At A, $\partial y/\partial t$ is given by

$$\frac{\partial y}{\partial t} = f(1 + x)$$

say, since $t = 2 - (1-x)$ where x is the ordinate of P. Hence

$$\frac{\partial y}{\partial x} = -f(1+x)$$

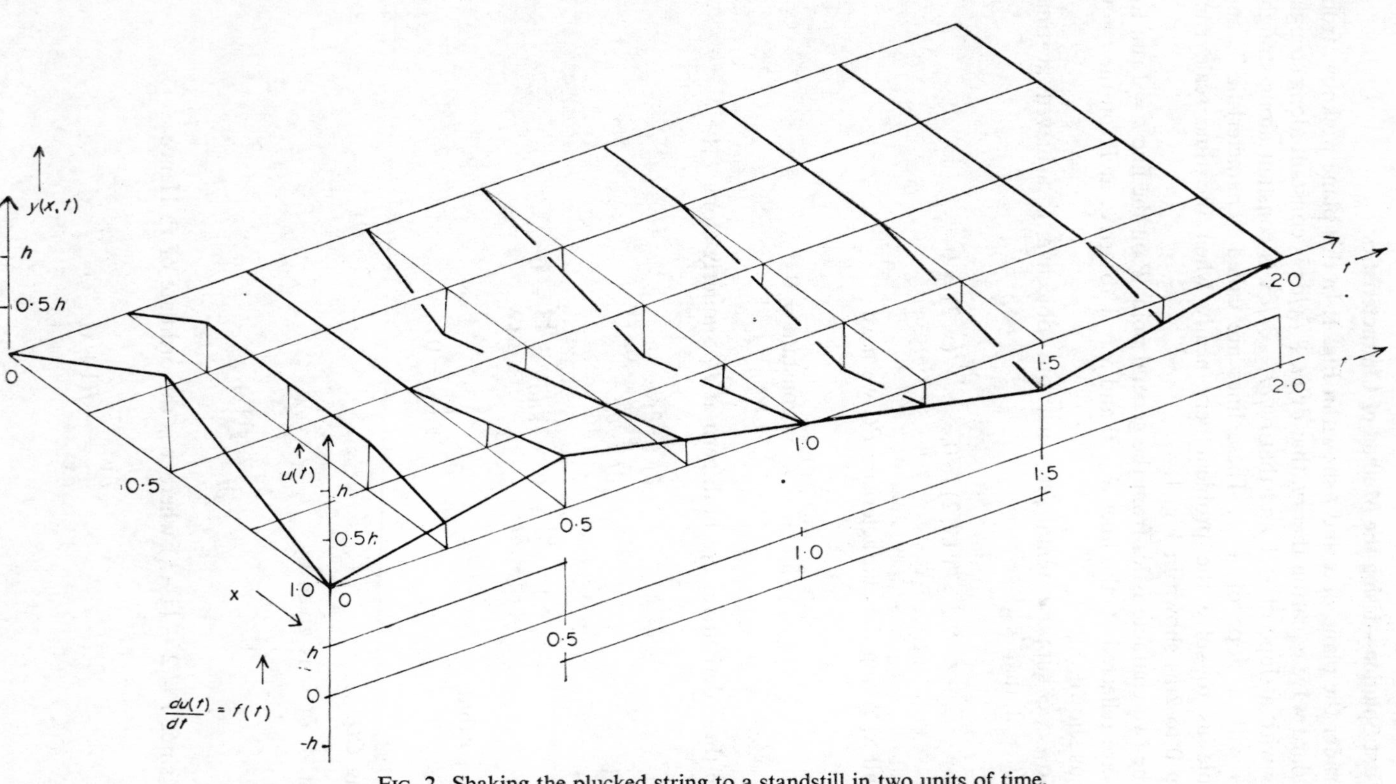

FIG. 2. Shaking the plucked string to a standstill in two units of time.

at A and along AB

$$\frac{\partial y}{\partial t} - \frac{\partial y}{\partial x} = 2f(1 + x).$$

At D, $\partial y/\partial t$ is given by

$$\frac{\partial y}{\partial t} = f(1 - x) \quad \text{so that} \quad \frac{\partial y}{\partial x} = -f(1 - x)$$

and consequently

$$\frac{\partial y}{\partial t} - \frac{\partial y}{\partial x} = 2f(1 - x)$$

along DP'.

At B, $\partial y/\partial t = 0$ so that

$$\frac{\partial y}{\partial x} = -2f(1 + x) \text{ and } \frac{\partial y}{\partial t} + \frac{\partial y}{\partial x} = -2f(1 + x)$$

along BP'.

Finally at P' we obtain the equations

$$\frac{\partial y}{\partial t}(x, 0) + \frac{\partial y}{\partial x}(x,0) = -2f(1 + x) \tag{3}$$

$$\frac{\partial y}{\partial t}(x, 0) - \frac{\partial y}{\partial x}(x, 0) = 2f(1 - x) \tag{4}$$

As x varies from 0 to 1 the control $f(t) = du/dt$ is determined by Eqn (4) for $0 \leqslant t \leqslant 1$ and by Eqn (3) for $1 \leqslant t \leqslant 2$ The control $u(t)$ is obtained as the time integral of $f(t)$ using an initial condition at $t = 0$ on $y(1, 0)$, which is known.

Figure 2 illustrates the control at $x = 1$ required to bring a plucked string to rest and various stages of the motion.

We notice that the control is uniquely determined. A time interval T less than 2 will yield in general an insoluble control problem, for the proposed characteristics diagram corresponding to Fig. 1 will then appear as shown in Fig. 3. The string has zero displacement and velocity within the triangle RST, as may be found by tracing characteristics back from RS. Considering the triangle CET we shall in general obtain conflicting values for $\partial y/\partial t$ and $\partial y/\partial x$ calculated by characteristics from the segment

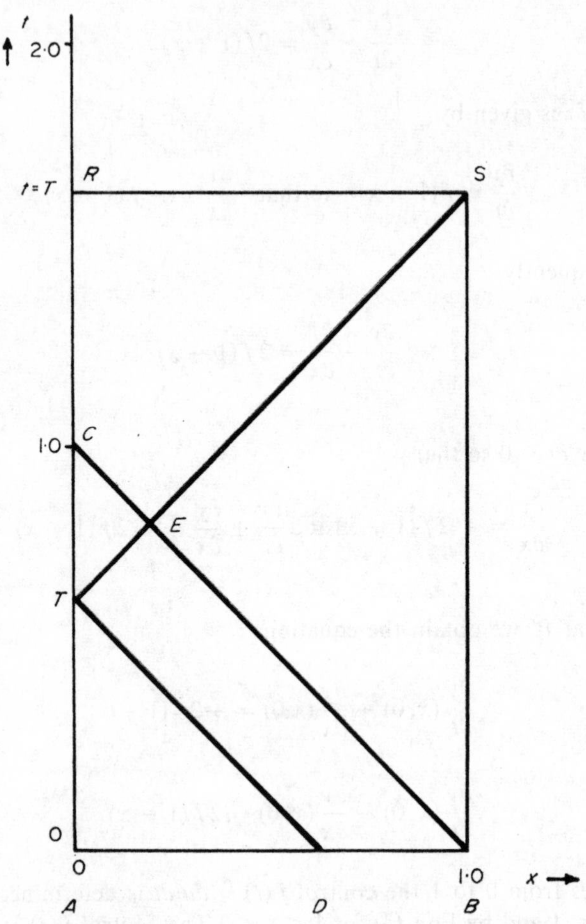

FIG. 3. Characteristics for $T < 2$.

DB, unless $\partial y/\partial t + \partial y/\partial x = 0$ at each point of DB, which is a very special initial state.

On the other hand for a time T greater than 2 a certain arbitrariness appears in the control problem. The characteristics diagram then appears as shown in Fig. 4. The problem may be made determinate if $\partial y/\partial x$ is given some prescribed but arbitrary value at $x = 0$ for $1 < t < T-1$.

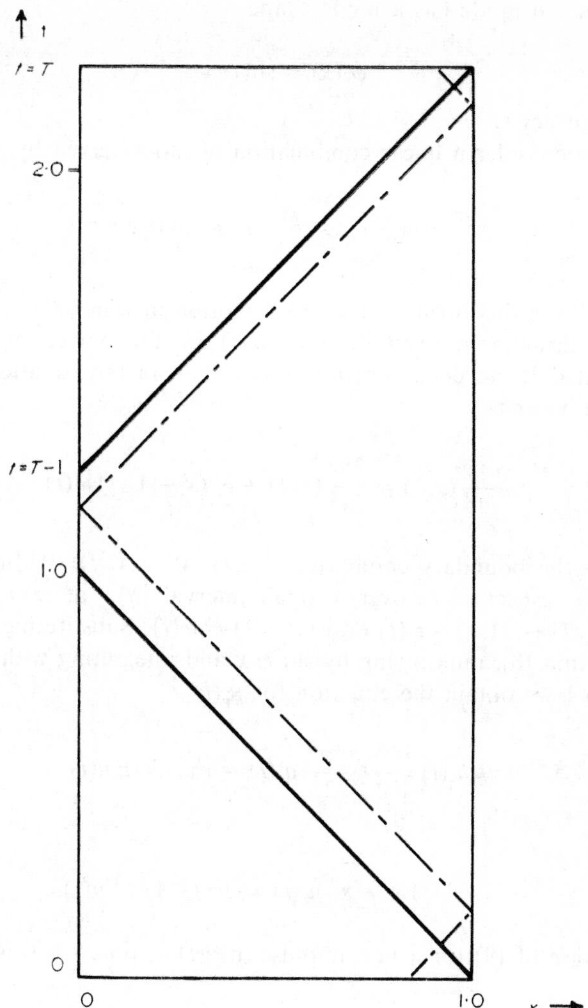

FIG. 4. Characteristics for $T > 2$.

4. Second Solution using Modal Analysis

Let us consider the modes of vibration of the string with boundary conditions $y = 0$ at $x = 0$ and $x = 1$. These modes of vibration are

$$\left.\begin{aligned} &y(x, t) = \sin r\pi x \sin r\pi t \\ &r = 1, 2, 3 \ldots \end{aligned}\right\} \tag{5}$$

so that the rth mode has a mode shape

$$\phi_r(x) = \sin r\pi x \tag{6}$$

and a frequency $r\pi$.

We now consider a linear combination of modes given by

$$y(x, t) = \sum_{r=1}^{\infty} \mu_r(t) \sin r\pi x \tag{7}$$

We shall use this form to describe a motion in which the $x = 1$ point is displaced through a given distance $u(t)$ by the device of including a differentiated Dirac delta function at $x = 1-$ in the equation of motion (1), which becomes

$$\frac{\partial^2 y}{\partial t^2}(x, t) = \frac{\partial^2 y}{\partial x^2}(x, t) + \delta'(x - 1 -)\, u(t) \tag{8}$$

subject to the boundary conditions $y(0, t) = 0$, $y(1, t) = 0$. Integrating (8) twice with respect to x over a small interval in x at $x = 1$ we obtain $y(1, t) - y(1-, t) = -u(t)$ or $y(1-, t) = u(t)$. Substituting the Fourier series (7) into (8), multiplying by $\sin r\pi x$ and integrating with respect to x from 0 to 1 we obtain the equation for $\mu_r(t)$

$$\tfrac{1}{2}\ddot{\mu}_r(t) = -\frac{r^2\pi^2}{2}\mu_r(t) - r\pi \cos r\pi\, u(t)$$

or

$$\ddot{\mu}_r(t) + r^2\pi^2 \mu_r(t) = 2r\pi(-1)^{r-1} u(t). \tag{9}$$

The response of (9) for a unit impulse in $u(t)$ at time $t = 0$ is

$$\mu_r(t) = 2(-1)^{r-1} \sin r\pi t \tag{10}$$

and so if the initial conditions on $\mu_r(t)$ and $\dot{\mu}_r(t)$ at $t = 0$ are $\mu_r(0)$ and $\dot{\mu}_r(0)$ respectively, the solution of (9) is

$$\mu_r(t) = 2(-1)^{r-1}\int_{\tau=0}^{t} \sin r\pi(t - \tau)\, u(\tau)d\tau + \mu_r(0)\cos r\pi t + \frac{\dot{\mu}_r(0)}{r\pi}\sin r\pi t. \tag{11}$$

Now the control $u(t)$ will be successful if we can find a T such that $\mu_r(T)$ and $\dot{\mu}_r(T)$ are both zero i.e.

$$\left.\begin{aligned} &2(-1)^{r-1}\int_{\tau=0}^{T} \sin r\pi(T-\tau)u(\tau)d\tau = -\mu_r(0)\cos r\pi T - \frac{\dot{\mu}_r(0)}{r\pi}\sin r\pi T \\ &\text{and} \\ &2(-1)^{r-1}\int_{\tau=0}^{T} \cos r\pi(T-\tau)u(\tau)d\tau = \mu_r(0)\sin r\pi T - \frac{\dot{\mu}_r(0)}{r\pi}\cos r\pi T \end{aligned}\right\} \quad (12)$$

This introduces classical problems in *closures and moments*, (see Section 7). In fact in the general case $u(t)$ can be found only when $T \geqslant 2$. When $T = 2$ and additionally we consider the plucked string of Fig. 2, in which a Fourier analysis of the shape yields

$$\begin{aligned} &\mu_r(0) = (-1)^{(r-1)/2}\, 8h/r^2\pi^2 && r = 1, 3, 5, \ldots \\ &\mu_r(0) = 0 && r = 2, 4, 6, \ldots \end{aligned} \qquad (13)$$

with

$$\dot{\mu}_r(0) = 0 \qquad r = 1, 2, 3, \ldots,$$

the solution to the moment problem may be shown to be satisfied by the Fourier series

$$u(t) = \sum_{r=1}^{\infty} \mu_r \sin r\pi t \qquad (0 \leqslant t \leqslant 2) \qquad (14)$$

since we obtain from the first of Eqns (12)

$$\mu_r . 2(-1)^{r-1}\int_{\tau=0}^{2} \sin r\pi(2-\tau)\sin r\pi\tau\, d\tau = -\mu_r(0)$$

and the second of Eqns (12) is satisfied identically.

We find

$$\mu_r = \tfrac{1}{2}(-1)^{r-1}\mu_r(0)$$

or

$$\begin{aligned} \mu_r &= (-1)^{(r-1)/2}\frac{4h}{r^2\pi^2} && r = 1, 3, 5, \ldots \\ &= 0 && r = 2, 4, 6 \ldots \end{aligned} \qquad (15)$$

This shows that $u(t)$ as given by (14) is precisely the wave form deduced earlier in Fig. 2.

5. Third Solution—Fourier Transforms

Let us take the Fourier transform of Eqn (1) subject to the boundary conditions (2) and initial conditions

$$\begin{aligned} y(x, -\tau) &= \phi(x) \\ \frac{\partial y}{\partial t}(x, -\tau) &= \dot{\phi}(x). \end{aligned} \tag{16}$$

We have taken the initial time as $t = -\tau$ instead of $t = 0$ since the subsequent theory concerns functions of time which are zero except in the interval $-\tau \leqslant t \leqslant \tau$, that is in an interval of time symmetrically disposed about $t = 0$.

The transformed equation becomes

$$-\omega^2 \bar{y}(x, i\omega) = \frac{d^2\bar{y}}{dx^2}(x, i\omega) + \left(i\omega\phi(x) + \dot{\phi}(x)\right) e^{i\omega\tau} \tag{17}$$

subject to the boundary conditions

$$\begin{aligned} \bar{y}(0, i\omega) &= 0 \\ \bar{y}(1, i\omega) &= \bar{u}(i\omega) \end{aligned} \tag{18}$$

where the bar denotes the Fourier transform.

Equation (17) is solved by the standard method of "variation of parameters" in which we assume

$$\bar{y}(x, i\omega) = A(x)\cos\omega x + B(x)\sin\omega x. \tag{19}$$

Now

$$\frac{d\bar{y}}{dx}(x, i\omega) = -\omega A(x)\sin\omega x + \omega B(x)\cos\omega x \tag{20}$$

provided that

$$\frac{dA(x)}{dx}\cos\omega x + \frac{dB(x)}{dx}\sin\omega x = 0. \tag{21}$$

Differentiating again

$$\begin{aligned} \frac{d^2\bar{y}}{dx^2}(x, i\omega) = {} & -\omega^2 A(x)\cos\omega x - \omega^2 B(x)\sin\omega x \\ & + \omega\left(-\frac{dA(x)}{dx}\sin\omega x + \frac{dB(x)}{dx}\cos\omega x\right) \end{aligned} \tag{22}$$

Substituting into (17) we obtain

$$-\frac{dA(x)}{dx}\sin\omega x+\frac{dB(x)}{dx}\cos\omega x=\left(-i\phi(x)-\frac{\dot{\phi}(x)}{\omega}\right)e^{i\omega\tau} \tag{23}$$

which, combined with (21), yields

$$\begin{aligned}\frac{dA(x)}{dx}&=\left(i\phi(x)+\frac{\dot{\phi}(x)}{\omega}\right)e^{i\omega\tau}\sin\omega x\\ \frac{dB(x)}{dx}&=-\left(i\phi(x)+\frac{\dot{\phi}(x)}{\omega}\right)e^{i\omega\tau}\cos\omega x.\end{aligned} \tag{24}$$

From the boundary conditions (18) we obtain

$$A(0)=0 \quad\text{and}\quad \bar{u}(i\omega)=A(1)\cos\omega+B(1)\sin\omega \tag{25}$$

so that (24) gives us

$$\begin{aligned}A(x)&=\int_0^x\left(i\phi(\xi)+\frac{\dot{\phi}(\xi)}{\omega}\right)e^{i\omega\tau}\sin\omega\xi\,d\xi\\ B(x)&=-\int_0^x\left(i\phi(\xi)+\frac{\dot{\phi}(\xi)}{\omega}\right)e^{i\omega\tau}\cos\omega\xi\,d\xi+C.\end{aligned} \tag{26}$$

The constant C is found from (25) which gives us

$$\begin{aligned}\bar{u}(i\omega)=&\left[\int_0^1\left(i\phi(\xi)+\frac{\dot{\phi}(\xi)}{\omega}\right)e^{i\omega\tau}\sin\omega\xi\,d\xi\right]\cos\omega\\ &-\left[\int_0^1\left(i\phi(\xi)+\frac{\dot{\phi}(\xi)}{\omega}\right)e^{i\omega\tau}\cos\omega\xi\,d\xi\right]\sin\omega+C\sin\omega.\end{aligned} \tag{27}$$

We obtain finally the following expression for $\bar{y}\,(x, i\omega)$:

$$\begin{aligned}\bar{y}(x,i\omega)=&\left\{\int_0^x\left[i\phi(\xi)+\frac{\dot{\phi}(\xi)}{\omega}\right]e^{i\omega\tau}\sin\omega\xi d\xi\right\}\cos\omega x\\ &+\left\{-\int_0^x\left[i\phi(\xi)+\frac{\dot{\phi}(\xi)}{\omega}\right]e^{i\omega\tau}\cos\omega\xi\,d\xi\right.\\ &+\frac{\bar{u}(i\omega)}{\sin\omega}-\left\{\int_0^1\left[i\phi(\xi)+\frac{\dot{\phi}(\xi)}{\omega}\right]e^{i\omega\tau}\sin\omega\xi\,d\xi\right\}\frac{\cos\omega}{\sin\omega}\\ &\left.+\int_0^1\left[i\phi(\xi)+\frac{\dot{\phi}(\xi)}{\omega}\right]e^{i\omega\tau}\cos\omega\xi\,d\xi\right\}\sin\omega x.\end{aligned} \tag{28}$$

We now seek to make $\bar{y}(x, i\omega)$ the Fourier transform of a time function which is zero except on an interval of time $-\tau \leqslant t \leqslant \tau$. For this to be the case $\bar{y}(x, z)$ must (i) have no poles in the complex z-plane, (ii) tend to infinity like $e^{\tau|z|}$ as $|z| \to \infty$, that is $\bar{y}(x, z)$ should be an *entire function* of of *order* 1 and *type* or *degree* τ. (See Section 7).

To remove the poles which will occur when $\sin \omega = 0$ we require

$$\bar{u}(i\omega) = \left\{ \int_0^1 \left[i\phi(\xi) + \frac{\dot{\phi}(\xi)}{\omega} \right] e^{i\omega\tau} \sin \omega\xi \, d\xi \right\} \cos \omega \tag{29}$$

when $\omega = 0, \pm r\pi, r = 1, 2, 3, \ldots$. We might suppose we could actually put $\bar{u}(i\omega)$ identically equal to the right-hand side of (29) but the resulting function would have degree $2 + \tau$.

The correct procedure is to make use of the *Lagrange interpolation formula* (see Section 7) so that

$$\bar{u}(i\omega) = \sum_{k=-\infty}^{+\infty} A_k \frac{\sin \omega}{(-1)^k(\omega - k\pi)} \tag{30}$$

where

$$A_k = \left\{ \int_0^1 \left[i\phi(\xi) + \frac{\dot{\phi}(\xi)}{k\pi} \right] e^{ik\pi\tau} \sin k\pi\xi \, d\xi \right\} \times (-1)^k \tag{31}$$

Now the inverse transform of

$$(-1)^k \frac{\sin \omega}{\omega - k\pi} \quad \text{is} \quad \frac{1}{2} e^{ik\pi t}, \quad \text{for} \quad -1 \leqslant t \leqslant 1,$$
$$0, \quad \text{for} \quad t < -1, t > 1.$$

Putting in the plucked string initial conditions (13) in which $\dot{\phi}(x) = 0$ and $\phi(x)$ may be represented by a Fourier series

$$\phi(x) = \sum_{r=1}^{\infty} b_r \sin r\pi x, \qquad 0 \leqslant x \leqslant 1 \tag{32}$$

where

$$b_r = 2 \int_0^1 \phi(\xi) \sin r\pi\xi \, d\xi = (-1)^{(r-1)/2} \frac{8h}{r^2\pi^2} \quad (r = 1, 3, 5 \ldots)$$
$$= 0 \quad (r = 2, 4, 6 \ldots)$$

and noting that we must fix $\tau = 1$ so that $\bar{u}(i\omega)$ and $\bar{y}(x, i\omega)$ both are of equal type or degree (equal to 1) we obtain the inverse transform of (30) as

$$u(t) = \sum_{k=0}^{\infty} A_k i \sin k\pi t, \qquad -1 \leqslant t \leqslant 1$$
$$= 0 \qquad t < -1, t > 1.$$

From (31) $A_k = \frac{1}{2}b_k i$ so that

$$\begin{aligned} u(t) &= -\tfrac{1}{2}\phi(t) \qquad & -1 \leqslant t \leqslant 1 \\ &= 0 & t < -1, t > 1 \end{aligned} \tag{33}$$

where $\phi(t)$ is the Fourier series defined in (32) as an odd function. (33) is precisely the solution for $u(t)$ obtained in Fig. 2 and in Eqn (15), allowing for the change of time origin.

6. A Fourth Attempt—Discretisation

Another method of solution which we might consider is to use a discrete "lumped mass" model of the heavy string on which point masses are spaced along a light string. A simple two-mass model is shown in Fig. 5.

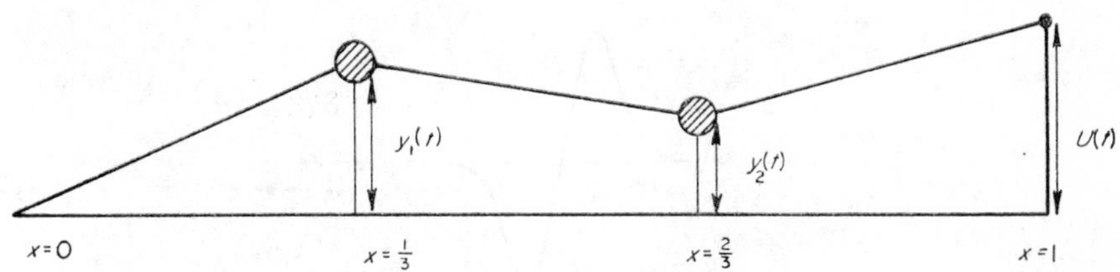

FIG. 5. A simple lumped mass model.

The equations of motion are

$$\begin{bmatrix} \ddot{y}_1 \\ \ddot{y}_2 \end{bmatrix} = k \begin{bmatrix} -2 & 1 \\ 1 & -2 \end{bmatrix} \begin{bmatrix} y_1 \\ y_2 \end{bmatrix} + k \begin{bmatrix} 0 \\ 1 \end{bmatrix} u(t) \tag{34}$$

or

$$\frac{d}{dt}\begin{bmatrix} y_1 \\ \dot{y}_1 \\ y_2 \\ \dot{y}_2 \end{bmatrix} = \begin{bmatrix} 0 & 1 & 0 & 0 \\ -2k & 0 & k & 0 \\ 0 & 0 & 0 & 1 \\ k & 0 & -2k & 0 \end{bmatrix} \begin{bmatrix} y_1 \\ \dot{y}_1 \\ y_2 \\ \dot{y}_2 \end{bmatrix} + \begin{bmatrix} 0 \\ 0 \\ 0 \\ k \end{bmatrix} u(t) \tag{35}$$

Equation (35) is completely controllable in the sense of Kalman and we deduce that with a suitably large $u(t)$ the system can be brought to rest in an infinitesimally short time. Specifically, if the initial displacements of the masses are $y_1(0)$ and $y_2(0)$ and the initial velocities are zero then if the Fourier transform of $u(t)$ is assumed to be the entire function $\bar{u}(i\omega) = a + bi\omega + c(i\omega)^2 + d(i\omega)^3$ (of order and type zero), the conditions

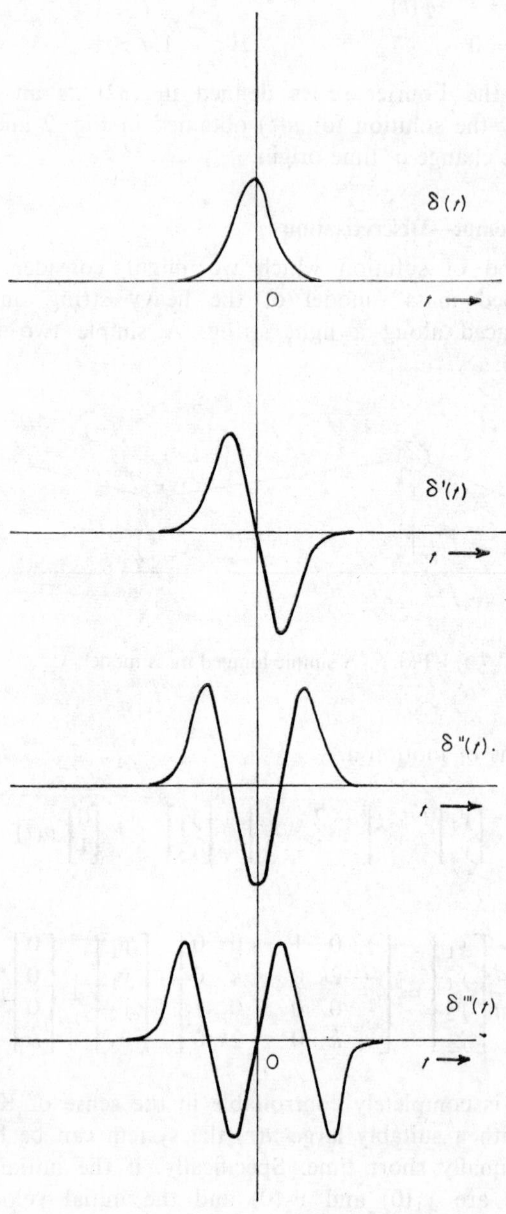

FIG. 6. An artist's impression of the delta function and its derivatives.

that $\bar{y}_1(i\omega)$ and $\bar{y}_2(i\omega)$ be entire functions (also of order and type zero) may be obtained by examining the solution vector

$$\begin{bmatrix} \bar{y}_1(i\omega) \\ \bar{y}_2(i\omega) \end{bmatrix} = \frac{1}{(-\omega^2 + k)(-\omega^2 + 3k)} \times \begin{bmatrix} i\omega(-\omega^2 + 2k)y_1(0) + ki\omega y_2(0) + k^2\bar{u}(i\omega) \\ ki\omega y_1(0) + i\omega(-\omega^2 + 2k)y_2(0) + k(-\omega^2 + 2k)\bar{u}(i\omega) \end{bmatrix} \tag{36}$$

The values of a, b, c, d are found by removing the poles which would otherwise appear in $\bar{y}_1(i\omega)$ and $\bar{y}_2(i\omega)$ when $\omega = \pm\sqrt{k}$ and $\pm\sqrt{3k}$ in (36). We find

$$a = c = 0,$$
$$b = \frac{-2y_1(0) - y_2(0)}{k}, \qquad d = \frac{-y_1(0)}{k^2} \tag{37}$$

This implies $u(t)$ is composed of a linear combination of the first and third derivatives of the delta function. These derivatives are sketched conceptually in Fig. 6.

Similar, and apparently erroneous, results will be obtained as the number of masses in the model is increased, for the model remains completely controllable and it can be brought to rest in infinitesimal time by a $u(t)$ composed of a linear combination of the delta function and its derivatives. There is no suggestion of the minimum time interval of $T = 2$ that characterised the exact solutions of Sections 3, 4 and 5.

7. New Concepts

The discussion in previous sections has led to a number of concepts which are not familiar to most control engineers and indeed many control theorists.

Closures and moments comprise an important topic in mathematics with principal authors Paley and Wiener [8], Ahiezer and Krein [9], Levinson [10].

A set of functions $\{f_n(x)\}$, $f_n(x) \in L_2$, is said to be closed over an interval (a, b) if

$$\int_a^b f(x) f_n(x) dx = 0 \qquad (n = 1, 2, 3, \ldots)$$

implies $f(x)$ vanishes except over a set of zero measure if $f(x) \in L_2$.

A theorem of Paley and Wiener, [8, p. 113], states a condition for the existence of a set $\{h_n(x)\}$ of functions orthogonal to the set $\{e^{i\lambda_n x}\}$ over the interval $(-\pi, \pi)$. This condition is that set $\{\lambda_n\}$ behaves like the natural numbers or more precisely

$$|\lambda_n - n| \leqslant K < 1/\pi^2. \tag{38}$$

A moment problem such as that in Eqns (12) may be solved by expressing $u(t)$ in the form $\sum b_r h_r(t)$. In this particular problem of Section 4 set $\{e^{i\lambda_n t'}\}$ is in fact $\{e^{in\pi t}\}$ and $\{h_n(t')\}$ is $\{e^{in\pi t}\}$ also; the interval in t' $(-\pi, \pi)$ becomes $(-1, 1)$ in t under a change of time scale $t' = \pi t$.

We note in passing the "L-moment problem" in Ahiezer and Krein [9, p. 60]. Here theorems are given for the existence of solutions of the moment problem subject to the restriction $|u(t)| < L$. These theorems are clearly going to be of importance in problems of "bang-bang" control.

Entire functions are functions $\phi(z)$ of the complex variable z that are holomorphic in the entire z-plane and consequently represented by a power series

$$\phi(z) = a_0 + a_1 z + a_2 z^2 + \ldots + a_n z^n + \ldots,$$

convergent everywhere.

The order of an entire function is the greatest lower bound of positive constants k such that

$$\max_{|z|=r} |\phi(z)| < e^{r^k}$$

is true for all sufficiently large $r > r_0(k)$.

The type or degree of an entire function of order ρ is the greatest lower bound of positive numbers A such that

$$\max_{|z|=r} |\phi(z)| < e^{Ar^\rho}$$

is true for all sufficiently large $r > r_0(A)$.

Again there is an extensive mathematical literature: Levin [11], Paley and Wiener [8], Boas [12].

A theorem of Paley and Wiener [8, p. 13], states that the Fourier transform of an L_2 function $f(t)$ which is zero outside the interval $-A \leqslant t \leqslant A$ is an entire function of order 1 and type σ where $\sigma \leqslant A$. This theorem is exploited frequently by Butkovskii [7], and here in Section 5.

Another theorem of Paley and Wiener [8, p. 115], using Lagrange interpolation is also used in Section 5. This states that given λ_n satisfying (38), $\lambda_0 = 0$, $\lambda_{-n} = -\lambda_n$, then the function

$$\phi(z) = \sum_{-\infty}^{+\infty} \frac{a_n G(z)}{G'(\lambda_n)(z - \lambda_n)}, \qquad \text{where } G(z) = z \prod_{1}^{\infty} \left(1 - \frac{z^2}{\lambda_n^{\,2}}\right)$$

is the Lagrange interpolation formula for a function $\phi(z)$ taking the values a_n when $z = \lambda_n$ and, provided

$$\sum_{-\infty}^{\infty} |a_n|^2 < \infty,$$

it is an entire function of order 1 and type π. In Section 5 we have

$$G(z) = \sin z = z \prod_{1}^{\infty} \left(1 - \frac{z^2}{n^2\pi^2}\right)$$

and so a transformation $z' = \pi z$ is necessary which will make $\phi(z')$ of order 1 and type 1: the inverse transform $u(t)$ in Eqn (33) consequently is zero outside the interval $-1 \leqslant t \leqslant 1$.

8. Conclusions

By studying an apparently simple problem by various techniques we have drawn attention to some areas of complex variable analysis which seem an essential background to current and future work on control of distributed parameter systems. These areas, namely entire function theory and the theory of moments are not too well known at present, and may well appear to some as "impossible things"!

9. Acknowledgements

It is a pleasure to acknowledge most useful discussions with Professor L. Markus and Dr. A. J. Pritchard during the preparation of this paper.

References

1. Wang, P. K. C., Control of distributed parameter systems. *In* "Advances in Control Systems" (ed. C. T. Leondes), Vol. 1, pp. 75–172, Academic Press, New York (1964).
2. Butkovskii, A. G., "Distributed Control Systems", American Elsevier, New York, 1969 (Translation from Russian, Moscow, 1965).
3. Lions, J. L., "Optimal Control of Systems Governed by Partial Differential Equations", Springer-Verlag, New York, 1971 (Translation from French, Paris, 1968).
4. Wang, P. K. C., Theory of stability and control for distributed parameter systems (a Bibliography). *Int. J. Control* **7** (1968), 101–116.
5. Hamza M. H. (Ed.), Proceedings IFAC Conference on Control of Distributed Parameter Systems, Banff, Canada, June 1971, Vols. 1 and 2, T.M.S. Services Box 3243, Postal Station B, Calgary, Alberta (1971).
6. Russell, D. L., Nonharmonic Fourier series in the control theory of distributed parameter systems. *J. Math. Appl.* **18** (1967), 542–560.

7. Butkovskii, A. G. and Poltavskii, L. N., Finite control of systems with distributed parameters. *Automation and Remote Control* **30** (1969), 23–33, 491–501.
8. Paley, R. E. A. C. and Wiener, N., "Fourier Transforms in the Complex Domain", Amer. Math. Soc. Colloquium Publications, Vol. 19, Providence (1934).
9. Ahiezer, N. I. and Krein, M., "Some Questions in the Theory of Moments", Amer. Math. Soc. Translations of Math Monographs, Vol. 2, Providence (1962). (Translation from Russian, Kharkov, 1938).
10. Levinson, N., "Gap and Density Theorems", Amer. Math. Soc. Colloquium Publications, Vol. 26, Providence, (1940).
11. Levin, B. Ya., "Distribution of Zeros of Entire Functions", Amer. Math. Soc. Translations of Math. Monographs, Vol. 5, Providence (1964) (Translation from Russian, Moscow, 1956).
12. Boas, R. P. Jr., "Entire Functions", Academic Press, New York (1954).

Discussion

BRERETON (*C.E.R.L.*). You gave a simple geometric way of constructing the boundary control and you depicted several more complicated ways of doing the same thing. Are there in fact generalizations of that?

PARKS. Oh yes, I think so, Of course, the wave equation has these characteristics which lend themselves to straightforward problems, but what other partial differential equations of interest are there? Well, particularly the heat equation, which although not having the characteristics set-up is suitable for modal methods and particularly for the entire function method.

In my paper I have been honest about the sources of all this. The characteristics business is well known, but has been particularly exploited in the control context by David Russell. He has actually turned the method of characteristics into a method for proving theorems in the theory of moments.

I mentioned that the entire function theory had been developed by Russian Academicians. In the control context all the work has been done by Butkovskii in a series of papers in *Automatika i Telemekhanika* in which he uses the theory for ordinary differential equations. He dealt with the partial differential equation case through the string example in the IFAC proceedings of the Banff Conference. It would be nice to think that the IFAC Conference remembered R.E.A.C. Paley*—they met forty years later in the very town where he died and I don't suppose anyone remembered. I've tried to put that right today!

MEES (*University of Cambridge*). Would it help if you carried out your experiment† with your eyes closed? The purpose of my question is that if you didn't know in advance how the system was going to behave and the only information you had of what was coming in to you as a controller was the local feel, then perhaps the knowledge could be used. Would it have been possible to bring the string to rest in any case? Maybe it would have taken a longer time.

PARKS. I'm not too sure; you are asking for some feedback control aren't you?

MEES. Yes, I suppose so. You know the state of the system only to the extent that

* Raymond Edward Alan Paley was killed by an avalanche at Banff on April 7th, 1933. Although only 26 years old he had already written over 20 important papers.

† The author gave a practical demonstration with a long rope at the end of the lecture.

you can reconstruct it from certain outputs: you cannot measure it directly. I imagine you could still bring the whole string to rest eventually though.

PARKS. Yes, I think you probably could.*

DIPROSE (*University of Bath*). Suppose you were substituted on the end of the string—you behave as if you were a continuing infinite string—and let the wave run itself off the string.

PARKS. Here it is getting reflected off the end.

DIPROSE. If you act as a normal reflector the waves will not come off through you. If you created forces of motion as if there was an infinite string behind you then the waves would run off down the "string" and that can be done without any knowledge other than what is at the end of the string.·

STOREY (*University of Loughborough*). Does not the paradox you mention arise because the control is not uniform?

PARKS. Well, it may be. I avoided saying much more about that. I was talking to Larry Markus, before he left for America, about this lecture and particularly about this paradox in the discrete model and his reply was that it had been explained. Someone has sorted out what is going on, although I haven't been able to find this. I'm not sure that Markus is absolutely correct about this, but someone in the audience may contradict me.

SÁNCHEZ (*University of California*). It's in some lectures notes of Greenspan.

PARKS. I'm not sure; I haven't found it myself. Anyway, Larry Markus says it's been done! However, there is a paradox and I can see in a way how one might get around it because you get an infinite series in s as the number of discrete masses builds up. One might be able to relate that to e^{-s}.

STOREY. With non-uniform control, states can be changed in an arbitrarily small time but not instantaneously. If, however, the control is uniform then instantaneous changes can be made.

PARKS. Can I leave you with this? The Laplace transform of the delayed delta function is $e^{-s\tau}$. Now

$$e^{-s\tau} = 1 - s\tau + \frac{s^2\tau^2}{2!} - \dots$$

an infinite series of terms and so a delta function at $t = \tau$ is equivalent to a sum of delta functions and derivatives at the *origin*,

$$\text{i.e.} \quad \delta(t-\tau) = \sum_{n=0}^{\infty} \frac{(-\tau)^n}{n!} \delta^n(t).$$

So if I'm right about how you can relate the two together, some paradox like this is involved.

* See postscript at the end of this discussion.

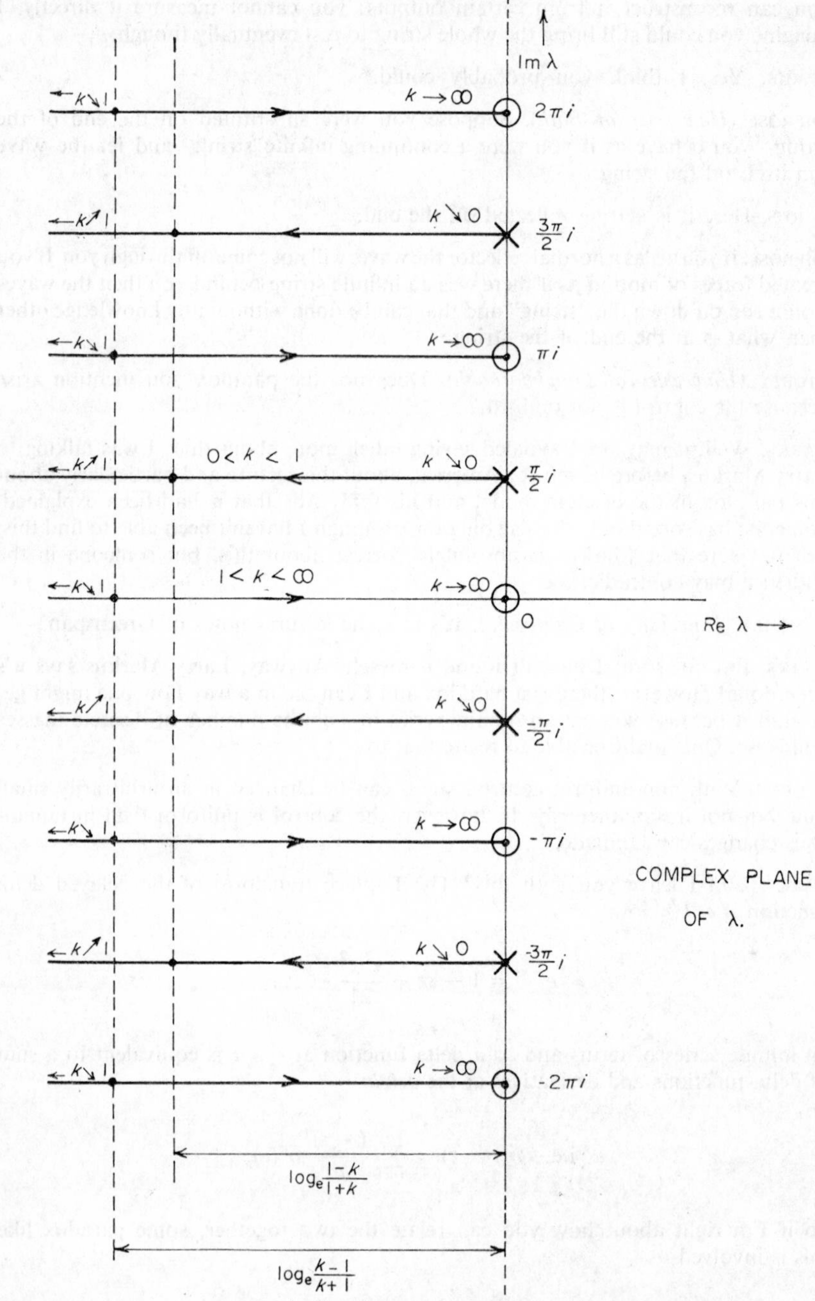

FIG. 7. Root locus diagram for λ when $\cosh \lambda + k \sinh \lambda = 0$.

Postscript

PARKS. Professor K. V. Diprose (private communication) points out that a massless dashpot placed at $x = 1$ on the string and tuned so that

$$\frac{\partial u}{\partial t} \equiv \left.\frac{\partial y}{\partial t}\right|_{x=1} = -\left.\frac{\partial y}{\partial x}\right|_{x=1}$$

will bring any initial configuration of the string to rest in 2 units of time, for examination of Fig. 1 reveals that

$$\frac{\partial u}{\partial t} \equiv \left.\frac{\partial y}{\partial t}\right|_{x=1} = -\left.\frac{\partial y}{\partial x}\right|_{x=1}$$

along the line $x=1$, $0 \leqslant t \leqslant 2$. This result was also observed by P. Quinn (reference 5, Vol. 1, Paper 12.6 p.5).

One might enquire what happens if the damper is not correctly tuned so that the boundary condition at $x=1$ is

$$k\left.\frac{\partial y}{\partial t}\right|_{x=1} = -\left.\frac{\partial y}{\partial x}\right|_{x=1}, k \neq 1.$$

Solving Eqn (1) subject to this boundary condition at $x=1$ and also $y(0, t)=0$ at $x=0$ one obtains

$$y(x, t) = (A\,e^{-\lambda x} + B\,e^{-\lambda x})e^{\lambda t}$$

where λ satisfies the transcendental equation

$$\cosh \lambda + k \sinh \lambda = 0$$

or

$$e^{2\lambda} = \frac{k-1}{k+1}.$$

This yields the unusual root locus diagram shown in Fig. 7. The time for the initial motion to decay to zero is infinite except for the special case $k=1$ when, as we have seen, it is 2 units of time. Clearly the damping will become very heavy as k approaches unity from above or below. A suitably arranged experiment might verify this effect, at least in an approximate quantitative way. The special case $k=1$ we shall call "Diprose damping". Professor J. L. Douce points out that the same phenomenon is observed in transmission line theory where it is known as "matched impedance", and Professor Sir James Lighthill reminds us that the phenomenon is mentioned in Lord Rayleigh's "*Theory of Sound*".

20. The Infinite Dimensional Riccati Equation

RUTH F. CURTAIN and A. J. PRITCHARD

Control Theory Centre,
University of Warwick, Coventry, England

1. Introduction

We consider the existence and uniqueness of solutions of the infinite dimensional Riccati equation, which arises from the solution of the infinite dimensional, linear, quadratic cost control problem. (see [6]). Both time-invariant and time-dependent linear dynamical systems are considered for a finite time interval.

The main difficulties in considering the infinite dimensional version of the classical control problem arise in two main areas—in the existence and uniqueness of solutions of the Hilbert space differential equation and in the interpretation of the infinite dimensional Riccati equation. We use the abstract evolution equation approach when considering the Hilbert space differential equation (see [1–5]), and in order to give sense to the Riccati equation, we define a mild inner product version which solves the problems of the domain of the differential operators.

Previously there have been two main approaches to this problem by Lions [7] and Lukes and Russell [8]. Lions uses a different approach to prove existence and uniqueness theorems for the Hilbert space differential equation, which is essentially a weakened coercivity condition on the differential operator. He establishes his Riccati equation results indirectly by considering a decoupled system of equations formed by introducing an adjoint system. Our final theorem includes his results for this particular system, although his methods provide more easily verifiable criteria. Lukes and Russell consider the time-invariant system, but include a special class of unbounded operators in the control term and in the cost functional. They prove the existence and uniqueness for the weak solution of an integrated version of the Riccati equation, using the abstract evolution equation approach.

2. Preliminaries

Let V, H be real Hilbert spaces, the control and state spaces respectively and let $T = [0, t_1]$ be a real, finite time interval. Consider the following abstract evolution equation on H.

$$\frac{dz}{dt} = A(t)\,z(t) \qquad z(0) = z_0 \tag{1.1}$$

where $A(t)$ is a closed, densely-defined operator on H, which generates an evolution operator $U(t, s)$† ensuring that (1.1) has the unique solution $z(t) = U(t, 0)z_0$.

Consider the inhomogeneous form of (1.1)

$$\frac{dz}{dt} = A(t)z(t) + f(t) \qquad z(0) = z_0. \tag{1.2}$$

Then for sufficiently smooth $f(t)$, (1.2) has the unique solution

$$z(t) = U(t, 0)z_0 + \int_0^t U(t, s)f(s)ds. \tag{1.3}$$

For $f(t)$ merely integrable, (1.3) is not a "strong" solution but it is a well defined H-valued function and we call it a "mild solution" of (1.2) as in Pritchard [6].

Consider now the infinite dimensional system

$$\frac{dz}{dt} = A(t)z(t) + B(t)\,u(t) \qquad z(0) = z_0 \tag{1.4}$$

with the cost functional

$$C(u) = \langle z(t_1), G\,z(t_1)\rangle + \int_0^{t_1} [\langle z(s), W(s)\,z(s)\rangle + \langle u(s), R(s)\,u(s)\rangle]\,ds \tag{1.5}$$

where $A(t)$ is as in (1.1); $z_0 \in H$; the admissible control $u \in L_2(T; V)$; $B(t) \in L(V, H)$; $G, W(t) \in L(H, H)$ and are self-adjoint and positive semi-definite; $R(t)$, $R(t)^{-1} \in L(V, V)$ and are self-adjoint and positive definite; $B(t)$, $W(t)$, $R(t)$, $R(t)^{-1}$ are all strongly continuous in t on T.

† See the Appendix of [9].

We prove existence and uniqueness of solutions $Q_\infty(t) \in L(H, H)$ for the associated inner product version of the Riccati equation.

$$\left\langle \left[\frac{dQ_\infty(t)}{dt} + A^*(t) Q_\infty(t) = Q_\infty(t) A(t) + W(t) \right.\right.$$
$$\left.\left. - Q_\infty(t) B(t) R^{-1}(t) B^*(t) Q_\infty(t) \right] y, x \right\rangle = 0 \tag{1.6}$$
$$Q_\infty(t_1) = G$$

(where x, y are arbitrary elements of $D = \bigcap_{t \in T} D(A(t))$)

As in Pritchard [6], we consider the sequence of control problems generated by a sequence $\{u_k(t)\}$ of admissible controls of the form $u_k(t) = -L_k(t) z(t)$, where $L_k(t) \in L(H, H)$ and is strongly continuous in t.

$$\frac{dz}{dt} = A_k(t) z(t) + B(t) \bar{u}(t) \tag{1.7}$$
$$z(0) = z_0$$

where $A_k(t) = A(t) - B(t) L_k(t)$ generates the evolution operator $U_k(t, s)$† and so for the admissible $\bar{u}(t)$, we can define the mild solution $z_k(t)$ of (1.7) to be

$$z_k(t) = U_k(t, 0) z_0 + \int_0^t U_k(t, s) B(s) \bar{u}(s)\, ds \tag{1.8}$$

and

$$L_0(t) = 0$$
$$L_k(t) = \Lambda_{k-1} R(t)^{-1} B^*(t) Q_{k-1}(t) + (I - \Lambda_{k-1}) L_{k-1}(t)$$
$$W_k(t) = W(t) + L_k^*(t) R(t) L_k(t) \tag{1.9}$$
$$Q_k(t) = U_k^*(t_1, t)\, G\, U_k(t_1, t) + \int_t^{t_1} U_k^*(s, t) W_k(s) U_k(s, t) ds$$

where $\{\Lambda_k\}$ is a sequence of bounded operators converging in the uniform topology to the identity I as $k \to \infty$ In [6], Pritchard shows that for the time-invariant case, under suitable conditions on the control system, the operators $Q_k(t)$, $L_k(t)$ and $W_k(t)$ converge in the strong topology. Here we extend the result to the time dependent case and prove that $Q_\infty(t)$, the strong limit of $Q_k(t)$ satisfies (1.6) and is unique. We consider two main classes of evolution equations, the time-invariant and the time-dependent cases.

† See the Appendix of [9].

3. The Riccati Equation

THEOREM 1 Existence theorem for the time invariant case. *Consider* (1.4) *and* (1.5) *under the following assumptions*

(i) $A(t) \equiv A$ *is the infinitesimal generator of the semigroup* $T(t)$.

(ii) $G, W(t)$ *are self-adjoint and positive semidefinite operators in* $L(H, H)$.

(iii) $R(t), R(t)^{-1} \in L(V, V)$ *and are self-adjoint and positive definite.*

(iv) $B(t) \in L(V, H), W(t), R(t)$ *and* $R(t)^{-1}$ *are strongly continuously differentiable in t on T.*

Then the associated inner product version of the Riccati equation (1.6) *has a solution.*

Proof. Consider the sequence of Eqns (1.7) to (1.9). From Pritchard [6], Theorem 3, Lemmas 5 and 6, we know that $\{Q_k(t)\}$, $\{L_k(t)\}$, and $\{W_k(t)\}$ are sequences of strongly continuous operators, uniformly bounded in norm in k and t which converge strongly to the strongly continuous operators $Q_\infty(t)$,

$$L_\infty(t) = R(t)^{-1} B(t) Q_\infty(t)$$

and

$$W_\infty(t) = W(t) + Q_\infty(t) B(t) L_\infty(t) \text{ as } k \to \infty$$

Also $U_k(t, s)$ converge strongly to $U_\infty(t, s)$, the evolution operator generated by $A_\infty(t) = A(t) - B(t) L_\infty(t)$ by a convergence result of [9]. If we consider

$$Q_k(t) = U_k^*(t_1, t)\, G\, U_k(t_1, t) + \int_t^{t_1} U_k^*(s, t)\, W_k(s)\, U_k(s, t)\, ds$$

then for $x, y \in D(A)$, we can take limits of $\langle Q_k(t)y, x\rangle$ in the strong sense as $k \to \infty$, since all the operators are uniformly bounded in norm, and obtain

$$\langle Q_\infty(t)y, x\rangle = \left\langle \left[U_\infty(t_1, t)\, G\, U_\infty(t_1, t) + \int_t^{t_1} U_\infty^*(s, t)\, W_\infty(s)\, U_\infty(s,t)\, ds \right] y, x \right\rangle. \tag{3.1}$$

This is a mild or integrated version of our "inner product" Riccati equation (1.6).

So far we have not needed to assume (iv). We assume (iv) to differentiate $\langle Q_\infty(t)\, y,x\rangle$ for arbitrary $x,y \in D(A)$. Now

$$\begin{aligned} \frac{d}{dt}\langle U_\infty^*(t_1, t)\, G\, U_\infty(t_1, t)\, y, x\rangle &= \frac{d}{dt}\langle G\, U_\infty(t_1, t)y, U_\infty(t_1, t)x\rangle \\ &= -\langle G\, U_\infty(t_1, t)\, A_\infty(t)y, U_\infty(t_1, t)x\rangle \\ &\quad - \langle G\, U_\infty(t_1, t)y, U_\infty(t_1, t)\, A_\infty(t)x\rangle \end{aligned}$$

using a property of the evolution operator and assumption (iv).

Also

$$\frac{d}{dt}\left\langle\left[\int_t^{t_1} U_\infty{}^*(s,t)\, W_\infty(s)\, U_\infty(s,t)\, ds\right] y, x\right\rangle$$

$$= \frac{d}{dt}\int_t^{t_1} \langle U_\infty{}^*(s,t)\, W_\infty(s)\, U_\infty(s,t)\, y, x\rangle\, ds$$

$$= -\langle U_\infty{}^*(t,t)\, W_\infty(t)\, U_\infty(t,t) y, x\rangle$$

$$+ \int_t^{t_1} \frac{\partial}{\partial t}\langle U_\infty{}^*(s,t)\, W_\infty(s)\, U_\infty(s,t)\, y, x\rangle\, ds$$

since all operators are strongly continuous in t and uniformly bounded in norm.

$$= -\langle W_\infty(t)\, y, x\rangle - \int_t^{t_1} [\langle U_\infty{}^*(s,t)\, W_\infty(s)\, U_\infty(s,t)\, A_\infty(t)\, y, x\rangle$$

$$+ \langle A_\infty{}^*(t)\, U_\infty{}^*(s,t)\, W_\infty(s)\, U_\infty(s,t)\, y, x\rangle]\, ds$$

$$= -\langle W_\infty(t) y, x\rangle - \left\langle \int_t^{t_1} U_\infty{}^*(s,t)\, W_\infty(s), U_\infty(s,t)\, ds\, A_\infty(t)\, y, x\right\rangle$$

$$- \left\langle A_\infty{}^*(t) \int_t^{t_1} U_\infty{}^*(s,t)\, W_\infty(s)\, U_\infty(s,t)\, ds\, y, x\right\rangle$$

taking $A_\infty(t)$, $A(t_\infty{}^*)$ outside the integral since they are closed operators and the integrands are all integrable.

The theorem now follows by substituting for $Q_\infty(t)$ from (2.1).

THEOREM 2. Existence theorem for the time dependent case. *Consider* (1.4) *and* (1.5) *under the following assumptions*

(i) $A(t)$ *is a closed, linear, densely-defined operator on* H *such that* (a) *the spectrum of* $-A(t)$ *is contained in the fixed sector* Σ:

$$|\arg \gamma| < \theta < \pi/2 \text{ and } \|(\gamma I - A(t))^{-1}\| \leqslant \frac{\varepsilon}{|\gamma|} \text{ for } \gamma \notin \Sigma.$$

(b) $A(t)^{-1} \in L(H,H)$ *and is Hölder continuously differentiable in* t *in the uniform operator topology.*

(c)

$$\left\| \frac{d}{dt}(\gamma I - A(t))^{-1} \right\| \leqslant \frac{\eta}{|\gamma|^{1-\rho}} \text{ for } \gamma \notin \Sigma, 0 \leqslant \rho < 1$$

(ii) *and* (iii) *as in Theorem* 1.

(iv) $B(t)$, $W(t)$, $R(t)$ *and* $R(t)^{-1}$ *are strongly continuous in t.*

Then the associated mild version of the Riccati equation, (1.6), *has a solution.*

Proof. Consider the sequence of Eqns (1.7) to (1.9). Since $B(t)\,L_k(t)$ is a strongly continuous, bounded operator for each k, $A_k(t) = A(t) - B(t)L_k(t)$ generates an evolution operator $U_k(t, s)$. So by the same argument as in theorem 1. (2.1) holds in this more general case also, using a convergence result proved in the appendix of [9]. We may differentiate $\langle Q_\infty(t)\,y, x\rangle$ as before, using the strong differentiability of $U_\infty(t, s)$ with respect to s to obtain the inner product version of the Riccati equation valid for all x and $y \in D = \bigcap_{t\in T} D(A(t))$

We note that we do not need strong differentiability of $B(t)$, $W(t)$, $R(t)$ $R(t)^{-1}$ in this case, because of the stronger assumptions on $A(t)$.

COROLLARY 1. *The results of Pritchard* [6] *for the finite time interval hold for time dependent systems with* $A(t)$ *satisfying the assumptions of Theorem* 2.

THEOREM 3. Uniqueness theorem. *Consider* (1.4) *and* (1.5) *under the assumptions of Theorem* 1 *or* 2. *Then the solution of the mild version of the Riccati equation* (1.6) *is unique.*

Proof. The proof uses Theorem 1 of Pritchard [6] and utilizes the fact that the control generated by the sequence minimizes the cost. Suppose that $Q_1(t)$ is strongly continuous in t and $\in L(H, H)$ and satisfies the integrated version of (1.6), i.e.

$$Q_1(t) = U_1{}^*(t_1, t)\,G\,U_1(t_1, t) + \int_t^{t_1} U_1{}^*(s, t)\,W_1(s)\,U_1(s, t)\,ds$$

where $U_1\ (t, s)$ is the evolution operator generated by

$$A(t) - B(t)\,R^{-1}(t)\,B^*(t)\,Q_1(t).$$

Let $Q_\infty(t)$ be the solution demonstrated in Theorem 1 (or 2). Consider the system equations.

$$\frac{dz}{dt} = A(t)\,z(t) + B(t)\,u(t) \tag{1.4}$$

$$z(0) = z_0$$

and the cost functional (1.5) as before.

Let $u_1 = \bar{u}_1 - R^{-1} B^* Q_1 z$ where

$$z(t) = U_1(t, 0)z_0 + \int_0^t U_1(t, s) B(s) \bar{u}_1(s) \, ds$$

and

$$u = \bar{u} - R^{-1} B^* Q_\infty z$$

where

$$z(t) = U_\infty(t, 0)z_0 + \int_0^t U_\infty(t, s) B(s) \bar{u}(s) \, ds.$$

Then by Theorem 1 of Pritchard [6], we have

$$\begin{aligned}\langle z(t), Q_1(t) z(t)\rangle = \langle z(t_1), Gz(t_1)\rangle + \int_t^{t_1} [\langle z(s), W_1(s) z(s)\rangle \\ - \langle z(s), Q_1(s) B(s) \bar{u}_1(s)\rangle - \langle Q_1(s) B(s) \bar{u}_1(s), z(s)\rangle] \, ds. \quad (2.2)\end{aligned}$$

The cost generated by the control u_1 is

$$\begin{aligned}C(u_1) &= \langle z(t)_1, G z(t_1)\rangle + \int_0^t [\langle z(s), W(s) z(s)\rangle + \langle u_1, R u_1\rangle] \, ds \quad \text{from (1.5)} \\ &= \langle z(t_1), G z(t_1)\rangle + \int_0^{t_1} \langle z(s), W_1(s) z(s)\rangle \, ds \\ &\quad + \int_0^{t_1} [\langle \bar{u}_1, R\bar{u}_1\rangle - \langle Q_1(s) B(s) \bar{u}_1, z\rangle - \langle z, Q_1 B\bar{u}_1\rangle] \, ds.\end{aligned}$$

Therefore

$$C(u_1) = \langle z_0, Q_1(0) z_0\rangle + \int_0^{t_1} \langle \bar{u}_1, R\bar{u}_1\rangle \, ds \quad \text{by (2.2)} \qquad (2.3)$$

Similarly,

$$C(u) = \langle z_0, Q_\infty(0) z_0\rangle + \int_0^{t_1} \langle \bar{u}, R\bar{u}\rangle \, ds \qquad (2.4)$$

Let $u_1{}^* = -R^{-1} B^* Q_1 z$ and $u^* = -R^{-1} B^* Q_\infty z$ i.e. the controls when $\bar{u}_1$ and $\bar{u}$ are zero, respectively.

Let

$$\bar{u} = -R^{-1}B^*(Q_1 - Q_\infty)z \quad \text{in (2.4)},$$

i.e.

$$u = -R^{-1}B^*Q_1 z = u_1{}^*.$$

$$C(u_1{}^*) = \langle z_0, Q_\infty(0)\, z_0\rangle + \int_0^{t_1} \langle \bar{u}, R\bar{u}\rangle\, ds$$

i.e.

$$C(u_1{}^*) = C(u^*) + \int_0^{t_1} \langle \bar{u}, R\bar{u}\rangle\, ds \quad \text{from (2.4).} \tag{2.5}$$

Let $\bar{u}_1 = -R^{-1}B^*(Q_\infty - Q_1)z$ in (2.3); therefore

$$C(u^*) = \langle z_0, Q_1(0)\, z_0\rangle + \int_0^{t_1} \langle \bar{u}_1, R\bar{u}_1\rangle\, ds$$

and

$$C(u^*) = C(u_1{}^*) + \int_0^{t_1} \langle \bar{u}_1, R\bar{u}_1\rangle\, ds. \tag{2.6}$$

Adding (2.5) and (2.6) we obtain

$$0 = \int_0^{t_1} \langle \bar{u}, R\bar{u}\rangle\, ds + \int_0^{t_1} \langle \bar{u}_1, R\,\bar{u}_1\rangle\, ds.$$

But $R(s)$ is positive and strongly continuous in t and so we deduce that $\bar{u} = \bar{u}_1 = 0$, i.e. $Q_1 = Q_\infty$ or there exists a unique strongly continuous solution of (1.6).

References

1. Hille, E. and Phillips, R. S., "Functional Analysis and Semi-groups", Amer. Math. Soc. Colloquium Publications, Vol. 31, revised edition, Providence (1957).
2. Phillips, R. S., Perturbation theory for semi-groups of linear operators. *Trans. Amer. Math. Soc.* **74** (1954), 199–221.
3. Balakrishnan A. V., Optimal control problems in Banach spaces. *SIAM J. Control Ser. A* **3** (1) (1965), 152–180.
4. Kato, T. and Tanabe, H., On the abstract evolution equation. *Osaka. Math J.* **14** (1962), 107–133.
5. Dunford, N. and Schwarz, J. T., "Linear Operators, Part 1: General Theory", Interscience, New York (1958).
6. Pritchard, A. J., "The Linear Quadratic Problem for Systems Described by Evolution Equations", Report 9, Control Theory Centre, University of Warwick. (1972).
7. Lions, J. L., "Optimal Control of Systems Governed by Partial Differential Equations", Springer-Verlag, New York (1970).

8. Lukes, D. L. and Russell, D. L., The quadratic criterion for distributed systems. *SIAM J. Control* 7 (1) (1969), 101–121.
9. Curtain, R. F. and Pritchard, A. J., "The Infinite Dimensional Riccati Equation". Report 10, Control Theory Centre, University of Warwick. (1972).

Discussion

GELDER (*Pilkington Bros. Ltd.*). You pointed out that you don't in general expect the u's to be smooth in optimal control. If you build up a sequence of controls presumably in general this sequence will tend to something which is less and less smooth.

PRITCHARD. I didn't say I didn't expect the optimal control to be smooth. I don't know *a priori.*

GELDER. In the case when you know *a priori* it isn't smooth, is it possible to construct a sequence of controls u_i satisfying some natural smoothness condition so that the iterates become less smooth?

PRITCHARD. I don't know; that is something I would have to think about. The way that we construct the optimal control there are various operators which are open to choice. It's very much like the Kleinman† iterative procedure. Maybe by progressively letting the smoothness of these operators decrease it may be possible to do what you are saying. But the other problem is that B, not just u, requires to be smooth, which cannot be handled by this approach.

†*Editors Note.* KLEINMAN, D. L. On the iterative technique for Riccati Equation Computations. *I.E.E.E. Trans. autom. Control,* Vol. AC–13 No. 1, Feb. 1968, pp. 114–115.

21. Discrete Time Control of Distributed Parameter Systems

M. J. E. Mayhew

Computer Science Department, Lanchester Polytechnic, Rugby, England

1. Introduction

The problem of designing optimal controllers for distributed parameter systems has been studied widely in recent years. Several methods have been proposed which involve sampling the state of the system continuously in space and time. A bibliography appears in [1]. More recently it has been shown that, for certain systems, it is sufficient to sample the state at discrete points in space [3, 7]. In [3] a transform technique is used to show that the system may be sampled at discrete points in time as well as in space.

In this paper a dynamic programming technique is adopted and the optimal control is obtained as an integral over the domain of the system at discrete points in time. This integral can then be approximated by a suitable quadrature formula to obtain a feedback control which depends on knowledge of the state at discrete points in space as well as time.

2. The Basic Equations

The system is defined on a fixed spatial domain Ω which is an open, simply connected, subset of an m-dimensional Euclidean space with boundary $\partial\Omega$. The mathematical model is assumed to be of the form

$$\frac{\partial u}{\partial t} = \mathscr{L}u \tag{2.1}$$

where $\mathscr{L}$ is a formal linear partial differential operator and $u = (u_1, u_2, \ldots, u_m)$. The state space U and the control space H of the system are Hilbert spaces. A is a closed linear operator, obtained from $\mathscr{L}$, defined on a dense subset $S \subseteq U$. It is assumed that A generates a strongly continuous semigroup of linear bounded operators T_t.

The control system is then

$$\frac{du}{dt} = Au + Bh, \qquad u(0) = u_0 \in S \tag{2.2}$$

where $B{:}H \to U$ is a linear bounded operator.

The system's equations are framed about some desired state and control laws have to be found to bring the system from a given state and minimise the quadratic performance index given by

$$C(u_0, T - t_0) = \int_{t_0}^{T} [\langle u, Qu\rangle + \langle h, Sh\rangle]\,dt + \langle u(T), Q_0 u(T)\rangle \tag{2.3}$$

where Q and Q_0 are self-adjoint, positive semi-definite and S is self adjoint, positive definite.

The system is sampled with a sampling period Δ. The time interval (t_0, T) is divided into N sub-intervals of duration Δ and on each sub-interval $(t_0 + n\Delta,\ t_0 + (n+1)\Delta)$ the control h is independent of time.

$$h(t) = h_n, \qquad n\Delta < t < (n+1)\Delta. \tag{2.4}$$

The optimal control which is continuous in time can be obtained by a dynamic programming technique [8]. A similar approach is followed in the discrete time case. This method has been used for ordinary differential equations [2].

If the optimal cost of controlling the state u_n at time $t_n = t_0 + n\Delta$ is denoted by C^* then

$$\begin{aligned} C^* &= \min_h \left\{ C(u_n, T - t_n) \right\} \\ &= \min_h \left\{ \int_{t_n}^{t_{n+1}} [\langle u, Qu\rangle + \langle h, Sh\rangle]\,dt + C^*(u_{n+1}, T - t_{n+1}) \right\}. \end{aligned} \tag{2.5}$$

The solution of Eqn (2.2) in the interval (t_n, t_{n+1}) will be written in the form

$$u(t) = T_t u_n + \int_{t_n}^{t} T_{t-s} B(s) h_n\, ds.$$

If $B(s)$ is sufficiently smooth [5], this will represent a true solution of (2.2). Otherwise the idea of a mild solution [6] will be adopted.

Since h_n is independent of time in the interval (t_n, t_{n+1}), this solution can be written

$$u(t) = T_t u_n + S_t h_n \tag{2.6}$$

where

$$S_t = \int_{t_n}^{t} T_{t-s} B(s)\, ds \tag{2.7}$$

and, in particular,

$$u_{n+1} = T_\Delta u_n + S_\Delta h_n. \tag{2.8}$$

If the optimal cost is assumed to be of the form

$$C^*(u_n, T - t_n) = \langle u_n, V_{N-n} u_n \rangle \tag{2.9}$$

then Eqn (2.5) becomes

$$C^*(u_n, T - t_n) = \min_{h_n} \left\{ \int_{t_n}^{t_{n+1}} [\langle u, Qu \rangle + \langle h_n, Sh_n \rangle]\, dt + \langle u_{n+1}, V_{N-n-1} u_{n+1} \rangle \right\}.$$

After substituting from Eqns (2.6), (2.7) and (2.8) this becomes

$$C^*(u_n, T - t_n) = \min_{h_n} [\langle u_n, C_1 u_n \rangle + \langle u_n, C_2 h_n \rangle + \langle C_2 h_n, u_n \rangle + \langle h_n, C_3 h_n \rangle] \tag{2.10}$$

where

$$C_1 = \int_{t_n}^{t_{n+1}} T_t^* Q T_t\, dt + T_\Delta^* V_{N-n-1} T_\Delta$$

$$C_2 = \int_{t_n}^{t_{n+1}} T_t^* Q S_t\, dt + T_\Delta^* V_{N-n-1} S_\Delta \tag{2.11}$$

$$C_3 = \int_{t_n}^{t_{n+1}} S_t^* Q S_t\, dt + S_\Delta^* V_{N-n-1} S_\Delta + \Delta S.$$

The minimum in Eqn (2.10) is attained when

$$h_n = -C_3^{-1} C_2^* u_n \tag{2.12}$$

and the minimum value is

$$C^*(u_n, T - t_n) = \langle u_n, V_{N-n} u_n \rangle = \langle u_n, (C_1 - C_2 C_3^{-1} C_2^*)\, u_n \rangle$$

or

$$V_{N-n} = C_1 - C_2 C_3^{-1} C_2^*. \tag{2.13}$$

This equation can be solved iteratively to obtain $V_1, V_2, \ldots, V_N$ starting with $V_0 = Q_0$.

If the time interval is infinite then the same procedure is applied with $V_0 = 0$. If there is a sequence of controls with finite cost then V_N is bounded above. Consequently the successive solutions form a monotone sequence

$$K > V_N > V_{N-1} > \dots > V_0$$

and the solutions converge.

It has been shown [7] that, for some systems, good results can be obtained in the continuous time case by replacing the equation corresponding to (2.12) by a Gaussian quadrature formula which only requires knowledge of the state at discrete points. The same technique can be adopted in the discrete time case, i.e. (2.12) has to be replaced by a corresponding numerical approximation

$$h_n(\mathbf{w}) = G^k(\mathbf{w}, \mathbf{v}^k)\, u_n(\mathbf{v}^k) \tag{2.14}$$

where G^k are derived from the numerical formula and $\underline{v}^k$ are discrete points in Ω. The following example will show an application of this technique.

3. An Example

The theory developed in Section 2 is applied to solve the heat equation in non-dimensional form

$$u_t = u_{xx}$$

with boundary control

$$\begin{aligned} u_x(0, t) &= h \\ u_x(1, t) &= 0 \\ u(x, 0) &= u_0(x) \end{aligned} \tag{3.1}$$

and performance index

$$C = \int_0^\infty \left[\int_0^1 u^2 dx + \lambda h^2 \right] dt \tag{3.2}$$

The system's equations are rewritten with homogeneous boundary conditions and the performance index in the form of Section 2, i.e.

$$\begin{aligned} u_t &= u_{xx} + h\delta(x) \\ u_x(0, t) &= U_x(1, t) = 0 \end{aligned} \tag{3.3}$$

and

$$C = \int_0^\infty \int_0^1 \int_0^1 (u(x)\delta(x - y)u(y) + \lambda\delta(x)\delta(y)h^2)\, dx\, dy\, dt. \tag{3.4}$$

The Green's function for this system is

$$A(t_0, t, x, y) = 1 + 2\sum_{n=1}^{\infty} e^{-n^2\pi^2(t-t_0)} \cos n\pi x \cos n\pi y \tag{3.5}$$

and the corresponding kernel of the operator S_t is

$$B(t_0, t, x, y) = (t - t_0)\delta(y) - 2\delta(y) \sum_{n=1}^{\infty} \frac{1}{n^2\pi^2}(e^{-n^2\pi^2(t-t_0)} - 1) \times \cos n\pi x \cos n\pi y. \tag{3.6}$$

These can then be substituted into Eqns (2.11) to give the respective kernels

$$C_1(z, w) = \Delta - \sum_{n=1}^{\infty} \frac{e^{-2n^2\pi^2\Delta} - 1}{n^2\pi^2} \cos n\pi z \cos n\pi w + \sum_{m=0}^{\infty}\sum_{n=0}^{\infty} a_m a_n \times v_{mn} \cos m\pi z \cos n\pi w.$$

$$C_2(z, w) = \left[\frac{\Delta^2}{2} - \sum_{n=1}^{\infty} \frac{(e^{-n^2\pi^2\Delta} - 1)^2}{n^4\pi^4} \cos n\pi z \cos n\pi w + \sum_{m=0}^{\infty}\sum_{n=0}^{\infty} a_m b_n \times v_{mn} \cos m\pi z \cos n\pi w\right] \delta(w) \tag{3.7}$$

$$C_3(z, w) = \left[\frac{\Delta^3}{3} + \Delta\lambda + \sum_{n=1}^{\infty} \frac{1}{n^6\pi^6}(-e^{-2n^2\pi^2\Delta} + 4e^{-n^2\pi^2\Delta} + 2n^2\pi^2\Delta - 3) \times \cos n\pi z \cos n\pi w + \sum_{m=0}^{\infty}\sum_{n=0}^{\infty} b_m b_n v_{mn} \cos m\pi z \cos n\pi w\right] \delta(z)\delta(w)$$

where

$$a_m = e^{-m^2\pi^2\Delta}$$

$$b_n = -\frac{1}{n^2\pi^2}(e^{-n^2\pi^2\Delta} - 1), \qquad n \neq 0$$

$$b_0 = \Delta.$$

Since $h_n(w)$ is constant it is possible to put

$$h_n = \int_0^1 k(v) u_n(v)\, dv$$

$$= \int_0^1 \sum_{i=0}^{\infty} k_i \cos i\pi v\, u_n(v)\, dv \tag{3.8}$$

and on substituting this into Eqn (2.12),

$$\frac{\Delta^2}{2} + \sum_{n=1}^{\infty} \frac{(e^{-n^2\pi^2\Delta} - 1)}{n^2\pi^2} \cos n\pi w + \sum_{m=0}^{\infty}\sum_{n=0}^{\infty} a_m v_{mn} b_n \cos m\pi w$$

$$= -\left\{\frac{\Delta^3}{3} + \Delta h + \sum_{n=1}^{\infty} (-e^{-2n^2\pi^2\Delta} + 4e^{-n^2\pi^2\Delta} + 2n^2\pi^2\Delta - 3)\right.$$

$$\left. + \sum_{m=0}^{\infty}\sum_{n=0}^{\infty} b_m v_{mn} b_n\right\} \sum_{r=0}^{\infty} k_r \cos r\pi w. \tag{3.9}$$

On equating coefficients, this leads to

$$(\mathbf{m} + AV\mathbf{b}) = -(n + \mathbf{b}^T V\mathbf{b})\,\mathbf{k} \tag{3.10}$$

where

$$\mathbf{m} = \left(\frac{\Delta^2}{2}, \frac{1}{\pi^4}(e^{-\pi^2\Delta} - 1)^2, \quad \frac{1}{2^4\pi^4}(e^{-2^2\pi^2\Delta} - 1)^2, \ldots\right)^T$$

$$n = \frac{\Delta^3}{3} + \Delta h + \sum_{n=1}^{\infty} (-e^{-2n^2\pi^2\Delta} + 4e^{-n^2\pi^2\Delta} + 2n^2\pi^2\Delta - 3)$$

$$A = \mathrm{diag}\,(a_0, a_1, \ldots)$$

and

$$\mathbf{b} = (b_0, b_1, \ldots).$$

Similarly, substituting into Eqn (2.10) gives

$$V(z, w) = \left[\Delta - \sum_{n=1}^{\infty} \left(\frac{e^{-2n^2\pi^2\Delta} - 1}{n^2\pi^2}\right) \cos n\pi z \cos n\pi w + \sum_{m=0}^{\infty}\sum_{n=0}^{\infty} a_m v_{mn}\right.$$

$$\left. \times\, a_n \cos m\pi z \cos n\pi w\right] + \left[\frac{\Delta^2}{2} - \sum_{n=1}^{\infty} \frac{(e^{-n^2\pi^2\Delta} - 1)^2}{n^4\pi^4} \cos n\pi w\right.$$

$$\left. + \sum_{m=0}^{\infty}\sum_{n=0}^{\infty} a_m v_{mn} b_n \cos m\pi w\right] k(z) + \left[\frac{\Delta^2}{2} + \sum_{n=1}^{\infty} \frac{(e^{-n^2\pi^2\Delta} - 1)^2}{n^4\pi^4} \cos n\pi z\right.$$

$$\left. + \sum_{m=0}^{\infty}\sum_{n=0}^{\infty} a_m v_{mn} b_n \cos n\pi z\right] k(w) + \left[\frac{\Delta^3}{3} + \Delta\lambda + \sum_{n=1}^{\infty} \frac{1}{n^6\pi^6}(-e^{-2n^2\pi^2\Delta}\right.$$

$$\left. + 4e^{-n^2\pi^2\Delta} + 2n^2\pi^2\Delta - 3) + \sum_{m=0}^{\infty}\sum_{n=0}^{\infty} b_m v_{mn} b_n\right] k(z)k(w)$$

which leads to the matrix equation

$$V = R + AVA + (\mathbf{m} + AV\mathbf{b})\,\mathbf{k}^T + \mathbf{k}(\mathbf{m} + AV\mathbf{b})^T + \mathbf{k}(n + \mathbf{b}^T V \mathbf{b})\,\mathbf{k}^T \tag{3.11}$$

or, using Eqn (3.10),

$$V = R + AVA + (\mathbf{m} + AV\mathbf{b})\,\mathbf{k}^T \tag{3.12}$$

The pair of Eqns (3.10) and (3.12) can be solved iteratively for the first M harmonics, starting with $V_0 = 0$.

The equations were solved on an ICL 1903A computer with $M = 10$. The cost of controlling initial states $u_0(x) = 1$ and $u_0(x) = \cos \pi x$ for different values of the sampling period Δ is shown in Figs 1 and 2. The optimal costs shown in the figures are optimal with continuous sampling in space and time.

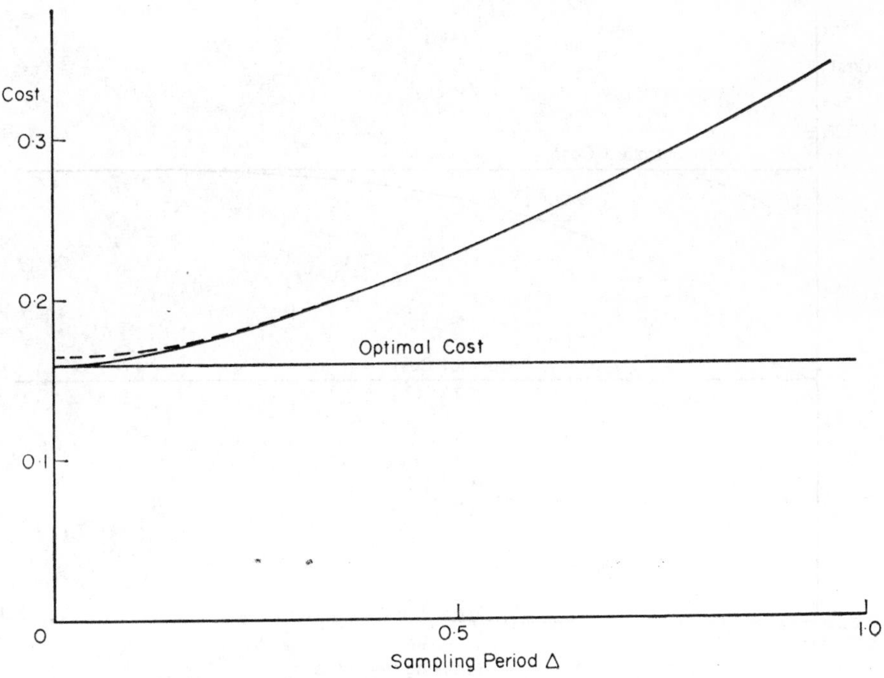

FIG. 1

The gains for feedback from discrete points were then calculated by substituting the appropriate values of $k(v)$ into the Gaussian quadrature formula [4]

$$\int_0^1 k(v)u_0(v)\,dv = 0{\cdot}17446[k(0{\cdot}07)u_0(0{\cdot}07) + k(0{\cdot}93)u_0(0{\cdot}93)]$$

$$+ 0{\cdot}32553[k(0{\cdot}33)u_0(0{\cdot}33) + k(0{\cdot}67)u_0(0{\cdot}67)]$$

$$= \sum_{i=1}^{4} g_i u_0(x^i) \qquad (3.13)$$

The values of g_1 and g_4 for different values of Δ are shown in Figs 3 and 4. In each case the gain converges to the corresponding gain computed for the continuous time case as $\Delta \to 0$.

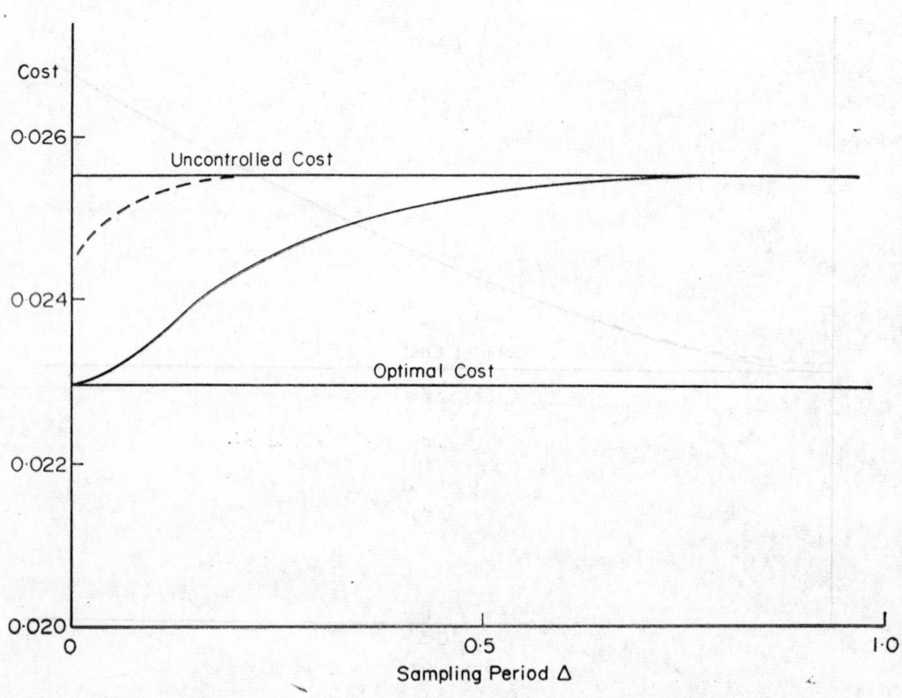

FIG. 2

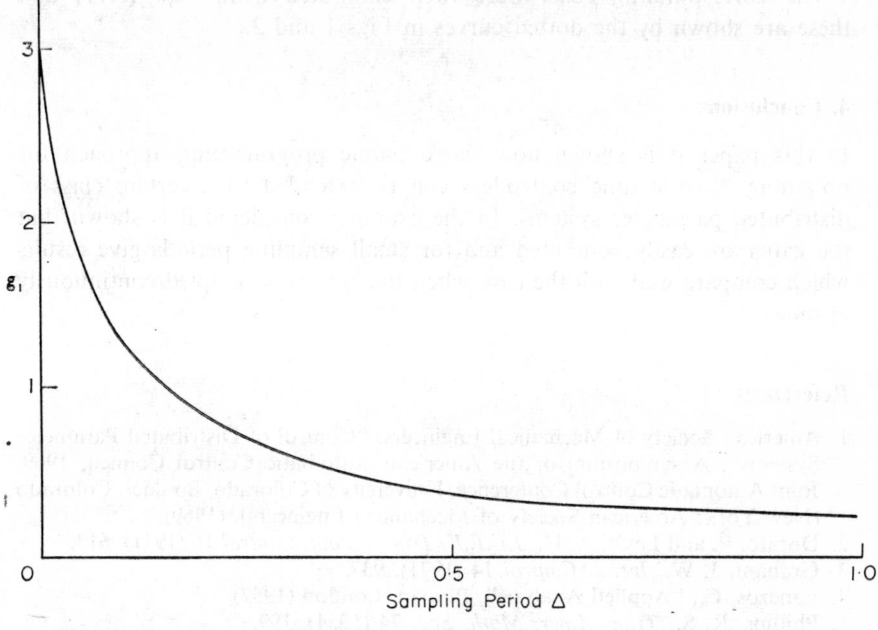

FIG. 3

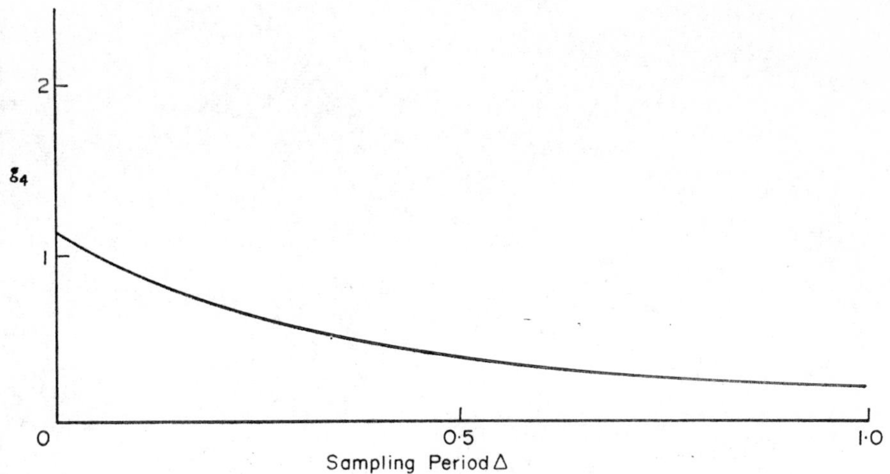

FIG. 4

The corresponding costs were then computed from Eqn (3.11) and these are shown by the dotted curves in Figs 1 and 2.

4. Conclusions

In this paper it is shown how the dynamic programming approach for obtaining discrete time controllers can be extended to a certain class of distributed parameter systems. In the example considered it is shown that the gains are easily computed and for small sampling periods give results which compare well with the case when the system is sampled continuously in time.

References

1. American Society of Mechanical Engineers, "Control of Distributed Parameter Systems", A symposium of the American Automatic Control Council, 1969, Joint Automatic Control Conference, University of Colorado, Boulder, Colorado (New York: American Society of Mechanical Engineers). (1969).
2. Dorato, P. and Levis, A. H., *I.E.E.E. Trans autom. Control* **16** (1971), 613.
3. Graham, J. W., *Int. J. Control* **14** (1971), 937.
4. Lanczos, C., "Applied Analysis", Pitman, London (1957).
5. Phillips, R. S., *Trans Amer. Math. Soc.* **74** (1954), 199.
6. Pritchard, A. J., Report No. 9. Control Theory Centre, University of Warwick. (1972).
7. Pritchard, A. J. and Mayhew, M. J. E., *Int. J. Control* **14** (1971), 619.
8. Wang, P. K. C., Control of distributed parameter systems. *In* "Advances in Control Systems" (ed. C. T. Leondes), Vol. 1. Academic Press, New York (1964) pp. 75–172.

22. A Maximum Principle for the Optimal Control of Distributed Parameter Systems

M. J. GREGSON

School of Mathematics Computing and Statistics, City of Leicester Polytechnic, England

Maximum principles for systems of ordinary differential equations with bounded controls were set up by Pontryagin [1] and others, and have since been considerably explored and exploited.

Similar maximum principles for a variety of systems of partial differential equations have been proposed, notably by Sage [2], Butkovskii [3], Egorov [4], and Degtyarev [5]. The maximum principle I shall describe has been inspired by these writers. Having established the principle, I shall discuss its use in the determination of controls which affect the characteristics of the partial differential equations.

Since a high-order partial differential equation may be reduced to a set of coupled first order equations (admittedly at some cost in terms of redundant solutions), I will begin by considering the following set of coupled first order equations:

$$\frac{\partial \phi_i}{\partial t} = f_i\,(x, t, \boldsymbol{\phi}, \boldsymbol{\phi}_x, \mathbf{u}), \qquad i = 1, 2, \ldots n \tag{1}$$

where

$\boldsymbol{\phi}(x, t) \in E^n$ is the set of state variables

$\mathbf{u}(x, t) \in$ some bounded subset of E^r is the set of distributed control variables.

The functions f_i are assumed to be differentiable in all their arguments. The domain of interest is $D \equiv \{0 \leqslant x \leqslant L, 0 \leqslant t \leqslant T\}$, and the functional to be minimised is

$$J = \iint_D G_1\,(x, t, \boldsymbol{\phi}, \boldsymbol{\phi}_x, \mathbf{u})\,dx\,dt + \int_0^T G_2(t, \phi(L, t))\,dt$$

$$+ \int_0^L G_3(x, \phi(x, T))dx \tag{2}$$

where G_1, G_2 and G_3 are also differentiable in their arguments (usually quadratic cost functions).

Boundary conditions are that $\boldsymbol{\phi}(x, 0)$ and $\boldsymbol{\phi}(0, t)$ are prescribed and that they satisfy the matching condition

$$\lim_{x\to 0} \boldsymbol{\phi}(x, 0) = \lim_{t\to 0} \boldsymbol{\phi}(0, t). \tag{3}$$

We introduce co-state variables $\psi_i(x, t)$, $\boldsymbol{\psi} \in E^n$, and the Hamiltonian H defined by

$$H = G_1 + \sum_{i=1}^{n} \psi_i f_i \tag{4}$$

so that

$$\phi_{it} = \frac{\partial H}{\partial \psi_i}. \tag{5}$$

The functions ψ_i satisfy the differential equations

$$\psi_{it} = -\frac{\partial H}{\partial \phi_i} + \frac{\partial}{\partial x}\left(\frac{\partial H}{\partial \phi_{ix}}\right), \quad i = 1, 2, \ldots, n \tag{6}$$

in the domain D, together with boundary conditions on the boundaries $x = L$ and $t = T$:

$$\psi_i(x, T) = \frac{\partial G_3}{\partial \phi_i(x, T)} \tag{7}$$

$$\left[\frac{\partial H}{\partial \phi_{ix}}\right]_{x=L} = -\frac{\partial G_2}{\partial \phi_i(L, t)}, \quad i = 1, 2, \ldots, n. \tag{8}$$

THEOREM. *The maximisation (minimisation) of J is achieved when H is maximised (minimised) with respect to the controls. That is, we choose u so that*

$$H^0 = \max_u (H) \qquad [\text{or} \min_u (H)]$$

where H^0 is the optimum value of H.

We assume that the set of contours Γ along which $\mathbf{u}$ changes discontinuously is a set of simple curves. $\boldsymbol{\phi}$ and $\boldsymbol{\psi}$ are continuous across Γ, but $\boldsymbol{\phi}_x$, $\boldsymbol{\phi}_t$, $\boldsymbol{\psi}_x$, $\boldsymbol{\psi}_t$ may not be.

Proof Consider small changes $\Delta\boldsymbol{\phi}$, $\Delta\boldsymbol{\psi}$, $\Delta\mathbf{u}$ which are admissible and compatible with the differential equations.

Then, to first order,

$$\Delta J = \iint_D \Delta G_1 \, dxdt + \int_0^T \Delta G_2 dt + \int_0^L \Delta G_3 dx \tag{9}$$

$$= \iint_D \Delta \left[H - \sum_{i=1}^n \psi_i f_i \right] dxdt + \int_0^T \Delta G_2 dt + \int_0^L \Delta G_3 dx \tag{10}$$

$$= \iint_D \left[\Delta H(x, t, \boldsymbol{\phi}, \boldsymbol{\phi}_x, \boldsymbol{\psi}, \mathbf{u}) - \sum_{i=1}^n \Delta \psi_i f_i - \sum_{i=1}^n \psi_i \Delta f_i \right] dxdt$$

$$+ \int_0^T \Delta G_2 dt + \int_0^L \Delta G_3 dx \tag{11}$$

$$= \iint_D \left[\sum_{i=1}^n \frac{\partial H}{\partial \phi_i} \Delta \phi_i + \sum_{i=1}^n \frac{\partial H}{\partial \phi_{ix}} \Delta \phi_{ix} + \sum_{i=1}^n \frac{\partial H}{\partial \psi_i} \Delta \psi_i + \Delta_{\mathbf{u}}(H) \right.$$

$$\left. - \sum_{i=1}^n \Delta \psi_i f_i - \sum_{i=1}^n \psi_i \Delta f_i \right] dxdt + \int_0^T \Delta G_2 dt + \int_0^L \Delta G_3 dx \tag{12}$$

$$= \iint_D \left[\sum_{i=1}^n \frac{\partial H}{\partial \phi_i} \Delta \phi_i + \sum_{i=1}^n \frac{\partial H}{\partial \phi_{ix}} \Delta \phi_{ix} + \Delta_{\mathbf{u}}(H) \right.$$

$$\left. - \sum_{i=1}^n \psi_i \Delta \left(\frac{\partial \phi_i}{\partial t} \right) \right] dxdt + \int_0^T \Delta G_2 dt + \int_0^L \Delta G_3 dx \tag{13}$$

where $\Delta_{\mathbf{u}}(H)$ is the variation in H due to the change $\Delta\mathbf{u}$ in $\mathbf{u}$. We integrate by parts the terms containing $\Delta\phi_{ix}$ and $\Delta(\partial\phi_i/\partial t)$, so that

$$\Delta J = \int_0^T \left[\sum_{i=1}^n \frac{\partial H}{\partial \phi_{ix}} \Delta \phi_i \right]_{x=0}^{x=L} dt - \int_0^L \left[\sum_{i=1}^n \psi_i \Delta \phi_i \right]_{t=0}^{t=T} dx$$

$$+ \iint_D \left[- \sum_{i=1}^n \frac{\partial}{\partial x} \left(\frac{\partial H}{\partial \phi_{ix}} \right) \Delta \phi_i + \sum_{i=1}^n \frac{\partial \psi_i}{\partial t} \Delta \phi_i + \sum_{i=1}^n \frac{\partial H}{\partial \phi_i} \Delta \phi_i \right.$$

$$\left. + \Delta_{\mathbf{u}}(H) \right] dxdt + \int_0^T \sum_{i=1}^n \frac{\partial G_2}{\partial \phi_i(L, t)} \Delta \phi_i(L, t) \, dt$$

$$+ \int_0^L \sum_{i=1}^n \frac{\partial G_3}{\partial \phi_i(x, T)} \Delta \phi_i(x, T) dx. \tag{14}$$

Now $[\Delta\boldsymbol{\phi}]_{x=0} = [\Delta\boldsymbol{\phi}]_{t=0} = 0$, since $\boldsymbol{\phi}(x, 0)$ and $\boldsymbol{\phi}(0, t)$ are prescribed. When the boundary conditions (7), (8) are incorporated into Eqn (14), we see that all the terms in ΔJ vanish except

$$\iint_D \Delta_{\mathbf{u}}(H)\,dxdt$$

so that

$$\Delta J = \iint_D \Delta_{\mathbf{u}}(H)\,dxdt. \tag{15}$$

Thus J is maximised (minimised) when H is maximised (minimised) with respect to the controls $\mathbf{u}(x, t)$, as stated in the theorem.

Note

Let p_i, $(i = 1, 2, \ldots, 3n)$ denote collectively the variables ϕ_i, ϕ_{ix}, ψ_i, $(i = 1, 2, \ldots, n)$.

Then in the expansion of ΔH,

$$\begin{aligned}\Delta H = H(\mathbf{p} + \Delta\mathbf{p}, \mathbf{u} + \Delta\mathbf{u}) - H(\mathbf{p}, \mathbf{u}) &= H(\mathbf{p} + \Delta\mathbf{p}, \mathbf{u} + \Delta\mathbf{u}) \\ &- H(\mathbf{p}, \mathbf{u} + \Delta\mathbf{u} + H(\mathbf{p}, \mathbf{u} + \Delta\mathbf{u}) - H(\mathbf{p}, \mathbf{u}) \\ &= \sum_{i=1}^{3n} \Delta_{pi} \frac{\partial H}{\partial p_i}(\mathbf{p}, \mathbf{u}) + \Delta_{\mathbf{u}}(H) + O(\Delta\mathbf{p} \,.\, \Delta\mathbf{u}).\end{aligned} \tag{16}$$

The second-order smallness of the term $O(\Delta\mathbf{p} \,.\, \Delta\mathbf{u})$ is ensured by the requirement that H be differentiable in $\mathbf{u}$ and $\mathbf{p}$.

We note that small changes $\Delta\mathbf{u}$ in $\mathbf{u}$ include changes of the following two possible kinds:

(i) those in which $\mathbf{u}(x, t)$ changes to $\mathbf{u}(x, t) + \varepsilon\gamma(x, t)$, where $\gamma(x, t)$ is bounded in D, the smallness of the change being small by reason of the smallness of ε.

(ii) those in which relatively large changes in $\mathbf{u}$ may occur over a subdomain δ of D, the change being small by reason of the smallness of δ. Here we have in mind variations of the switching curve in a "bang-bang" control.

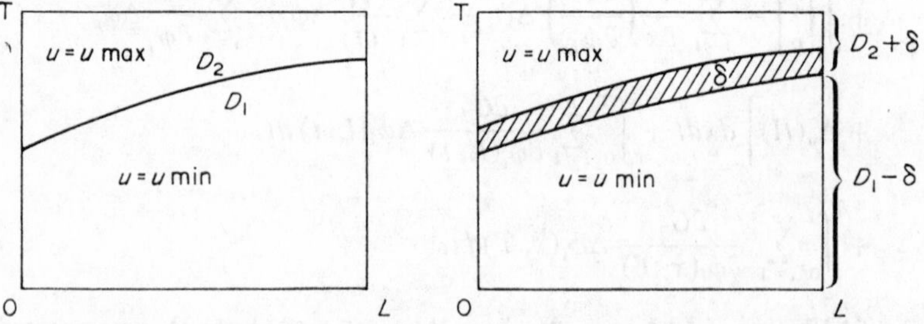

FIG. 1. Variation of the switching curve in a "bang-bang" control.

Extension 1: To Boundary Controls (i.e. Boundary Values of $\boldsymbol{\phi}$) which are Free

When $\boldsymbol{\phi}(x, 0)$ and $\boldsymbol{\phi}(0, t)$ are free, and J is of the form

$$J = \iint_D G_1(x, t, \boldsymbol{\phi}, \boldsymbol{\phi}_x, \mathbf{u})\, dxdt + \int_0^T G_2 dt + \int_0^L G_3 dx \tag{17}$$

where

$$G_2 = G_2'[t, \boldsymbol{\phi}(0, t)] + G_2''[t, \boldsymbol{\phi}(L, t)] \tag{18}$$

$$G_3 = G_3'[x, \boldsymbol{\phi}(x, 0)] + G_3''[x, \boldsymbol{\phi}(x, T)] \tag{19}$$

all these being differentiable in their arguments, then we have

$$\begin{aligned}\Delta J = &\int_0^T \left[\sum_{i=1}^n \frac{\partial H}{\partial \phi_{ix}} \Delta\phi_i\right]_{x=0}^{x=L} dt - \int_0^L \left[\sum_{i=1}^n \psi_i \Delta\phi_i\right]_{t=0}^{t=T} dx \\ &+ \iint_D \Delta_{\mathbf{u}}(H)\, dxdt + \int_0^T \left\{\sum_{i=1}^n \frac{\partial G_2'}{\partial \phi_i(0, t)} \Delta\phi_i(0, t) + \sum_{i=1}^n \frac{\partial G_2''}{\partial \phi_i(L, t)} \right. \\ &\left. \times \Delta\phi_i(L,t)\right\} dt + \int_0^L \left\{\sum_{i=1}^n \frac{\partial G_3'}{\partial \phi_i(x, 0)} \Delta\phi_i(x, 0) + \sum_{i=1}^n \frac{\partial G_3''}{\partial \phi_i(x,T)} \right. \\ &\left. \times \Delta\phi_i(x, T)\right\} dx \end{aligned} \tag{20}$$

whence

$$\psi_i(x, T) = \frac{\partial G_3''}{\partial \phi_i(x, T)} \tag{21}$$

$$\psi_i(x, 0) = -\frac{\partial G_3'}{\partial \phi_i(x, 0)} \tag{22}$$

$$\left[\frac{\partial H}{\partial \phi_{ix}}\right]_{x=L} = -\frac{\partial G_2''}{\partial \phi_i(L, t)} \tag{23}$$

$$\left[\frac{\partial H}{\partial \phi_{ix}}\right]_{x=0} = \frac{\partial G_2'}{\partial \phi_i(0, t)} \tag{24}$$

that is, two extra sets of conditions for ψ_i, $i = 1, 2, \ldots n$.

The values of $\boldsymbol{\phi}(x, 0)$ and $\boldsymbol{\phi}(0, t)$ which result when all the equations for state and co-state variables are solved may be considered as boundary controls whose values may be set in order to achieve a desired optimal result.

In practice one may have mixed sets of boundary conditions with perhaps $\phi_i(0, t)$ given for certain values of i only and $\phi_i(L, t)$ given for others, and likewise for $\phi_i(x, 0)$ and $\phi_i(x, T)$. Wherever there is a boundary on which one of the ϕ_i's is free, there will be a condition of the above type for the corresponding ψ_i.

Extension 2: To a System of Higher Order Equations

We may extend the order of the terms on the right-hand sides of the equations so that we have a system of the form

$$\frac{\partial \phi_i}{\partial t} = f_1\left(x, t, \boldsymbol{\phi}, \frac{\partial \boldsymbol{\phi}}{\partial x}, \ldots, \frac{\partial^k \boldsymbol{\phi}}{\partial x^k}, \mathbf{u}\right), \quad i = 1, 2, \ldots, n \tag{25}$$

with

$$J = \iint_D G_1\left(x, t, \boldsymbol{\phi}, \frac{\partial \boldsymbol{\phi}}{\partial x}, \ldots, \frac{\partial^k \boldsymbol{\phi}}{\partial x^k}, \mathbf{u}\right) dx dt + \int_0^T G_2\left(t, \boldsymbol{\phi}, \ldots, \frac{\partial^{k-1} \boldsymbol{\phi}}{\partial x^{k-1}}\right) dt + \int_0^L G_3(x, \boldsymbol{\phi})\, dx \tag{26}$$

the boundary conditions being that

$$\phi_i(x, 0), \phi_i(0, t), \frac{\partial \phi_i}{\partial x}(0, t), \ldots, \frac{\partial^{k-1} \phi_i}{\partial x^{k-1}}(0, t)$$

are prescribed, $i = 1, 2, \ldots, n$.

The Hamilton H, as before, is

$$H = G_1 + \sum_{i=1}^{n} \psi_i f_i \tag{27}$$

with the co-state variables satisfying

$$\frac{\partial \psi_i}{\partial t} = -\frac{\partial H}{\partial \phi_i} - \sum_{r=1}^{k} (-1)^r \frac{\partial^r}{\partial x^r}\left(\frac{\partial H}{\partial[\partial^r p_i/\partial x^r]}\right), \quad i = 1, 2, \ldots, n \tag{28}$$

in the domain D, together with the boundary conditions

$$\text{(1)}\quad [\psi_i]_{t=T} = \frac{\partial G_3}{\partial \phi_i(x, T)} \tag{29}$$

$$\text{(2)}\quad \left[\sum_{r=\alpha}^{k} (-1)^{r=\alpha} \frac{\partial^{r-\alpha}}{\partial x^{r-\alpha}}\left(\frac{\partial H}{\partial(\partial^r \phi_i/\partial x^r)}\right)\right]_{x=L} = -\frac{\partial G_2}{\partial(\partial^{\alpha-1}\phi_i/\partial x^{\alpha-1})} \tag{30}$$

for $\alpha = 1, 2, \ldots, k$ and $i = 1, 2, \ldots, n$.

Other combinations of boundary conditions are also possible. For example, certain of

$$\phi_i, \frac{\partial \phi_i}{\partial x}, \ldots, \frac{\partial^{k-1}\phi_i}{\partial x^{k-1}}$$

may be given at $x = 0$ and/or $x = L$, and there will then be corresponding conditions for ψ_i and its derivatives, as in Extension 1.

Extension 3: To a Larger Number of Independent Variables

Let us suppose we have the system

$$\frac{\partial \phi_i}{\partial t} = f_i\left(t, x_1, x_2, \ldots x_m, \boldsymbol{\phi}, \frac{\partial \boldsymbol{\phi}}{\partial x_1}, \ldots, \frac{\partial \boldsymbol{\phi}}{\partial x_m}, \mathbf{u}(t, x_1, \ldots x_m)\right),$$

$$i = 1, 2, \ldots, n \qquad (31)$$

where $\mathbf{x} = \{x_1, x_2, \ldots, x_m\}$ belongs to a certain closed set Ω, boundary $\partial\Omega$ and $\tau \equiv \{0 \leqslant t \leqslant T\}$ is a fixed time interval.

Boundary conditions are

(1) $\boldsymbol{\phi}(0, x_1, \ldots, x_m)$ given on Ω at $t = 0$

(2) $\boldsymbol{\phi}(t, x_1, \ldots, x_m)$ given on $\partial\Omega$ for $[0 \leqslant t \leqslant T]$.

The domain D is now $\Omega x \tau$, and the cost functional is

$$J = \int_D G_1\left(\mathbf{x}, t, \boldsymbol{\phi}, \frac{\partial \boldsymbol{\phi}}{\partial x_1}, \frac{\partial \boldsymbol{\phi}}{\partial x_2}, \ldots, \frac{\partial \boldsymbol{\phi}}{\partial x_m}, \mathbf{u}\right) d\mathbf{x}dt + \int_\Omega G_2(\mathbf{x}, \boldsymbol{\phi}(\mathbf{x}, T))\, d\mathbf{x}. \quad (32)$$

The theory is largely unchanged, with

$$H = G_1 + \sum_{i=1}^{n} \psi_i f_i \,. \qquad (33)$$

The co-state variables ψ_i satisfy the differential equations

$$\frac{\partial \psi_i}{\partial t} = -\frac{\partial H}{\partial \phi_i} + \sum_{j=1}^{m} \frac{\partial}{\partial x_j}\left(\frac{\partial H}{\partial[\partial \phi_i/\partial x_j]}\right), \qquad i = 1, 2, \ldots, n \qquad (34)$$

with the boundary conditions

$$[\psi_i]_{t=T} = \frac{\partial G_2}{\partial \phi_i(x, T)}, \qquad i = 1, 2, \ldots, n. \qquad (35)$$

Extension 4: Extensions 2 and 3 may be combined

The results are given in [2], Chapter 7, Section 7.1.

Example 1: *A simple example*

Consider the linear parabolic partial differential equation of heat conduction

$$\frac{\partial \phi}{\partial t} = \frac{\partial^2 \phi}{\partial x^2} + u(x,t), \qquad D \equiv \{0 \leqslant t \leqslant T, 0 \leqslant x \leqslant L\} \tag{36}$$

with initial condition

$$\phi(x,0) = \phi_0(x) \tag{37}$$

and boundary conditions

$$\left.\frac{\partial \phi}{\partial x}\right|_{x=0} = \left.\frac{\partial \phi}{\partial x}\right|_{x=L} = 0 \tag{38}$$

We seek the unrestricted control $u(x, t)$ *which minimises*

$$J = \tfrac{1}{2}\int_0^T \int_0^L \{\phi^2(x,t) + u^2(x,t)\}\,dxdt. \tag{39}$$

Now

$$H = \tfrac{1}{2}(\phi^2 + u^2) + \psi\left(\frac{\partial^2 \phi}{\partial x^2} + u\right) \tag{40}$$

and ψ satisfies

$$\frac{\partial \psi}{\partial t} = -\frac{\partial H}{\partial \phi} - \frac{\partial^2}{\partial x^2}\left(\frac{\partial H}{\partial(\partial^2\phi/\partial x^2)}\right) = -\phi - \frac{\partial^2 \psi}{\partial x^2} \tag{41}$$

with the boundary conditions

$$[\psi]_{t=T} = \frac{\partial G_3}{\partial \phi(x,T)} = 0 \tag{42}$$

$$\left[\frac{\partial \psi}{\partial x}\right]_{x=0} = \left[\frac{\partial \psi}{\partial x}\right]_{x=L} = 0 \tag{43}$$

H is minimised when $u = -\psi$.

This system can be solved completely, given that $\phi_0(x)$ is known.

Example 2: *An example in which the choice of the control* $u(x, t)$ *affects the characteristics of the partial differential equations.*

Consider the first order transport equation

$$\phi_t = -u\phi_x = f \tag{44}$$

where $D \equiv \{0 \leqslant t \leqslant T, 0 \leqslant x \leqslant 2\pi\}$, the control $u(x, t)$ satisfies $\frac{1}{2} \leqslant u \leqslant 1$, and the boundary conditions are

$$\phi = 0 \quad \text{on} \quad x = 0 \tag{45}$$

$$\phi = \sin x \quad \text{on} \quad t = 0. \tag{46}$$

We seek to minimise the quadratic functional

$$J = \int_0^{2\pi} \phi^2(x, T)dx = \int_0^{2\pi} G_3\,[\phi(x, T)]dx \tag{47}$$

The Hamiltonian is

$$H = \sum_{i=1}^{n} \psi_i f_i = -\,u\phi_x\psi \tag{48}$$

J is minimised when H is minimised with respect to u, so that for

$$\phi_x\psi > 0, u = u_{\max} = 1 \tag{49}$$

$$\phi_x\psi < 0, u = u_{\min} = \tfrac{1}{2}, \tag{50}$$

i.e. the control is "bang-bang".

ψ satisfies

$$\frac{\partial\psi}{\partial t} = -\frac{\partial H}{\partial\phi} + \frac{\partial}{\partial x}\left(\frac{\partial H}{\partial\phi_x}\right) = -\,u\psi_x \tag{51}$$

that is, the same equation as for ϕ. Hence, in particular, characteristics for ψ are the same as those for ϕ.

Boundary conditions for ψ are that on $t = T$,

$$\psi(x, T) = \frac{\partial G_3}{\partial\phi(x, T)} = 2\phi(x, T) \tag{52}$$

while on $x = 2\pi$,

$$\left[\frac{\partial H}{\partial\phi_x}\right]_{x=2\pi} = -\frac{\partial G_2}{\partial\phi(2\pi, t)}, \tag{53}$$

that is,

$$\psi(x, 2\pi) = 0. \tag{54}$$

The characteristics of the equations are the lines

$$x - ut = \text{const}$$

along which ϕ and ψ are constant. The slope of these lines is $1/u$.

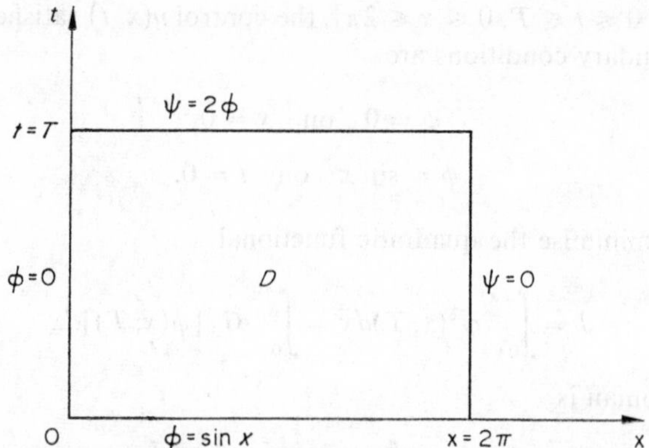

FIG. 2. Boundary conditions for Example 2.

The domain D will be subdivided into regions of differing u.

The following argument establishes the optimal control and shows that in some parts of the domain D at least, the control, and the state and co-state functions are non-unique. Values of $\phi(x, t)$ are determined by transporting boundary values of ϕ forward (i.e. in the direction of increasing t) along characteristics, while values of $\psi(x, t)$ are determined by transporting boundary values of ψ back along characteristics.

Consider the neighbourhood of the point $(\pi/2, 0)$ for a small value of the end time T. The boundary condition

$$\psi(x, T) = 2\phi(x, T)$$

shows that

$$\operatorname{sgn}\{\psi(x, t)\} = \operatorname{sgn}\{\phi(x, t)\}$$

and since $\phi = \sin(x - ut)$, which is positive in the neighbourhood of $t = 0$, $\psi(x, t)$ will also be positive. Since

$$\frac{\partial \phi}{\partial x}\left(\frac{\pi}{2}, 0\right) = 0, \quad \text{so that} \quad \frac{\partial \phi}{\partial x}$$

changes sign, we may expect a change of control along some contour emanating from the point $(\pi/2, 0)$ with $u = 1$ on the left where $\phi_x \psi > 0$ and $u = \frac{1}{2}$ on the right where $\phi_x \psi < 0$. The requirement that $\phi(x, t)$ be continuous shows this contour to be a straight line of slope 4/3. Such a contour will be called a "switching curve", although it is not a curve in phase space, as in the ordinary differential equation case.

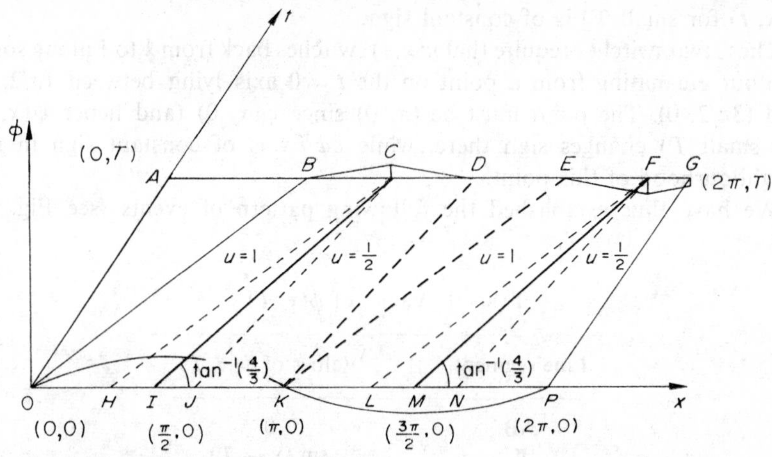

FIG. 3. The evolution of the switching curves and of the state function in Example 2.

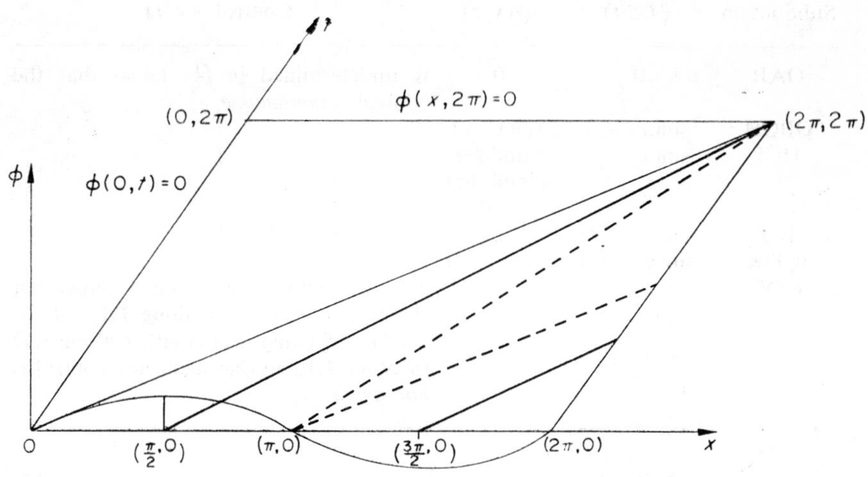

FIG. 4. The case $T = 2\pi$.

In a similar way, one would expect a switching curve emanating from $(3\pi/2, 0)$ where $\partial\phi/\partial x\ (x, 0)$ again changes sign, while $\phi(x, 0)$ (and hence $\psi(x, t)$ for small T) is of constant sign.

These two switches require that $u(x, t)$ switches back from $\frac{1}{2}$ to 1 along some contour emanating from a point on the $t = 0$ axis lying between $(\pi/2, 0)$ and $(3\pi/2, 0)$. The point must be $(\pi, 0)$ since $\phi(x, 0)$ (and hence $\psi(x, t)$ for small T) changes sign there, while $\partial\phi/\partial x$ is of constant sign in the neighbourhood of this point.

We have thus established the following pattern of events (see Fig. 3).

TABLE 1. Values of $\phi(x, T)$

Line segment	Values of $\phi(x, T)$
AB	0
BC	$\sin(x - T)$
CD	$\sin(x - \frac{1}{2}T)$
DE	0
EF	$\sin(x - T)$
FG	$\sin(x - \frac{1}{2}T)$

TABLE 2. Subdomains of D

Subdomain	$\phi(x, t)$	$\psi(x, t)$	Control $u(x, t)$
OAB	0	0	is undetermined in $(\frac{1}{2}, 1)$ so that the control is *non-unique*.
OBCH	$\sin(x - t)$	$2\sin(x - t)$	1
HCI	$\sin(x - t)$	is undetermined, but > 0.	1
ICJ	$\sin(x - \frac{1}{2}t)$	,,	$\frac{1}{2}$
JCDK	$\sin(x - \frac{1}{2}t)$	$2\sin(x - \frac{1}{2}t)$	$\frac{1}{2}$
KDE	0	0	The switching curve is undetermined, but the contribution to J along DE will be zero if switching occurs either along KD or along KE, so that again the control is *non-unique*.

Similar results hold for the remainder of D. We note that for $T \geqslant 2\pi$, $\phi(x, T)$ can be controlled to zero (see Fig. 4).

Conclusion

In the optimal control of distributed parameter systems, it is clear that the use of distributed controls which affect the characteristics of the partial differential equations will lead to non-unique controls and also to regions of the solution domain in which the state and co-state functions are not uniquely determined.

References

1. Pontryagin, L. S., *et al.*, "The Mathematical Theory of Optimal Processes", J. Wiley, New York, (1962).
2. Sage, A. P., "Optimum Systems Control". Prentice Hall, London (1968).
3. Butkovskii, A. G., "Distributed Control Systems". Elsevier, New York (1969).
4. Egorov, A. I., Necessary optimality conditions for distributed parameter systems. *SIAM J. Control* **5** (1967), 352.
5. Degtyarev, G. L. and Sirazetdinov, T. K., Optimal control of one dimensional processes with distributed parameters. *Aut. and Remote Control* (Nov. 1967), 1642–1650.

Discussion

SARGENT (*Imperial College*). Can you say anything about the possibility of singular trajectories in linear, partial differential systems?

GREGSON. I found the difficulties bad enough with something very simple. Although we have a certain kind of singular situation here, I have not investigated as yet whether the ideas of the theory of singular trajectories for ordinary differential equations are going to be helpful or not.

SARGENT. There are two papers by Jackson† in the International Journal of Control in which he discusses singular segments in hyperbolic systems, but I believe these are non-linear.

† Jackson, R., Optimisation Problems in a class of systems described by Hyperbolic Partial Differential Equations. Part 1. Variational Theory. *Int. J. Control,* **4,** No. 2, 127–136 (1966). Part II. A Maximum Principle *Int. J. Control,* **4,** No. 6, 585–598 (1966).

23. Some Topics in Algebraic Systems Theory: A Survey

STEPHEN BARNETT

School of Mathematics, University of Bradford, Yorkshire, England

1. Introduction

During the past quarter of a century a large amount of research effort has been devoted to the study of linear control systems theory. The aim of this paper is to present a concise review of some of the major areas in this field, the emphasis being on matrix methods and techniques. It is of course impossible in a limited space to do more than make brief comments, and the book [1.2] should be consulted for an account written in the same spirit but giving further details and references. In order to complement this text rather than merely give a condensed version of its contents, an attempt has been made to highlight material which does not feature prominently in the book or which has appeared since its completion.

The field is now sufficiently well established for there to be several good textbooks giving accounts of basic material, for example [1.4], [1.5]. For definitions of fundamental concepts of systems theory [1.6] and [1.7] should be consulted. An excellent survey of linear multivariable feedback theory has been given by MacFarlane [1.10], including a very extensive bibliography. A complete issue of the *I.E.E.E. Transactions on Automatic Control* [1.9] has recently been devoted to valuable review material on linear control systems, and a very full bibliography on linear optimal control theory is included [1.11]. Other up to date surveys have been given at the Fifth IFAC Congress [1.1], [1.8].

The contents of the paper are briefly as follows. In Section 2 we discuss criteria for determining asymptotic stability in terms of the coefficients of the characteristic polynomial, it being appropriate to begin with this problem since historically it was one of the first to be studied mathematically [see the now famous paper by Maxwell in [1.3]). In Section 3 the application of Liapunov stability theory to systems in vector-matrix form is outlined.

Some aspects of the concept of relative primeness of two polynomial matrices in relation to the minimal realization problem are explained in Section 4. In Section 5 we are concerned with the matrix equations of Riccati type which arise when the control vector is chosen so as to minimize a quadratic performance index, and the two problems of positive-realness and spectral factorization associated with rational matrices are discussed in Section 6. The paper closes with some remarks on other topics which, for reasons of space, could not be dealt with in any detail.

A select bibliography is given, and again for conciseness it is intended to be consulted in conjunction with [1.2].

2. Polynomials and Stability

In a paper on the governing of steam engines Maxwell (1868) studied special cases of the following problem: if

$$a(\lambda) = \lambda^n + a_1\lambda^{n-1} + \dots + a_n \tag{1}$$

is a polynomial associated with the nth order linear differential equation

$$a(D)z(t) = 0 \tag{2}$$

where $D \equiv d/dt$, under what conditions on the a_i is the null solution of (2) asymptotically stable, i.e. $\lim_{t \to \infty} z(t) = 0$? Algebraically, this is equivalent to $a(\lambda)$ having all its zeros in the left half plane Γ_1: $\text{Re}(\lambda) < 0$. The first general solution was published by Hermite in 1854, but earlier work had been done by Cauchy. A well-known solution for the case when the a_i are real is due to Hurwitz and for this reason a polynomial having all its zeros in Γ_1 is often called *Hurwitzian* (it is convenient to use this terminology even if the a_i are complex). A very considerable amount of published work has appeared on this problem, continuing up to the present day, but accounts of various criteria are readily available and it would be tedious to reproduce details here (see [1.2] for references; the paper by Bateman in [1.3] contains some interesting remarks on the history of the problem). Inter-relationships between many of the theorems are known, and continue to be discovered (e.g. [2.16]). It is worth remarking however that if the a_i are given numerically then the most convenient method for computational purposes is that of Routh [2.21].

The Hurwitz and other determinantal criteria for a real polynomial involve minors of order up to n, but it seems to be less well appreciated that there exist theorems involving determinants of about half the orders. Indeed, the Hermite result when applied to real polynomials involves determinants of order not more than $\frac{1}{2}(n+1)$ [see [2.18], p. 93). To discuss other reductions requires the concept of *resultant*: this is a scalar function of two polynomials which is nonzero if and only if they are relatively prime. Determinantal expressions for resultants are associated with Sylvester and Bézout. If the polynomials have a nontrivial greatest common divisor (*g.c.d.*) then this can be obtained using *subresultants*, which are symmetrically located principal minors of Sylvester's determinant. Illuminating accounts of the way in which these resultants are involved in the stability problem have been given by Householder [2.11], [2.12]. A reduced order determinantal stability criterion based on the use of Bézout's resultant was given by Liénard and Chipart in 1914 and subsequently rediscovered by several authors (see [2.7]). Incidentally, Liénard and Chipart are better known for their theorem which makes use of the fact that for real polynomials a necessary condition for asymptotic stability is $a_i > 0$, all i, and then only about half the Hurwitz determinants need to be calculated. This result was rediscovered by Fuller [2.10] who in the course of his argument exposed the rather neglected fact that the Hurwitz determinants are subresultants associated with the odd and even parts of $a(\lambda)$. This explains why reductions to half order determinants are possible.

Another way of looking at the g.c.d. problem is to introduce the *companion* matrix associated with (1):

$$C = \begin{bmatrix} 0 & 1 & 0 & \cdot & \cdot & 0 \\ 0 & 0 & 1 & \cdot & \cdot & \cdot \\ \cdot & \cdot & \cdot & \cdot & \cdot & 1 \\ -a_n & -a_{n-1} & -a_{n-2} & \cdot & \cdot & -a_1 \end{bmatrix} \tag{3}$$

this having the property that

$$\det(\lambda I_n - C) = a(\lambda)$$

(I_n denotes the unit matrix of order n). If

$$b(\lambda) = b_0\lambda^m + b_1\lambda^{m-1} + \ldots + b_m$$

is any other polynomial then it is easy to see that det $b(C)$ forms a resultant for $a(\lambda)$ and $b(\lambda)$, this fact being first recorded by MacDuffee [2.19] and also rediscovered [2.1], [2.17]. If $m < n$ it is also easy to show that the rows of $b(C)$ are successively

$$\left.\begin{aligned} r = [b_m, b_{m-1}, \ldots, b_0, 0, \ldots, 0] \\ rC, rC^2, \ldots, rC^{n-1} \end{aligned}\right\} \tag{4}$$

and if $m \geqslant n$ application of the Cayley–Hamilton theorem reduces $b(C)$ to a matrix polynomial in C of degree not exceeding $n - 1$. Note that $b(C)$ has order n compared with $m + n$ for Sylvester's determinant, and that $b(C)$ is recognised from (4) to be the observability matrix for the pair $[C, r]$. Further relationships between polynomials and controllability and observability can be developed (see [2.8]). In view of the connection between resultants and stability criteria it is not surprising that the companion matrix approach can be applied to the stability problem, and it turns out that for a real polynomial the largest order of determinant involved is $\frac{1}{2}n$ or $\frac{1}{2}(n - 1)$ according as n is even or odd [2.4], [2.5]. Determinantal expressions for Sturm sequences have been known for a considerable time [2.33], and the companion matrix method can also be used to calculate these sequences and related Routh arrays [2.5]. Links have been established between companion matrices and Bezoutians [2.6], Sylvester's resultant [2.3] and Hermite's theorem for complex polynomials. Various other problems in the qualitative study of polynomials, including g.c.d. determination, can be dealt with rather nicely using appropriate companion matrices (see [1.2]).

If (1) represents the characteristic polynomial associated with a linear difference equation with constant coefficients

$$a(E)z_k = 0 \tag{5}$$

where $Ez_k = z_{k+1}$, then the necessary and sufficient condition for asymptotic stability is that all the zeros of $a(\lambda)$ have modulus less than unity, i.e. lie within the unit disc $\Gamma_2 : |\lambda| < 1$, in which case $a(\lambda)$ is said to be *convergent*. Of course this stability problem can be transformed into the previous one simply by applying the transformation

$$\lambda \to \frac{\lambda + 1}{\lambda - 1} \tag{6}$$

which sends Γ_2 into Γ_1. The coefficients in the polynomial

$$(\lambda - 1)^n a\left(\frac{\lambda + 1}{\lambda - 1}\right)$$

can be obtained from a direct formula [2.26] or by matrix multiplication [2.9], [2.22]. Incidentally, the bilinear transformation

$$\lambda \to \frac{\alpha\lambda + \beta}{\gamma\lambda + \delta}$$

can also be expressed in matrix terms so that, for example, more general circles in the complex plane can be dealt with [2.7]. Again, there are several well-known determinantal criteria which can be found in the standard sources. One of these theorems is due to Schur and Cohn, and the companion matrix approach can be applied to it [2.2], giving the condition that the leading principal minors of the matrix

$$\bar{a}_n C^n + \bar{a}_{n-1} C^{n-1} + \ldots + I_n \tag{7}$$

(the bar denotes complex conjugate) be positive, as necessary and sufficient for $a(\lambda)$ to be convergent. Equation (7) illustrates that the resultant occurring in the discrete-time case is associated with the polynomials $a(\lambda)$ and $\lambda^n \bar{a}(1/\lambda)$ which both have degree n, which suggests that it is unlikely that a direct reduction to a half-order determinant formulation exists. Of course the half-order theorems for the continuous-time case can be used after (6) has been applied. For computational purposes a tabular scheme of Jury [2.13] is most efficient, and its relationship to the Schur–Cohn theorem has been demonstrated [2.23].

Other regions in the complex plane of interest in control theory for which stability criteria have been given include the shifted unit disc $|\lambda + 1| < 1$, which arises in the theory of sampled-data systems using ζ-transforms [2.32]; the region to the left of a line through the origin inclined at $\theta\,(> \frac{1}{2}\pi)$ to the positive real axis [2.27], [2.31], this being involved in relative stability; and an infinite sector [2.34]. Hurwitz polynomials also arise in the study of functions which are positive-real with respect to Γ_1 [2.29] or Γ_2 [2.30]. Other interesting applications include evaluation of time weighted performance indices using Hurwitz determinants [2.24], [2.25] and inversion of continued fractions using Routh arrays [2.28]. Despite the large number of previously published papers in the field covered by this section [see [2.20]] the fund of new results seems almost inexhaustible, and some further references to recent work can be found in [2.8].

We have indicated that the approaches using resultants or companion matrices provide a unified setting for most of the various stability criteria. It should also be remarked that Jury [2.14], [2.15] has proposed the term *inners* to refer to symmetrically situated principal submatrices of any matrix formed by deleting rows and columns numbered $1, n$; $1, 2, n-1, n$; etc. When applied to Sylvester's resultant array the determinants of the inners are precisely the classical subresultants, but the inners also arise in unit disc and other criteria, and thus provide yet another framework within which the various results can be expressed.

3. Linear Liapunov Theory

The Russian mathematician Liapunov published details of his so-called "direct" method of stability analysis in 1892, but it is only comparatively recently (within the past 15–20 years) that his work has become appreciated in the West. We shall consider here only the application to linear continuous or discrete-time systems which can be written

$$\dot{x} = Ax \tag{8}$$

or

$$x_{k+1} = Ax_k \tag{9}$$

where A is a constant $n \times n$ matrix. By a suitable choice of the components of the state vector x, (2) and (5) can be written in the forms (8) and (9) respectively, in which cases A takes the companion form (3), and $a(\lambda)$ coincides with the characteristic polynomial of A:

$$p(\lambda) = \det(\lambda I_n - A).$$

Generally the condition for asymptotic stability of (8) [or (9)] is that $p(\lambda)$ be respectively a Hurwitz (or convergent) polynomial and A is called a *stability* (or *convergent*) matrix. The appropriate Liapunov function is a Hermitian form x^*Px which leads to the linear matrix equations

$$A^*P + PA = -Q \tag{10}$$

$$A^*PA - P = -Q. \tag{11}$$

For asymptotic stability of (8) [(9)], Eqn (10) [(11)] must have a positive definite Hermitian solution P for arbitrary positive definite Hermitian Q [3.4]; when A is real conjugate transpose in (10) and (11) reduces to transpose. The *Liapunov matrix equations* (10) and (11) represent $\frac{1}{2}n(n+1)$ linear equations for the elements of P and a number of efficient solution methods are available [3.13], [3.18]. In particular, if A is in companion form then an effective procedure enables the rows of P to be calculated in succession [3.12]. An interesting scheme for (10) when A is real based on transformation into upper Hessenberg form is proposed in [3.6]. In practice however if the elements of A are known numerically then it must be admitted that the stability problem can be satisfactorily solved by direct calculation of the characteristic roots of A using a standard computer algorithm.

Although the Liapunov approach provides a different route to the solution of the polynomial problem discussed in the preceding section, the relationship with the classical results has been demonstrated for the Routh–Hurwitz case [3.14], the Hermite theorem [3.16] and the Schur–Cohn criterion [3.8], [3.15]. Links with the companion matrix method have also been discovered [3.5]. For example, if A in (11) is taken to be C^T, where C is defined in (3). then consideration of the rows p_i of P in (11) leads to

$$p_n[\bar{a}_n(C^T)^n + \bar{a}_{n-1}(C^T)^{n-1} + \ldots + I_n] = q \tag{12}$$

where q is easily obtained in terms of C and Q, and there is a simple recurrence relation for the other p_i. It is clear from (12) that in this case the solution P of (11) is expressible in terms of the inverse transpose of the matrix occurring in (7), which arose from the companion matrix version of the Schur–Cohn theorem. A similar argument can be applied to (10).

It should be noted that (11) can be obtained from (10) by the matrix transformation corresponding to (6), namely

$$A \to (A + I)(A - I)^{-1}.$$

In fact both (10) and (11) are special cases of a general matrix equation [3.7] which enables regions in the complex λ-plane more general than Γ_1 or Γ_2 (for example, ellipses) to be dealt with. A good survey of linear matrix equations has been given by Lancaster [3.9] with subsequent material presented in [3.19]. The use of Kronecker products to solve both algebraic and differential linear matrix equations is described in [3.1].

A distinct advantage of using Liapunov's method for stability analysis of linear systems in state space form is that it is thereby possible to deal with the elements of A themselves rather than the coefficients in $p(\lambda)$. Furthermore the Liapunov approach provides a very valuable tool for the study of linear system properties, including evaluation of system functionals [3.10], [3.11], with applications to optimal linear control problems [3.2]; synthesis of asymptotically stable linear systems [3.3]; and investigation of sensitivity properties [3.17].

4. Realization Theory and Polynomial Matrices

Consider now the standard state space representation of a constant linear control system

$$\dot{x} = Ax + Bu \tag{13}$$

$$y = Cx \tag{14}$$

where A, B, C are constant matrices, u is the m-vector of controlled variables and y is the output r-vector. Taking Laplace transforms of

(13) and (14) and assuming zero initial conditions leads to

$$\bar{y}(\lambda) = G(\lambda)\bar{u}(\lambda)$$

where $\mathscr{L}\{x(t)\} = \bar{x}(\lambda)$ and the rational matrix

$$G(\lambda) = C(\lambda I_n - A)^{-1} B \tag{15}$$

is the *transfer function matrix*. More generally, if the differential equations describing the system are not first order then after Laplace transformation we obtain

$$T(\lambda)\bar{w}(\lambda) = U(\lambda)\bar{u}(\lambda), \qquad \bar{y}(\lambda) = V(\lambda)\bar{w}(\lambda) \tag{16}$$

where $w(t)$ is the state vector and T, U, V are polynomial matrices having dimensions $p \times p$, $p \times m$ and $r \times p$ respectively and $\det T(\lambda) \not\equiv 0$. The expression for G is now

$$G(\lambda) = VT^{-1}U. \tag{17}$$

Note that a polynomial matrix can be added to the expression in (17(with little additional complication. We shall also assume that $G(\lambda)$ is *proper*, i.e. $G(\lambda) \to 0$ as $\lambda \to \infty$. The *order* of the system in (16) is the degree of $\det\ T(\lambda)$, and the *minimal realization* problem is: given $G(\lambda)$, to find triples $\{T, U, V\}$ satisfying (17) having least possible order. Generally in practice a state space realization $\{A, B, C\}$ is required satisfying (15), in which case the minimality condition implies that A must have smallest possible dimensions. A state space realization is minimal if and only if the pairs $[A, B]$ and $[A, C]$ are respectively controllable and observable, and all such minimal realizations are related by the similarity transformation of the state space $\hat{x} = Tx$ [2.17]. There are a very large number of published methods for computing minimal realizations. Many of these involve determination of the Smith–McMillan form of $G(\lambda)$ (e.g. [4.6]) or evaluation of the transition matrix. Further details are given in [1.2], and a useful collection of references can be found in [4.4], this paper also containing a proposed new procedure. The most efficient algorithms seem to be those which avoid polynomial manipulation, for example [4.7]. Methods also exist for converting polynomial realizations into state-space form [4.5]. It is also possible to compute merely the order of a minimal realization [4.1], [4.3], [4.8] or to obtain an upper bound for the least order [4.10]. A succinct review of realization theory (including the case when A, B, C have time varying elements) has been given by Silverman [4.12]; in both this paper and [4.5] the usefulness of Hankel matrices is demonstrated.

An interesting interpretation of the minimal realization problem has been developed by Rosenbrock [4.9] and partially rediscovered [4.13]. This relies on the following standard definition: if

$$T(\lambda) = D(\lambda)T_1(\lambda), \quad U(\lambda) = D(\lambda)U_1(\lambda)$$

where T_1, U_1, D are polynomial matrices, then $D(\lambda)$ is a *common left divisor* of T and U, and if all such divisors have determinants independent of λ then T and U are *relatively left prime*. An obviously similar statement applies for right primeness of T and V. A crucial theorem states that $\{T, U, V\}$ is minimal if and only if T, U and T, V are relatively left and right prime respectively. When T, U, V take the forms in (15) then the usual criteria for controllability and observability of $\{A, B, C\}$ can be deduced. A basic theorem for scalar polynomials $t(\lambda)$, $v(\lambda)$ states that they are relatively prime if and only if the scalar equation

$$t(\lambda)\alpha(\lambda) + v(\lambda)\beta(\lambda) = 1$$

has a unique solution for α and β with their degrees less than those of v, t respectively. Several different generalizations of this to polynomial matrices have been made—see [1.2] and [2.8]. If $T(\lambda)$ is assumed to be *regular*, i.e. having the matrix of leading coefficients nonsingular, then it has recently been shown [4.11] that a generalization of Sylvester's resultant (referred to in Section 2) can be developed, giving a necessary and sufficient condition for T and V to be prime. The assumption on T can in fact be partially relaxed [4.14]. There are some other interesting relationships with the scalar case—for example, if it is assumed that the g.c.d. of T and V is regular then it turns out [4.2] to be possible to determine this g.c.d. from Rowe's theorem using a generalized form of companion matrix, and this corresponds precisely to the scalar result in [2.1]. Further discussion of these aspects can be found in [2.8].

5. Optimal Linear Control and Matrix Riccati Equations

Consider the system described by

$$\dot{x}(t) = A(t)x(t) + B(t)u(t) \tag{18}$$

where the elements of A and B are real continuous functions of t, and $x(0) = x_0$. If it is desired to choose the control vector $u(t)$ so as to minimise the *quadratic performance index*

$$x^T(\tau)Mx(\tau) + \int_0^\tau [x^T(t)Q(t)x(t) + u^T(t)R(t)u(t)]dt \tag{19}$$

then the optimal control is *linear* in x:

$$u(t) = -R^{-1}B^TPx \tag{20}$$

and the symmetric matrix $P(t)$ is the solution of

$$\dot{P} = PBR^{-1}B^TP - A^TP - PA - Q \tag{21}$$

subject to $P(\tau) = M$. In (19) M and $Q(t)$ are symmetric positive semi-definite and $R(t)$ is symmetric positive definite. Equation (21) is often called a matrix *Riccati* equation because it can be regarded as a generalization of the scalar equation studied by Riccati in the eighteenth century. Several properties of this scalar equation can be extended to the general matrix case and details can be found in [1.2]. We shall restrict ourselves here to the equation arising in optimal control, and point out that the situation described by (18) and (19) is known as the optimal *state regulator*, since the aim is to "regulate" the state $x(t)$ so as to return it to equilibrium after an initial disturbance. Of course Riccati equations have many applications in other branches of control theory [5.20].

It is known [5.4] that (21) has a unique solution $\Pi(t)$ satisfying the given boundary condition and that the minimum value of (19) is $x_0^T\,\Pi(0)x_0$. This solution $\Pi(t)$ can be expressed in terms of the transition matrix $\Phi(t)$ defined by

$$\dot{\Phi} = H\Phi, \qquad \Phi(\tau) = I_{2n}$$

where

$$H(t) = \begin{bmatrix} A & -BR^{-1}B^T \\ -Q & -A^T \end{bmatrix}. \tag{22}$$

This provides a way of computing $\Pi(t)$ as an alternative to standard numerical techniques for solving (21).

When the final time τ in (19) tends to infinity we may set $M = 0$, and if (18) is controllable then

$$\lim_{\tau\to\infty} \Pi\,(t)$$

exists and is a solution of (21). If from now on we also assume that A, B, Q and R are constant matrices then the theory can be developed further. The limiting solution of (21) is time invariant and is the solution of the so-called *algebraic* matrix Riccati equation

$$PBR^{-1}B^TP - A^TP - PA - Q = 0. \tag{23}$$

Define a matrix Q_1 having the same rank as Q and such that $Q = Q_1{}^TQ_1$.

Then provided the triple $[A, B, Q_1]$ is controllable and observable the optimal control is given by (20) with P the unique positive definite solution of (23) [5.4]. The resulting closed loop system

$$\dot{x} = [A - BR^{-1}B^TP]x$$

is easily shown to be asymptotically stable, since a suitable Liapunov function is $x^T Px$. A well-known theorem [5.21] states that provided the pair $[A, B]$ is controllable then real linear feedback $u = Kx$ exists such that the characteristic roots of the closed loop matrix $A + BK$ can be made equal to those of an arbitrary real matrix. Thus (23) provides a way of generating matrices K which have the highly desirable property of producing an asymptotically stable system.

The positive-definite solution of (23) can be obtained in terms of the matrix H (now time-invariant) in (22). If it is assumed that the Jordan form of H is diagonal then [5.13]

$$P = [k_1, k_2, \ldots, k_n]\,[h_1, h_2, \ldots, h_n]^{-1}$$

where h_i, k_i denote the upper and lower halves of the ith characteristic vector of H corresponding to a root with negative real part. If the Jordan form of H is not diagonal then a similar expression for P involves the generalized characteristic vectors of H [5.10].

The conditions of controllability and observability can be replaced by the weaker ones of stabilizability and detectability [5.22]. The real pair $[A, B]$ is *stabilizable* if there exists a real matrix K such that $A + BK$ is a stability matrix, and dually $[A, Q_1]$ is *detectable* if $[A^T, Q_1{}^T]$ is stabilizable. A fundamental theorem [5.8] is that (23) possesses a unique symmetric positive semidefinite solution P and $A - BR^{-1}B^TP$ is a stability matrix if and only if the triple $[A, B, Q_1]$ is stabilizable and detectable. If the condition of detectability is relaxed then (23) does not have a unique positive semidefinite solution, and the problem of finding all such solutions is solved in [5.7] under the assumption that H has no purely imaginary characteristic roots. This generalizes results given in [5.10]. Willems [5.20] has also classified all the solutions of (23) using a different approach. Other methods of solving (23), including an iterative scheme [5.5], [5.16] which relies on repeatedly solving an equation of the form (10), are described in [1.2]. A different iterative scheme has been suggested recently [5.3] based on solution of a quadratic matrix difference equation, and a frequency domain representation of (23) is derived in [5.9]. Other material can be found in the comprehensive textbook [5.1].

For the time invariant problem when the upper limit τ in the performance integral (19) is finite, it has recently been demonstrated [5.15], [5.17], [5.18], [5.19] how the solution of (21) can be written in terms of the steady state solution of (23) and an expression which decays exponentially as $\tau \to \infty$. The last of these papers also gives an explicit expression for the optimal trajectory $x(t)$.

A disadvantage of the Riccati equation method for generating linear feedback control is that no way is known of determining directly the relationship between the weighting matrices Q and R and the feedback matrix in (20). However, for the time-invariant regulator problem with a single control variable a necessary and sufficient condition has been given [5.6] for Q to be replaceable by a diagonal matrix, thus reducing the number of parameters to be adjusted.

Sensitivity aspects of the optimal linear control problem have been studied (e.g. [5.2], [5.12]). In particular if θ is a parameter then it has been suggested (e.g. [5.14]) that adding a term $w^T S w$ to the integrand in (19), with $w = \partial x / \partial \theta$ and S positive definite, will lead to a linear control which makes the system less sensitive to small changes in θ. However, it is now known [5.11] that a solution to this new optimization problem of the form

$$u = K_1 x + K_2 w,$$

with K_1 and K_2 independent of x, does not exist.

6. Positive-realness and Spectral Factorization

We now give brief details of two problems associated with rational matrices. Firstly, an $n \times n$ rational matrix $Z(\lambda) = [z_{ij}(\lambda)]$ is *positive-real* (*p.r.*) if for $\mathrm{Re}(\lambda) > 0$

$$z_{ij} \text{ is analytic, all } i, j \tag{24a}$$

$$\bar{Z}(\lambda) = Z(\bar{\lambda}) \tag{24b}$$

$$Z^T(\bar{\lambda}) + Z(\lambda) \text{ is positive semidefinite Hermitian.} \tag{24c}$$

Condition (24b) implies that all the coefficients in z_{ij} are real. We can assume that $Z(\lambda)$ tends to a finite matrix as $\lambda \to \infty$. The concept of positive-realness has important applications in network theory [6.7] and elsewhere. Let $Z(\lambda) = Z_1(\lambda)/z(\lambda)$ and

$$[Z(\lambda) + I_n]^{-1} = Z_2(\lambda)/w(\lambda)$$

where Z_1 and Z_2 are polynomial matrices, z and w are scalar polynomials and any common factors have been removed. Then [2.29] necessary and sufficient conditions for Z to be p.r. are that (i) $w(\lambda)$ is a Hurwitz polynomial and (ii) the Hermitian matrix $z(-i\omega)Z_1(i\omega) + z(i\omega)Z_1^T(-i\omega)$

is positive semidefinite for all real $\omega \geqslant 0$ (here $i^2 = -1$). Although other criteria have been given (see [1.2]) this result is useful since not only can $z(\lambda)$ be dealt with as discussed in Section 2 but it is also shown in [2.29] how a Routh array can be used to test (ii). If the conditions (24) hold for $|\lambda| > 1$ then $Z(\lambda)$ is *circle* p.r. and Siljak [2.30] has also given algorithms for this case.

There are important connections between p.r. matrices and realization theory. For example, if $Z(\lambda)$ is p.r., proper and has no poles on $\text{Re}(\lambda) = 0$, then in any minimal realization $\{A, B, C\}$ of Z, A is a stability matrix; conversely, if $G(\lambda)$ in (15) is square then it is p.r. provided A is a stability matrix and $G^T(-i\omega) + G(i\omega)$ is positive semidefinite Hermitian for all real ω. Other results, which include the use of Liapunov matrix equations, can be found in [6.1].

A real rational matrix $Y(\lambda)$ is *parahermitian* if $Y(\lambda) = Y^T(-\lambda)$, and the *spectral factorization* problem is then defined as follows: given that $Y(i\omega)$ is positive semidefinite, find a rational matrix $W(\lambda)$ such that

$$Y(\lambda) = W^T(-\lambda)W(\lambda). \tag{25}$$

This problem has many applications [6.9] and a feedback interpretation is given in [6.6].

A basic paper is due to Youla [6.10] whose method for obtaining $W(\lambda)$ relies on finding the Smith–McMillan form of $Y(\lambda)$. A procedure which avoids this was introduced by Davis [6.4] and subsequently extended and improved [6.3], [6.5], [6.8].

It is known [6.2] that solution of the spectral factorization problem is closely related to algebraic Riccati equations. If $Y(\lambda) = Z(\lambda) + Z^T(-\lambda)$, where $Z(\lambda)$ is p.r. having a minimal realization $\{A, B, C\}$, then a solution of (25) can be obtained in terms of this realization, the matrix

$$\left[\lim_{\lambda \to \infty} Y(\lambda)\right]^{\frac{1}{2}}$$

and the solution of an equation of the form (23). Willems [6.9] gives further material on the relationship with Riccati equations.

7. Additional Topics

Of considerable importance for theoretical developments is the use of *canonical forms* for time-invariant systems described by (13) and (14). For the single variable control variable case, provided $[A, B]$ is controllable then A can be transformed into companion form (3) and B into $[0, 0, \ldots, 0, 1]^T$; a dual result holds with respect to observability. For the multivariable case canonical forms are not unique, and extensions of the scalar control result can be found in [7.11] and [7.13], the former including systems which are not controllable. A different approach puts A into Jordan form [7.9], and one of the few books containing material on canonical forms is [1.5]. The Luenberger-type form [7.13] has been applied to realization and decoupling problems [7.28] and to solution of the algebraic Riccati equation [7.10]. Canonical forms and feedback invariants are studied in [7.22]. A canonical form for time varying systems has also been given [7.18].

There is no point in attempting to review the general field of multivariable linear system design in view of the admirable and extensive survey [1.10]. However, the idea of arbitrary allocation of closed-loop characteristic roots by linear state feedback has already been referred to in connection with the regulator problem in Section 5, and this forms the basis of *modal control.* A basic paper is [7.21], the Luenberger canonical form is used in [7.1] and an attractively simple procedure is given in [7.20]. An expository account is contained in [7.23], and other work can be found in [7.7] and [7.17]. If only the output (14) is measurable, then linear output feedback $u = Ky$ can fix $\max(r, m)$ of the characteristic roots of the closed-loop matrix $A + BKC$ provided the triple $[A, B, C]$ is controllable and observable and B and C have full rank [7.5], [7.6]. One suggested procedure for choosing K is in [7.8], and the problem of stabilization by output feedback is discussed in [7.12]. Alternatively, if the system is observable then it is possible to contruct an approximation to $x(t)$ using a system termed a *state observer* (or *estimator*). Details can be found in [1.5] and the review paper [7.14].

The idea of generalized inverse of a rectangular matrix or singular square matrix is often useful and in addition to Chapter 6 of [1.2] two textbooks [7.3], [7.19] between them cover much known material. Of more specialized interest is the concept of infinite matrices [7.15].

All the preceding material in this survey has been concerned with the concrete aspects of algebraic linear systems theory provided by matrix representations. A body of work is now building up on the applications of abstract algebra, and an exposition based on automata theory is given in [7.2]. An account of Kalman's module theory approach is contained in [1.7], and Wonham's remarks [7.25] on this book are worth reading. Wonham's own work here is also interesting [7.16], [7.26], [7.27] and in particular his study of appropriate subspaces instead of linear transformations [7.24] seems potentially powerful. Algebraic properties of classes of nonlinear systems are also beginning to be studied [7.4].

References

1.1. Anderson, B. D. O., Linear multivariable control systems. Survey paper, *5th IFAC Congress* Paris (June, 1972).

1.2. Barnett, S., "Matrices in Control Theory". Van Nostrand Reinhold, London (1971).

1.3. Bellman, R. and Kalaba, R., "Selected Papers on Mathematical Trends in Control Theory", Vol. 1. Dover, New York (1964).

1.4. Brockett, R. W., "Finite Dimensional Linear Systems". John Wiley, New York (1970).

1.5. Chen, C. T., "Introduction to Linear System Theory". Holt, Rinehart and Winston, New York (1970).

1.6. Desoer, C. A., "Notes for a Second Course on Linear Systems". Van Nostrand Reinhold, New York (1970).

1.7. Kalman, R. E., Falb, P. L. and Arbib, M. A., "Topics in Mathematical System Theory". McGraw-Hill, New York (1969).

1.8. Kwakernaak, H., "The state of the art in linear systems", Commentator paper, Session 35, *5th IFAC Congress*, Paris (June, 1972).

1.9. *Special Issue on the Linear-Quadratic-Gaussian Problem, I.E.E.E. Trans. autom. Control* **AC-16** (6) (1971).

1.10. MacFarlane, A. G. J., Linear multivariable feedback theory: A survey. *In* "Proceedings of the 2nd IFAC Symposium on Multivariable Technical Control Systems". (H. Schwarz, Ed.), Vol 4. North-Holland, Amsterdam (1971). *Also in: Automatica* **8** (1972), 455–492.

1.11. Mendel, J. M. and Gieseking, D. L., Bibliography on the Linear-Quadratic-Gaussian problem. *I.E.E.E. Trans. autom. Control* **AC-16** (1971), 847–869.

Section 2

2.1. Barnett, S., Greatest common divisor of two polynomials. *Linear Algebra and Applications* **3** (1970), 7–9.

2.2. Barnett, S., Number of zeros of a complex polynomial inside the unit circle. *Electronics Lett.* **6** (1970), 164–165.

2.3. Barnett, S., Greatest common divisor of several polynomials. *Proc. Camb. phil. Soc. math. phys. Sci.* **70** (1971), 263–268.

2.4. Barnett, S., A new formulation of the Liénard–Chipart stability criterion. *Proc. Camb. phil. Soc. math. phys. Sci.* **70** (1971), 269–274.

2.5. Barnett, S., A new formulation of the theorems of Hurwitz, Routh and Sturm. *J. Inst. Maths. Applics.* **8** (1971), 240–250.

2.6. Barnett, S., A note on the Bezoutian matrix. *SIAM J. Appl. Math.* **22** (1972), 84–86.

2.7. Barnett, S., Some applications of matrices to location of zeros of polynomials. *Int. J. Control* **17** (1973), 823–831.

2.8. Barnett, S., Matrices, polynomials and linear time-invariant systems. *I.E.E.E. Trans. autom. Control* **AC-18** (1973), 1–10.

2.9. Duffin, R. J., Algorithms for classical stability problems. *SIAM Rev.* **11** (1969), 196–213.

2.10. Fuller, A. T., Stability criteria for linear systems and realizability criteria for RC networks. *Proc. Camb. phil. Soc. math. phys. Sci.* **53** (1957) 878–896.

2.11. Householder, A. S., Bigradients and the problem of Routh and Hurwitz. *SIAM Rev.* **10** (1968), 56–66.

2.12. Householder, A. S., Bezoutiants, elimination and localization. *SIAM Rev.* **12** (1970), 73–78.

2.13. Jury, E. I., "Theory of *z*-transforms". John Wiley, New York (1964).

2.14. Jury, E. I., Inners approach to some problems of system theory. *I.E.E.E. Trans. autom. Control* **AC-16** (1971), 233–240.

2.15. Jury, E. I., Stability, root clustering and inners, Paper 35.2, *5th IFAC Congress*, Paris (June, 1972).

2.16. Jury, E. I. and Anderson, B. D. O., Some remarks on simplified stability criteria for continuous linear systems. *I.E.E.E. Trans. autom. Control* **AC-17** (1972), 371–372.

2.17. Kalman, R. E., Mathematical description of linear dynamical systems. *SIAM J. Control* **1** (1963), 152–192.

2.18. Lehnigk, S. H., "Stability Theorems for Linear Motions", Prentice Hall, Englewood Cliffs, N.J. (1966).

2.19. MacDuffee, C. C., Some applications of matrices in the theory of equations. *Amer. Math. Mon..* **57** (1950), 154–161.

2.20. Marden, M., "Geometry of Polynomials", American Mathematical Society, Providence (1966).

2.21. Pace, I. S. and Barnett, S., Numerical comparison of root-location algorithms for constant linear systems. This Volume pp. 373–392

2.22. Power, H. M., The mechanics of the bilinear transformation. *I.E.E.E. Trans. education* **E-10** (1967), 114–116.

2.23. Power, H. M., Note on the Hermite–Fujiwara theorem. *Electronics Lett.* **6** (1970), 39–40.

2.24. Puri, N. N. and Weygandt, C. N., Calculation of quadratic moments of high-order linear systems via Routh canonical transformations. *I.E.E.E. Trans. appl. Ind.* **83** (1964), 428–433.

2.25. Ramar, K. and Ramaswami, B., A new method of evaluating quadratic and time-weighted quadratic form performance indices. *Int. J. Control* **13** (1971), 83–96.

2.26. Samuelson, P. A., Conditions that the roots of a polynomial be less than unity in absolute value. *Ann. math. Statist.* **12** (1941), 360–364.

2.27. Sherman, S., Generalized Routh–Hurwitz discriminant. *Phil. mag.* **37** (1946), 537–551.

2.28. Shieh, L. S., Schneider, W. P. and Williams, D. R., A chain of factored matrices for Routh array inversion and continued fraction inversion. *Int. J. Control* **13** (1971), 691–703.

2.29. Siljak, D. D., New algebraic criteria for positive realness. *J. Franklin Inst.* **291** (1971), 109–120.

2.30. Siljak, D. D., Algebraic criteria for positive realness relative to the unit circle. *Rept. No.* NGR 05–017–010–721, University of Santa Clara, California (1972). *Also in:* Proc. 2nd Intern. Symp. on Network Theory, Herceg–Novi, Yugoslavia (1972), 151–158.

2.31. Stojic, M. R. and Siljak, D. D., Generalization of Hurwitz, Nyquist and Mikhailov stability criteria. *I.E.E.E. Trans. autom. Control* **AC-10** (1965), 250–254.

2.32. Tschauner, J., A general formulation of the stability constraints for sampled-data control systems. *Proc. I.E.E.E.* **51** (1963), 619–620.

2.33. Van Vleck, E. B., On the determination of the series of Sturm's functions by the calculation of a single determinant. *Ann. Math.* (2), **1** (1899), 1–13.

2.34. Williams, J. The distribution of the roots of a complex polynomial equation. *A.R.C. Tech. Rpt. No.* 2238, Ministry of Supply, London (1946).

Section 3

3.1. Barnett, S., Matrix differential equations and Kronecker products. *SIAM J. Appl. Math.* **24** (1973), 1–5.
3.2. Barnett, S. and Storey, C., On the general functional matrix for a linear system. *I.E.E.E. Trans. autom. Control* **AC-12** (1967), 436–438.
3.3. Barnett, S. and Storey, C., Analysis and synthesis of stability matrices. *J. Diff. Eqns.* **3** (1967), 414–422.
3.4. Barnett, S. and Storey, C., "Matrix Methods in Stability Theory", Nelson, London (1970).
3.5. Datta, B., "Quadratic Forms, Matrix Equations and the Matrix Eigenvalue Problem", Ph.D. Dissertation, University of Ottawa, Canada (1972).
3.6. Howland, J. L. and Senez, J. A., A constructive method for the solution of the stability problem. *Numer. Math.* **16** (1970), 1–7.
3.7. Howland, J. L., Matrix equations and the separation of matrix eigenvalues. *J. Math. Anal. Appl.* **33** (1971), 683–691.
3.8. Kalman, R. E., On the Hermite–Fujiwara theorem in stability theory. *Quart. Appl. Math.* **23** (1965), 279–282.
3.9. Lancaster, P., Explicit solutions of linear matrix equations. *SIAM Rev.* **12** (1970), 544–566.
3.10. MacFarlane, A. G. J., The calculation of functionals of the time and frequency response of a linear constant coefficient dynamical system. *Q. J. Mech. Appl. Math.* **16** (1963), 259–271.
3.11 MacFarlane, A. G. J., Functional-matrix theory for the general linear network. *Proc. I.E.E.* **112** (1965),763–770.
3.12. Molinari, B. P., Algebraic solution of matrix linear equations in control theory. *Proc. I.E.E.* **116** (1969), 1748–1754.
3.13. Pace, I. S. and Barnett, S., Comparison of numerical methods for solving Liapunov matrix equations. *Int. J. Control* **15** (1972), 907–915.
3.14. Parks, P. C., A new proof of the Routh–Hurwitz stability criterion using the second method of Liapunov. *Proc. Camb. phil. Soc. math. phys. Sci.* **58** (1962), 694–702.
3.15. Parks, P. C., Liapunov and the Schur–Cohn stability criterion. *I.E.E.E. Trans. autom. Control* **AC-9** (1964), 121.
3.16. Ralston, A., A symmetric matrix formulation of the Hurwitz–Routh stability criterion. *I.R.E. Trans. autom. Control* **AC-7** (1962), 50–51.
3.17. Shane, B. A. and Barnett, S., Insensitivity of constant linear systems to finite variations in parameters. This Volume pp. 393–404.
3.18. Smith, P. G., Numerical solution of the matrix equation $AX + XA^t + B = 0$. *I.E.E.E. Trans. autom. Control* **AC-16** (1971), 278–279.
3.19. Wimmer,H. and Ziebur,A.D., Solving the matrix equation $\sum_{p=1}^{r} f_p(A)Xg_p(B) = C$. *SIAM Rev.* **14** (1972), 318–323.

Section 4

4.1. Barnett, S., Relationship between two methods for calculating the least order of a transfer function matrix. *Int. J. Control* **15** (1972), 509–512.

4.2. Barnett, S., Regular greatest common divisor of two polynomial matrices. *Proc. Camb. phil. Soc. math. phys. Sci.* **72** (1972), 161–165.

4.3. Gilbert, E. G., Controllability and observability in multivariable control systems. *SIAM J. Control* **1** (1963), 128–150.

4.4. Giorgi, C. G. and Isidori, A., A new algorithm for the irreducible realization of a rational matrix. *In* "Proc. 9th Annual Allerton Conference on Circuit and System Theory", University of Illinois, Urbana, pp. 884–893 (1971).

4.5. Gueguen, C. J., An algebraic algorithm for reducing a differential system to a state form. Paper 1.2.2. *In* "Proc. 2nd IFAC Symposium on Multivariable Technical Control Systems (H. Schwarz, ed.), Vol. 1. North-Holland, Amsterdam (1971).

4.6. Heymann, M. and Thorpe, J. A., Transfer equivalence of linear dynamical systems. *SIAM J. Control* **8** (1970), 9–40.

4.7. Munro, N. and McLeod, R. S., Minimal realization of transfer function matrices using the system matrix. *Proc. I.E.E.* **118** (1971), 1298–1301.

4.8. Rosenbrock, H. H., Efficient computation of least order of a given transfer function matrix. *Electronics. Lett.* **3** (1967), 413–414.

4.9. Rosenbrock, H. H., "State-space and Multivariable Theory", Nelson, London (1970).

4.10. Roveda, C and Schmid, R., An upper bound on the dimension of minimal realizations of linear time-invariant dynamical systems. *I.E.E.E. Trans. autom. Control* **AC-15** (1970), 639–644.

4.11. Rowe, A., The generalised resultant matrix. *J. Inst. Maths. Applics.* **9** (1972), 390–396.

4.12. Silverman, L. M., Realization of linear dynamical systems. *I.E.E.E. Trans. autom. Control* **AC-16** (1971), 554–567.

4.13. Wolovich, W. A., The determination of state-space representations for linear multivariable systems. Paper 1.2.3. *In* "Proc. 2nd IFAC Symposium on Multivariable Technical Control Systems (H. Schwarz, ed.), Vol. 1. North-Holland, Amsterdam (1971).

4.14. Wolovich, W. A., A frequency domain approach to state feedback and estimation. Paper presented at the I.E.E.E. Conference on Decision and Control, Miami, Florida (1971).

Section 5

5.1. Anderson, B. D. .O. and Moore, J. B., "Linear Optimal Control", Prentice-Hall, Englewood Cliffs, N.J. (1971).

5.2. Barnett, S., Sensitivity of optimal linear systems to small variations in parameters. *Int. J. Control* **4** (1966), 41–48.

5.3. Hitz, K. L. and Anderson, B. D. O., Iterative method of computing the limiting solution of the matrix Riccati differential equation. *Proc. I.E.E.* **119** (1972), 1402–1406.

5.4. Kalman, R. E., Contributions to the theory of optimal control. *Bol. Soc. mat. Mex.* **5** (1960), 102–119.

5.5. Kleinman, D. L., On an iterative technique for Riccati equation computations. *I.E.E.E. Trans. autom. Control* **AC-13** (1968), 114–115.

5.6. Kreindler, E. and Hedrick, J. K., On the equivalence of quadratic loss functions. *Int. J. Control* **11** (1970), 213–222.

5.7. Kucera, V., On nonnegative definite solutions to matrix quadratic equations. Paper 29.4. *Proc. 5th IFAC Congress,* Paris (June, 1972). *Also in: Automatica* **8** (1972), 413–423.
5.8. Kucera, V., A contribution to matrix quadratic equations. *I.E.E.E. Trans. autom. Control* **AC-17** (1972), 344–347.
5.9. MacFarlane, A. G. J., Return-difference and return-ratio matrices and their use in analysis and design of multivariable feedback control systems. *Proc. I.E.E.* **117** (1970), 2037–2049.
5.10. Martensson, K., On the matrix Riccati equation. *Information Sciences* **3** (1971), 17–49.
5.11. Newmann, M. M., On attempts to reduce the sensitivity of the optimal linear regulator to a parameter change. *Int. J. Control* **11** (1970), 1079–1084.
5.12. Pagurek, B., Sensitivity of the performance of optimal control systems to plant parameter variations. *Int. J. Control* **1** (1965), 33–45.
5.13. Potter, J. E., Matrix quadratic solutions. *SIAM J. Appl. Math.* **14** (1966), 496–501.
5.14. Tuel, W. G., Lee, I. and DeRusso, P. M., Synthesis of optimal control systems with sensitivity constraints. Paper 24. *In* "Proc. 3rd IFAC Congress", Vol. 1, Book 2 (1966).
5.15. Vaughan, D. R., A negative exponential solution for the matrix Riccati equation. *I.E.E.E. Trans. autom. Control* **AC-14** (1969), 72–75.
5.16. Vit, K., Iterative solution of the Riccati equation. *I.E.E.E. Trans. autom. Control* **AC-17** (1972), 258–259.
5.17. Walter, O. H. D., Formulas for linear optimal control over a finite or infinite interval. *Electronics Lett.* **6** (1970), 632–633.
5.18. Walter, O. H. D., Eigenvector scaling in a solution of the matrix Riccati equation. *I.E.E.E. Trans. autom. Control* **AC-15** (1970), 486–487.
5.19. Walter, O. H. D., Real-matrix solutions for the linear optimal regulator. *Proc. I.E.E.* **119** (1972), 621–624.
5.20. Willems, J. C., Least squares stationary optimal control and the algebraic Riccati equation. *I.E.E.E. Trans. autom. Control* **AC-16** (1971), 621–634.
5.21. Wonham, W. M., On pole assignment in multi-input controllable linear systems. *I.E.E.E. Trans. autom. Control* **AC-12** (1967), 660–665.
5.22. Wonham, W. M., On a matrix Riccati equation of stochastic control. *SIAM J. Control* **6** (1968), 681–697.

Section 6

6.1. Anderson, B. D. O., A system theory criterion for positive real matrices. *SIAM J. Control* **5** (1967), 171–182.
6.2. Anderson, B. D. O., An algebraic solution to the spectral factorization problem. *I.E.E.E. Trans. autom. Control* **AC-12** (1967), 410–414.
6.3. Csaki, F. and Fischer, P., On the spectrum factorization. *Acta. tech. Acad. Sci. Hung.* **58** (1967), 145–168.
6.4. Davis, M. C., Factoring the spectral matrix. *I.E.E.E. Trans. autom. Control* **AC-8** (1963), 296–305.
6.5. Kavanagh, R. J., Spectral matrix factorization. *Electronics Lett.* **4** (1968), 527–528.
6.6. MacFarlane, A. G. J., Return-difference matrix properties for optimal stationary Kalman–Bucy filter. *Proc. I.E.E.* **118** (1971), 373–376.

6.7. Newcomb, R. W., "Linear Multiport Synthesis", McGraw-Hill, New York (1966).

6.8. Riddle, A. C. and Anderson, B. D. O., Spectral factorization–computational aspects. *I.E.E.E. Trans. autom. Control* **AC-11** (1966), 764–765.

6.9. Willems, J. C., Stationary covariance generation via the algebraic Riccati equation. *In* "Proc. 4th UKAC Control Convention" Manchester, 1971; IEE Conference Publication 78.

6.10. Youla, D. C., On the factorization of rational matrices. *I.R.E. Trans. Information Theory* **IT-7** (1961), 172–189.

Section 7

7.1. Anderson, B. D. O. and Luenberger, D. G., Design of multivariable feedback systems. *Proc. I.E.E.* **114** (1967), 395–399.

7.2. Arbib, M. A. and Zeiger, H. P., On the relevance of abstract algebra to control theory. *Automatica* **5** (1969), 589–606.

7.3. Boullion, T. L. and Odell, P. L., "Generalized Inverse Matrices", Wiley-Interscience, New York (1971).

7.4. Brockett, R. W., System theory on group manifolds and coset spaces. *SIAM J. Control* **10** (1972), 265–284.

7.5. Davison, E. J., On pole assignment in linear systems with incomplete state feedback. *I.E.E.E. Trans. autom. Control* **AC-15** (1970), 348–351.

7.6. Davison, E. J. and Chatterjee, R., A note on pole assignment in linear systems with incomplete state feedback. *I.E.E.E. Trans. autom. Control* **AC-16** (1971), 98–99.

7.7. Fallside, F. and Seraji, H., Direct design procedure for multivariable feedback systems. *Proc. I.E.E.* **118** (1971), 797–801.

7.8. Fallside, F. and Seraji, H., Pole-shifting procedure for multivariable systems using output feedback. *Proc. I.E.E.* **118** (1971), 1648–1654.

7.9. Heymann, M., A unique canonical form for multivariable linear systems. *Int. J. Control* **12** (1970), 913–927.

7.10. Howerton, R. D. and Hammond, J. L., A new computational solution of the linear optimal regulator problem. *I.E.E.E. Trans. autom. Control* **AC-16** (1971), 645–651.

7.11. Johnson, C. D., A unified canonical form for controllable and uncontrollable linear dynamical systems. *Int. J. Control* **13** (1971), 497–517.

7.12. Johnson, C. D., Stabilization of linear dynamical systems with output feedback. Paper 29.3, *5th IFAC Congress*, Paris, (June, 1972).

7.13. Luenberger, D. G., Canonical forms for linear multivariable systems. *I.E.E.E. Trans. autom. Control* **AC-12** (1967), 290–293.

7.14. Luenberger, D. G., An introduction to observers. *I.E.E.E. Trans. autom. Control* **AC-16** (1971), 596–602.

7.15. Makhlouf, M. A., An algebra for the analysis of linear stationary and non-stationary multivariable systems using the time-domain infinite matrices. *Int. J. Control* **15** (1972), 29–64.

7.16. Morse, A. S. and Wonham, W. M., Decoupling and pole assignment by dynamic compensation. *SIAM J. Control* **8** (1970), 317–337.

7.17. Porter, B. and Crossley T. R., "Modal Control", Taylor and Francis, London (1972).

7.18. Ramar, K. and Ramaswami, B., Transformation of time-variable multi-input systems to a canonical form. *I.E.E.E. Trans. autom. Control* **AC-16** (1971), 371–374.

7.19. Rao, C. R. and Mitra, S. K., "Generalized Inverse of Matrices and Its Applications", John Wiley, New York (1971).

7.20. Retallack, D. G. and MacFarlane, A. G. J., Pole-shifting techniques for multivariable feedback systems. *Proc. I.E.E.* **117** (1970), 1037–1038.

7.21. Simon, J. D. and Mitter, S. K., A theory of modal control. *Information and Control* **13** (1968), 316–353.

7.22. Wang, S. H. and Davison, E. J., Canonical forms of linear multivariable systems. *Control Systems Rpt. No.* 7203, University of Toronto, Canada (1972).

7.23. Willems, J. C. and Mitter, S. K., Controllability, observability, pole allocation and state reconstruction. *I.E.E.E. Trans. autom. Control* **AC-16** (1971), 582–595.

7.24. Wonham, W. M., Algebraic methods in linear multivariable control. Paper 6-A3, *Preprints J.A.C.C*, St. Louis, Missouri, 1971.

7.25. Wonham, W. M., Review of [1.7]. *I.E.E.E. Trans. autom. Control* **AC-17** (1972), 182–183.

7.26. Wonham, W. M. and Morse, A. S., Decoupling and pole assignment in linear multivariable systems: A geometric approach. *SIAM J. Control* **8** (1970), 1–18.

7.27. Wonham, W. M. and Morse, A. S., Feedback invariants of linear multivariable systems. *Automatica* **8** (1972), 93–100.

7.28. Wolovich, W. A. and Falb, P. L., On the structure of multivariable systems. *SIAM J. Control* **7** (1969), 437–450.

Discussion

MEES (*University of Cambridge*). What is the theory that you mentioned about a transfer function matrix which is an irrational transfer function rather than a rational one? You didn't start off with state-space representation.

BARNETT. How do you mean, irrational?

MEES. If you have a time-delay or something like that.

BARNETT. I don't think there's been much work done on that yet. It didn't really come within my field. I am not aware of anything that has been published on this.*

PARKS (*University of Warwick*). One of the things that is apparent in your lecture is the amazing amount of theory that is being used ranging from work done in the last century to very recent results. This raises great problems for educators who have to teach courses at the undergraduate and postgraduate stage. What, in your opinion on your own topic of matrices, ought now to be taught? Because many of these topics go beyond what is taught in undergraduate courses, what do you think the minimum tool kit of matrices ought to be?

* *Editors Note*. See Willems, J. L., Stability Theory of Dynamical Systems. Chap. 3. Section 7.2. Nelson, 1970.

BARNETT. I certainly believe that not enough is taught to undergraduates about matrix theory. A lot of them don't even do any matrix analysis; they stick to matrix algebra. They don't even get on to the ideas of functions of matrices and so on.

PARKS. It is a real problem in Control—it doesn't seem to arise in, say, Fluid Mechanics. The tool kit there is the same as it has been for many years and new research and problems are just variations of the old theory and they aren't using new techniques, so the problem doesn't seem to arise there in the same way.

BARNETT. I think really that is the point. There is so much material, one could almost spend a full year teaching undergraduates the matrix theory they would need! There are some American books which cover this fairly well at undergraduate level. I can think of one by Cullen which covers quite a lot on matrix analysis which you don't find in the English books. The latter seem to stop at the ideas of quadratic forms and characteristic vectors!

PARKS. It's fairly clear we should do something on polynomial matrices.

BARNETT. Oh yes, I think so, and this can be connected with ideas of functions of a matrix. But perhaps we shouldn't get too carried away with our own specialities!

FULLER (*University of Cambridge*). Would you say that the algebraic stability criteria is just about buttoned up now?

BARNETT. No, I wouldn't like to say that.

WALSH (*University of York*). I read Routh's original paper and it struck me that he almost had the Nyquist criterion and that was years before Nyquist. Do you agree with this view and why was it such a long time between Routh and the discovery of the Nyquist test?

BARNETT. I have not read Routh's original paper, but in this field there is quite a history of workers being unaware of earlier published results.

PARKS. If I could come in on this story. The Routh–Hurwitz criterion comes from the fact that you are doing effectively a Nyquist diagram in going round the origin N times so that you have got to cut the real and imaginary axis alternately. So if you put real and imaginary parts to zero, the roots of those two equations have got to alternate. (One is going round the origin and not round the point —1.) Nyquist's particular contribution was the feedback case where you go round, or not go round, the —1 point. Of course, he was an electrical engineer and was particularly interested in this because it made a very nice experimental method.

BARNETT. Yes, Routh was only interested, I think, in the left half-plane problem without control.

24. Multivariable Circle Theorems*

H. H. Rosenbrock

Control Systems Centre, University of Manchester
Institute of Science and Technology, Manchester, England

1. Introduction

Let a linear, time-invariant system have the equations

$$\begin{aligned} \dot{x} &= Ax + Bu \\ y &= Cx \end{aligned} \tag{1}$$

where x, u, y are real vectors, of dimension n, m, m respectively. We do not assume in general that this system has least order. The resulting $m \times m$ transfer function matrix is

$$G(s) = C(sI_n - A)^{-1}B \tag{2}$$

Now let feedback be applied to the system (1) according to the equation

$$u = -Fy \tag{3}$$

where

$$F = \text{diag}\,[f_i(t, y)] \tag{4}$$

and

$$0 < \alpha_i \leqslant f_i(t, y) \leqslant \beta_i, \qquad \alpha_i < \beta_i, \qquad i = 1, 2, \ldots, m \tag{5}$$

The resulting closed-loop system obeys the nonlinear time-dependent equation

$$\dot{x} = (A - BFC)\,x \tag{6}$$

We assume F to be such that this equation has a unique solution, passing through any $x(t_0) = c$, which can be extended to all times t. Our problem is to find conditions for the stability of (6).

When $m = 1$ and $F(t, y) = F(y)$, the above problem is a form of the extensively-studied Lur'e problem [1, 2], for which Popov's criterion [3, 4] and the circle theorem [4] are available: the latter permits F to depend

* This paper was presented by the editor of these proceedings.

explicitly on t. When $m > 1$, no simple result corresponding to the circle criterion has previously been known. In this paper such a result will be obtained by placing further restrictions on $G(s)$.

2. Stability

We start from the following theorem, which we regard as known. It does not appear in precisely this form in the literature, but can readily be constructed from known results. The outline of a proof is given in the Appendix.

THEOREM 1. *Let* (1) *have least order, and let*

$$[\beta^{-1} + G(s)]^{-1}[\alpha^{-1} + G(s)] \tag{7}$$

$$\left[\text{resp } [\alpha^{-1} + G(s)][\beta^{-1} + G(s)]^{-1}\right] \tag{8}$$

$$\left[\text{resp } [I_m + G(s)\alpha]^{-1}[I_m + G(s)\beta]\right] \tag{9}$$

$$\left[\text{resp } [I_m + \beta G(s)][I_m + \alpha G(s)]^{-1}\right] \tag{10}$$

exist and be positive real [5], *where* $\alpha = \text{diag}(\alpha_i)$ *and* $\beta = \text{diag}(\beta_i)$. *Then the system* (6) *is stable in the sense of Liapunov.*

The desired results are obtained from Theorem 1 by a sequence of further theorems. Of these 2 is the key result and is believed to be new, while 3 and 4 are slightly adapted from standard results [5].

THEOREM 2. *Let* $\hat{Z}$ *be a complex* $m \times m$ *matrix, and let real constants* $\theta_i > 0$ *exist such that*

$$|\hat{z}_{ii}|\theta_i^{-1} - \sum_{j=1,\ j\neq i}^{m} |\hat{z}_{ij}|\theta_j^{-1} > 1, \qquad i = 1, 2, \ldots, m \tag{11}$$

$$\left[\text{resp } \theta_i^{-1}|\hat{z}_{ii}| - \sum_{j=1, j\neq i}^{m} \theta_j^{-1}|\hat{z}_{ji}| > 1, i = 1, 2, \ldots, m\right] \tag{12}$$

Then $\hat{Z}$ *has an inverse* Z *satisfying*

$$\|Z\| = \|Z^*\| < \sqrt{\left(\sum_i \theta_i^{-2}\right)} \tag{13}$$

where the norm is subordinate to the Euclidean vector norm, and the star denotes the complex conjugate transpose.

Proof. Write $\theta = \text{diag}(\theta_i)$, which is non-singular. The inverse of $\hat{Z}\theta^{-1}$ [resp $\theta^{-1}\hat{Z}$] exists by Gershgorin's theorem, and so therefore does the inverse of $\hat{Z}$. Because $\|Z\|^2$ is the largest eigenvalue of Z^*Z, which has the same eigenvalues as ZZ^*, it follows that $\|Z\| = \|Z^*\|$. If a is a complex m-vector we also have

$$\|Z\| = \max_a \frac{\|Za\|}{\|a\|} \tag{14}$$

where $\|Za\|$, $\|a\|$ are the Euclidean norms. Write $Za = \theta^{-1}b$, so that

$$\|Z\| = \max_b \frac{\|\theta^{-1}b\|}{\|\hat{Z}\theta^{-1}b\|} \tag{15}$$

$$= 1\bigg/\left\{\min_b \frac{\|\hat{Z}\theta^{-1}b\|}{\|\theta^{-1}b\|}\right\} \tag{16}$$

If

$$\|b\|_c = \max_i |b_i|$$

is the cubic norm of b, we have

$$\min_b \frac{\|\hat{Z}\theta^{-1}b\|}{\|\theta^{-1}b\|} = \min_b \left\{\frac{\|\hat{Z}\theta^{-1}b\|}{\|b\|_c} \cdot \frac{\|b\|_c}{\|\theta^{-1}b\|}\right\} \tag{17}$$

$$\geqslant \left\{\min_b \frac{\|\hat{Z}\theta^{-1}b\|}{\|b\|_c}\right\} \left\{\min_b \frac{\|b\|_c}{\|\theta^{-1}b\|}\right\} \tag{18}$$

$$\geqslant \left\{\min_b \frac{\|\hat{Z}\theta^{-1}b\|_c}{\|b\|_c}\right\}\bigg/\left\{\max_b \frac{\|\theta^{-1}b\|}{\|b\|_c}\right\} \tag{19}$$

Here (18) follows from (17) by relaxing the constraint that b should be the same in both factors within braces in (17), and (19) uses the fact that for any complex m-vector v, the Euclidean norm $\|v\|$ and the cubic norm $\|v\|_c$ satisfy $\|v\| \geqslant \|v\|_c$.

Now we have

$$\|\hat{Z}\theta^{-1}b\|_c = \max_i \left|\sum_{j=1}^{m} \hat{z}_{ij}\theta_j^{-1}b_j\right| \tag{20}$$

$$= \max_i \left|\hat{z}_{ii}\theta_i^{-1}b_i + \sum_{j=1, j\neq i}^{m} \hat{z}_{ij}\theta_j^{-1}b_j\right| \tag{21}$$

$$\geqslant \max_i \left\{|\hat{z}_{ii}\theta_i^{-1}b_i| - \sum_{j=1, j\neq i}^{m} |\hat{z}_{ij}\theta_j^{-1}b_j|\right\} \tag{22}$$

$$> \max_i \left\{|b_i| + \sum_{j=1, j\neq i}^{m} |\hat{z}_{ij}\theta_j^{-1}b_i| - \sum_{j=1, j\neq i}^{m} |\hat{z}_{ij}\theta_j^{-1}b_j|\right\} \tag{23}$$

on using (11). Because $\|\mu b\| = \mu\|b\|$ for any norm and any real $\mu > 0$, we may put $\|b\|_c = 1$ in (19) without affecting the result. In (23), we then have some k for which $|b_k| = 1$, and $|b_j| \leqslant 1$ for $j \neq k$. Evaluating the quantity within braces for $i = k$, we obtain

$$1 + \sum_{j=1, j\neq k}^{m} |\hat{z}_{kj}\theta_j^{-1}| - \sum_{j=1, j\neq k}^{m} |\hat{z}_{kj}\theta_j^{-1}|\,|b_j| \geqslant 1 \tag{24}$$

whence the maximum over i is certainly not less than 1. That is, $\|\hat{Z}\theta^{-1}b\|_c > 1$ when $\|b\|_c = 1$.

Similarly,

$$\|\theta^{-1}b\|^2 = \sum_{i=1}^{m} |\theta_i^{-1}b_i|^2 \tag{25}$$

and if $\|b\|_c = 1$, the maximum of this expression is clearly achieved when $|b_i| = 1, i = 1, 2, \ldots, m$, and is

$$\sum_{i=1}^{m} \theta_i^{-2}.$$

On comparing (16) and (19), the results thus established give the first part of the theorem. For the second part of the theorem, we notice that if (12) is true for $\hat{Z}$, then (11) is true for $Q = \hat{Z}^*$, because

$$\theta_i^{-1}|\hat{z}_{ii}| - \sum_{j=1, j\neq i}^{m} \theta_j^{-1}|\hat{z}_{ji}| = \theta_i^{-1}|q_{ii}| - \sum_{j=1, j\neq i}^{m} \theta_j^{-1}|\bar{q}_{ij}|$$

$$= |q_{ii}|\theta_i^{-1} - \sum_{j=1, j\neq i}^{m} |q_{ij}|\theta_j^{-1} \tag{26}$$

THEOREM 3. *With the same norm as in Theorem* 2, *let* $\|Z\| < 1$. *Then*

$$Y = (I_m + Z)(I_m - Z)^{-1} = (I_m - Z)^{-1}(I_m + Z)$$

is finite, and $Y + Y^*$ *is positive definite.*

Proof. The two expressions for Y are equal whenever they exist because $(I_m + Z)$ and $(I_m - Z)^{-1}$ commute. Because $\|Z\| < 1$, it follows that Z is finite. If $I_m - Z$ were singular, then $\lambda = 1$ would be an eigenvalue of Z. There would then be an eigenvector v such that $Zv = v$, giving $\|Zv\| = \|v\|$, and so contradicting the assumption that $\|Z\| < 1$. Hence $I_m - Z$ is non-singular and Y is finite. Now we have to show that if a is any complex m-vector,

$$a^*(Y + Y^*)\,a > 0, \qquad a \neq 0. \tag{27}$$

From the definition of Y we obtain

$$a^*(Y + Y^*)\,a$$

$$= a^*[(I_m + Z)(I_m - Z)^{-1} + (I_m - Z^*)^{-1}(I_m + Z^*)]\,a \quad (28)$$

$$= b^*[(I_m - Z^*)(I_m + Z) + (I_m + Z^*)(I_m - Z)]\,b \quad (29)$$

$$= 2b^*[I_m - Z^*Z]\,b \quad (30)$$

$$= 2[\|b\|^2 - \|Zb\|^2] \quad (31)$$

$$\geqslant 2\|b\|^2\,[1 - \|Z\|^2] > 0 \qquad \text{if } b \neq 0 \quad (32)$$

where

$$b = (I_m - Z)^{-1}a \quad (33)$$

and is non-zero if a is non-zero. This completes the proof.

THEOREM 4. *Let the rational $m \times m$ matrix $Y(s)$ have no pole inside or on a closed contour C. Let $Y(s) + Y^*(s)$ be positive definite for all s on C. Then $Y(s) + Y^*(s)$ is positive definite for all s inside or on C.*

Proof. Let a be any non-zero complex m-vector. Then

$$\phi(s) = a^*Y(s)\,a = \sum_{i,j=1}^{m} \bar{a}_i y_{ij}(s)\,a_j \quad (34)$$

is a rational function of s which has no pole inside or on C. By the maximum-modulus theorem, $|e^{-\phi(s)}|$ therefore achieves its greatest value on C. But on C,

$$|e^{-\phi(s)}| = \exp\{-\tfrac{1}{2}a^*[Y(s) + Y^*(s)]\,a\} < 1 \quad (35)$$

whence it follows that for any s_0 inside C,

$$|e^{-\phi(s_0)}| < 1 \quad (36)$$

$$\operatorname{Re}\phi(s_0) > 0 \quad (37)$$

$$a^*[Y(s_0) + Y^*(s_0)]\,a > 0 \quad (38)$$

This is true for any non-zero a, whence $Y(s_0) + Y^*(s_0)$ is positive definite. Since s_0 is any point inside C, this completes the proof.

Now define matrices ζ, η by

$$\zeta = \operatorname{diag}(1/2\alpha_i + 1/2\beta_i), \qquad \eta = \operatorname{diag}(1/2\alpha_i - 1/2\beta_i) \quad (39)$$

where α_i, β_i are defined by (5). Also let D be a contour consisting of the imaginary axis from $s = -iR$ to $s = iR$, together with a semi-circle of radius R in the right half-plane. The contour D is indented into the left half-plane to avoid any imaginary poles of $g_{ii}(s)$, and R is chosen large enough to ensure that all poles of $g_{ii}(s)$ lying in the closed right half-plane are inside D, $i = 1, 2, \ldots, m$.

THEOREM 5. *Let there exist real constants* $\theta_i > 0$, *satisfying*

$$\sum_{i=1}^{m} \theta_i^{-2} \leqslant 1,$$

such that for each s *on* D' *having* $R' \geqslant R$

$$|\zeta_i + g_{ii}(s)| - \sum_{j=1, j\neq i}^{m} |g_{ij}(s)| > \theta_i \eta_i, \qquad i = 1, 2, \ldots, m \tag{40}$$

$$\left[\text{resp}\, |\zeta_i + g_{ii}(s)| - \sum_{j=1, j\neq i}^{m} |g_{ji}(s)| > \theta_i \eta_i, \qquad i = 1, 2, \ldots, m\right]. \tag{41}$$

Let (1) *have least order, and let the linear time-invariant closed-loop system defined by* (1) *and*

$$u = -\beta y \tag{42}$$

be asymptotically stable. Then the nonlinear time-dependent closed-loop system defined by (1), (3) *is stable in the sense of Liapunov.*

Proof. By the choice of D, $g_{ii}(s)$ is finite for all s on D' having $R' \geqslant R$, and by (40) [resp (41)] so also is $g_{ij}(s), i, j = 1, 2, \ldots, m$. Make the substitutions $u_1 = \theta^{-1} u, y_1 = \theta^{-1} y$ in (1) and (3), where $\theta = \operatorname{diag}(\theta_i)$. This replaces G by $Q = \theta^{-1} G \theta$ and F by $\theta F \theta^{-1} = F$. Now (40) gives

$$1 < |\eta_i^{-1} \theta_i^{-1} [\zeta_i + g_{ii}(s)]| - \sum_{j=1, j\neq i}^{m} |\eta_i^{-1} \theta_i^{-1} g_{ij}(s)|$$

$$= |\eta_i^{-1} [\zeta_i + q_{ii}(s)]\, \theta_i^{-1}| - \sum_{j=1, j\neq i}^{m} |\eta_i^{-1} q_{ij} \theta_j^{-1}| \tag{43}$$

and by Theorem 2

$$\|[\zeta + Q(s)]^{-1} \eta\| < 1. \tag{44}$$

Then by Theorem 3

$$Y(s) = \{I_m - [\zeta + Q(s)]^{-1} \eta\}^{-1} \{I_m + [\zeta + Q(s)]^{-1} \eta\} \tag{45}$$

$$= [\zeta + Q(s) - \eta]^{-1} [\zeta + Q(s) + \eta] \tag{46}$$

$$= [\beta^{-1} + Q(s)]^{-1} [\alpha^{-1} + Q(s)] \tag{47}$$

is finite and $Y + Y^*$ is positive definite. This is true for all s on D. Moreover, Y has no pole in the closed right half-plane, for if it did (47) shows that $\theta^{-1}[I_m + \beta G(s)]^{-1}\beta\alpha^{-1}\theta$ or $\theta^{-1}[I_m + \beta G(s)]^{-1}\beta G(s)\theta$ would have a pole there. In either case the system defined by (1) and (42) could not be asymptotically stable. Provided that the indentations of D into the left half-plane are sufficiently small, Y therefore has no pole inside or on D. By Theorem 4, $Y + Y^*$ is positive definite for all s inside or on D. As R can be increased indefinitely, $Y + Y^*$ is positive definite everywhere in the closed right half-plane.

We have now established that Y satisfies the following conditions:

(i) Y is a rational function of s with real coefficients

(ii) Y has no pole in the closed right half-plane

(iii) $Y + Y^*$ is positive definite everywhere in the closed right half-plane.

It therefore follows that Y is positive real, and so the condition relating to (7) in Theorem 1 is satisfied for the system giving $Q(s)$. But apart from a change in notation, this is the system (1), (3), and the first part of the theorem is proved.

For the second part, make the substitutions $u_1 = \theta u$, $y_1 = \theta y$ in (1) and (3), which replaces G by $Q = \theta G \theta^{-1}$ and F by $\theta^{-1}F\theta = F$. Then (41) gives

$$1 < |[\zeta_i + g_{ii}(s)]\,\theta_i^{-1}\eta_i^{-1}| - \sum_{j=1, j\neq i}^{m} |g_{ji}(s)\theta_i^{-1}\eta_i^{-1}|$$

$$= |\theta_i^{-1}[\zeta_i + q_{ii}(s)]\eta_i^{-1}| - \sum_{j=1, j\neq i}^{m} |\theta_j^{-1}\hat{q}_{ji}(s)\eta_i^{-1}| \tag{48}$$

and as above we obtain

$$\|\eta[\zeta + Q(s)]^{-1}\| < 1 \tag{49}$$

$$Y(s) = \{I_m + \eta[\zeta + Q(s)]^{-1}\}\{I_m - \eta[\zeta + Q(s)]^{-1}\}^{-1} \tag{50}$$

$$= [\zeta + Q(s) + \eta]\,[\zeta + Q(s) - \eta]^{-1} \tag{51}$$

$$= [\alpha^{-1} + Q(s)]\,[\beta^{-1} + Q(s)]^{-1} \tag{52}$$

which has the same properties as before. The result then follows on verifying the conditions of Theorem 1, using (8).

We now assume that the $m \times m$ matrix $G(s)$ has an inverse $G^{-1}(s) = \hat{G}(s)$. This condition ensures that the system is controllable(f), and will normally be satisfied. Use of the inverse transfer function matrix G is convenient because [6] when all loops except loop i, in a linear time-invariant closed-loop system, are tightly controlled, the transfer function seen in loop i approximates $\hat{g}_{ii}$. We also define matrices γ, δ by

$$\gamma = \operatorname{diag}\left(\frac{\beta_i}{2} + \frac{\alpha_i}{2}\right), \qquad \delta = \operatorname{diag}\left(\frac{\beta_i}{2} - \frac{\alpha_i}{2}\right). \tag{53}$$

The contour D consists as before of the imaginary axis from $s = -iR$ to $s = iR$, and of a semi-circle of radius R in the right half-plane. The contour is indented into the left half-plane to avoid any imaginary pole of $\hat{g}_{ii}(s)$, while R is chosen large enough to ensure that every pole of $\hat{g}_{ii}(s)$ lying in the closed right half-plane is inside D, $i = 1, 2, \ldots, m$.

THEOREM 6. *Let there exist real constants $\theta_i > 0$, satisfying*

$$\sum_{i=1}^{m} \theta_i^{-2} \leqslant 1,$$

such that for each s on D' having $R' \geqslant R$

$$|\gamma_i + \hat{g}_{ii}(s)| - \sum_{j=1, j\neq i}^{m} |\hat{g}_{ij}(s)| > \theta_i \delta_i, \qquad i = 1, 2, \ldots, m \tag{54}$$

$$\left[\text{resp } |\gamma_i + \hat{g}_{ii}(s)| - \sum_{j=1, j\neq i}^{m} |\hat{g}_{ji}(s)| > \theta_i \delta_i, \qquad i = 1, 2, \ldots, m\right] \tag{55}$$

Let (1) *have least order, and let the linear time-invariant system defined by* (1) *and*

$$u = -\alpha y \tag{56}$$

by asymptotically stable. Then the nonlinear time-dependent closed-loop system defined by (1), (3) *is stable in the sense of Liapunov.*

Proof. By the choice of D, $\hat{g}_{ii}(s)$ is finite for all s on D' having $R' \geqslant R$, and by (54) [resp (55)] so also is $\hat{g}_{ij}(s)$, $i, j = 1, 2, \ldots, m$. Make the substitutions $u_1 = \theta^{-1}u$, $y_1 = \theta^{-1}y$ in (1) and (3). Then $\hat{G}$ is replaced by $\hat{Q} = \theta^{-1}\hat{G}\theta$ and F by $\theta F \theta^{-1} = F$. Then rearrange (54) to give

$$\begin{aligned} 1 &> |\delta_i^{-1}\theta_i^{-1}[\gamma_i + \hat{g}_{ii}(s)]| - \sum_{j=1, j\neq i}^{m} |\delta_i^{-1}\theta_i^{-1}\hat{g}_{ij}(s)| \\ &= |\delta_i^{-1}[\gamma_i + \hat{q}_{ii}(s)]\theta_i^{-1}| - \sum_{j=1, j\neq i}^{m} |\delta_i^{-1}\hat{q}_{ij}(s)\theta_j^{-1}| \end{aligned} \tag{57}$$

and from Theorem 2 obtain

$$\|[\gamma + Q(s)]^{-1}\delta\| < 1. \tag{58}$$

Then

$$Y(s) = \{I_m - [\gamma + \hat{Q}(s)]^{-1}\delta\}^{-1}\{I_m + [\gamma + \hat{Q}(s)]^{-1}\delta\} \tag{59}$$

$$= [\gamma + \hat{Q}(s) - \delta]^{-1}[\gamma + \hat{Q}(s) + \delta] \tag{60}$$

$$= [\hat{Q}(s) + \alpha]^{-1}[\hat{Q}(s) + \beta] \tag{61}$$

$$= [I_m + Q(s)\alpha]^{-1}[I_m + Q(s)\beta] \tag{62}$$

which agrees with (9) and by Theorem 3 is finite, while $Y + Y^*$ is positive definite. Moreover, Y has no pole in the closed right half-plane, for if it did, (62) shows that

$$\theta^{-1}[I_m + G(s)\alpha]^{-1}\theta \quad \text{or} \quad \theta^{-1}[I_m + G(s)\alpha]^{-1}G(s)\alpha(\alpha^{-1}\beta\theta)$$

would have a pole there. In either case the system defined by (1) and (56) could not be asymptotically stable. Provided that the indentations of D into the left half-plane are sufficiently small, Y therefore has no pole inside or on D. By Theorem 4, $Y + Y^*$ is positive definite for all s inside or on D, and so for all s in the closed right half-plane. As in Theorem 5, this establishes that Y is positive real. All the conditions of Theorem 1 are satisfied, and the first part of the theorem is proved.

For the second part, substitute $u_1 = \theta u, y_1 = \theta y$ in (1) and (3), so replacing $\hat{G}$ by $\hat{Q} = \theta\hat{G}\theta^{-1}$ and F by $\theta^{-1}F\theta = F$. From (55) obtain

$$1 > |\gamma_i + \hat{g}_{ii}]\theta_i^{-1}\delta^{-1}| - \sum_{j=1, j\neq i}^{m} |\hat{g}_{ji}\theta_i^{-1}\delta_i^{-1}|$$

$$= |\theta_i^{-1}[\gamma_i + \hat{q}_{ii}]\delta_i^{-1}| - \sum_{j=1, j\neq i}^{m} |\theta_j^{-1}\hat{q}_{ji}\delta_i^{-1}| \tag{63}$$

and from Theorems 2 and 3

$$\|\delta[\gamma + \hat{Q}(s)]^{-1}\| < 1 \tag{64}$$

$$Y(s) = \{I_m + \delta[\gamma + \hat{Q}(s)]^{-1}\}\{I_m - \delta[\gamma + \hat{Q}(s)]^{-1}\}^{-1} \tag{65}$$

$$= [\gamma + \hat{Q}(s) + \delta][\gamma + \hat{Q}(s) - \delta]^{-1} \tag{66}$$

$$= [\hat{Q}(s) + \beta][\hat{Q}(s) + \alpha]^{-1} \tag{67}$$

$$= [I_m + \beta Q(s)][I_m + \alpha Q(s)]^{-1} \tag{68}$$

which is finite and positive definite everywhere on D. The result then follows as before, using (10) in Theorem 1.

3. Graphical Interpretation

Theorems 5 and 6 have a simple graphical interpretation, which we now state. First we give some preliminary definitions. Let the system (1) have p_0 poles in the closed right half-plane (that is, let A have p_0 eigenvalues there). Let the contour D be indented into the left half-plane to avoid any imaginary poles of $g_{ii}(s)$, and let R be chosen so that all poles of $g_{ii}(s)$ lying in the closed right half-plane are inside D, $i = 1, 2, \ldots, m$. For each s on D write

$$d_i(s) = \sum_{j=1, j\neq i}^{m} |g_{ij}(s)|, \qquad i = 1, 2, \ldots, m \tag{69}$$

$$d_i'(s) = \sum_{j=1, j\neq i}^{m} |g_{ji}(s)|, \qquad i = 1, 2, \ldots, m \tag{70}$$

Call a disc with centre $g_{ii}(s)$ and radius $d_i(s)$ [resp $d'_i(s)$] an ith Gershgorin row [resp column] disc. Call the band swept out by the ith Gershgorin row [resp column] disc, as s goes round D, the ith Gershgorin row [resp column] band. Let a disc be constructed with centre at $[-(1/2\alpha_i + 1/2\beta_i), 0]$ and radius $\theta_i(1/2\alpha_i - 1/2\beta_i)$; call this the ith critical disc (see Figure 1). The θ_i are any real positive numbers satisfying

$$\sum_{i=1}^{m} \theta_i^{-2} \leqslant 1$$

so that there is freedom to reduce one critical disc at the expense of another.

THEOREM 5a. *Let* (1) *have least order, and let the ith Gershgorin row [resp column] band have no point in common with the ith critical disc. Let this band go* N_{ci} *times clockwise around the ith critical disc,* $i = 1, 2, \ldots, m$, *as s goes once clockwise round D. Let these conditions remain satisfied for all* D' *having* $R' \geqslant R$. *If*

$$\sum_{i=1}^{m} N_{ci} = -p_0 \tag{71}$$

the closed-loop system defined by (1), (3) *is stable in the sense of Liapunov.*

Proof. Consider the system defined by (1) and (42). Because $(-1/\beta_i, 0)$ is in the ith critical disc, the conditions of the theorem show that for all s on D

$$|1/\beta_i + g_{ii}(s)| - \sum_{j=1, j\neq i}^{m} |g_{ij}(s)| > 0, \qquad i = 1, 2, \ldots, m \tag{72}$$

$$\left[\text{resp } |1/\beta_i + g_{ii}(s)| - \sum_{j=1, j\neq i}^{m} |g_{ji}(s)| > 0, \qquad i = 1, 2, \ldots, m\right]. \tag{73}$$

By (72) [resp (73)] and the definition of D, no $g_{ij}(s)$ has a pole on D, and so $|\beta^{-1} + G(s)|$ has no pole there. Also by (72) [resp (73)] no $1/\beta_i + g_{ii}(s)$ has a zero on D, and by Gershgorin's theorem $|\beta^{-1} + G(s)|$ has no zero there. As R may be increased indefinitely, it follows that every finite pole and zero of $|\beta^{-1} + G(s)|$ lying in the closed right half-plane is inside D. Let $1/\beta_i + g_{ii}(s)$ map D into Γ_i, which encircles the origin n_i times clockwise as s goes once clockwise around D. Let $|\beta^{-1} + G(s)|$ map D into Γ, which goes n times clockwise round the origin as s goes once clockwise round D, Then by a known theorem [6] we find

$$n = \sum_{i=1}^{m} n_i. \tag{74}$$

But the number n_i of encirclements of the origin by Γ_i is the number of encirclements of $(-1/\beta_i, 0)$ by the map of D under $g_{ii}(s)$, and this is N_{ci}. It is known that the system defined by (1), (42) is asymptotically stable if $n = -p_0$, and by what has been proved this is ensured by (71).

Moreover, the centre of the ith critical disc is $\zeta_i = 1/2\alpha_i + 1/2\beta_i$ and its radius is $\theta_i\eta_i = \theta_i(1/2\alpha_i - 1/2\beta_i)$. The conditions of the theorem ensure that (40) [resp (41)] is true, and remains true as R is increased. All the conditions of Theorem 5 are satisfied, and the proof is complete.

For the corresponding generalisation of Theorem 6, we suppose D to be suitably indented, and R suitably chosen, so that every pole of $\hat{g}_{ii}(s)$ lying in the closed right half-plane is inside D, $i = 1, 2, \ldots, m$. The radii of the Gershgorin row and column discs are now respectively

$$\hat{d}_i(s) = \sum_{j=1, j\neq i}^{m} |\hat{g}_{ij}(s)| \tag{75}$$

$$\hat{d}_i'(s) = \sum_{j=1, j\neq i}^{m} |\hat{g}_{ji}(s)| \tag{76}$$

The Gershgorin row and column bands are defined in the corresponding way. The ith critical disc now has its centre at $[-\frac{1}{2}(\beta_i + \alpha_i), 0]$ and has radius $\frac{1}{2}\theta_i(\beta_i - \alpha_i)$. The θ_i again satisfy

$$\sum_{i=1}^{m} \theta_i^{-2} \leqslant 1.$$

THEOREM 6a. *Let* (1) *have least order, and let the ith Gershgorin row* [*resp column*] *band have no point in common with the origin or with the ith critical disc,* $i = 1, 2, \ldots, m$. *Let this band go* $\hat{N}_{0i}$ *times clockwise round the origin, and* $\hat{N}_{ci}$ *times clockwise round the ith critical disc,* $i = 1, 2, \ldots, m$, *as s goes once clockwise round D. Let these conditions remain satisfied for all D′ having R′ ⩾ R. Then if*

$$\sum_{i=1}^{m}\left(\hat{N}_{0i} - \hat{N}_{ci}\right) = p_0 \tag{77}$$

the system defined by (1), (3) *is stable in the sense of Liapunov.*

Proof. Consider the system (1), (56). By the conditions of the theorem, for all s on D,

$$|\hat{g}_{ii}(s)| - \sum_{j=1, j\neq i}^{m} |\hat{g}_{ij}(s)| > 0, \qquad i = 1, 2, \ldots, m \tag{78}$$

$$\left[\text{resp } |\hat{g}_{ii}(s)| - \sum_{j=1, j\neq i}^{m} |\hat{g}_{ji}(s)| > 0, \quad i = 1, 2, \ldots, m\right]. \tag{79}$$

Hence, by the definition of D, no $\hat{g}_{ii}(s)$ nor $|\hat{G}(s)|$ has a pole on D. Also no $\hat{g}_{ii}(s)$ nor (by Gershgorin's Theorem) $|\hat{G}(s)|$ has a zero on D. As R can be increased indefinitely, every finite pole and zero of $|\hat{G}(s)|$ lying in the closed right half-plane is inside D. Let the map of D under $\hat{g}_{ii}(s)$ go $\hat{n}_{0i}$ times clockwise round the origin, and the map under $|\hat{G}(s)|$ go $\hat{n}_0$ times clockwise round the origin, as s goes once clockwise round D. Then [6]

$$\hat{n}_0 = \sum_{i=1}^{m} \hat{n}_{0i}. \tag{80}$$

In a similar way every finite pole and zero of $|\alpha + G(s)|$ lying in the closed right half-plane is inside D. Let the map of D under $\alpha_i + \hat{g}_{ii}(s)$ go $\hat{n}_{ci}$ times clockwise round the origin, and the map under $|\alpha + \hat{G}(s)|$ go $\hat{n}_c$ times clockwise round the origin, as s goes once clockwise round D. Then [6]

$$\hat{n}_c = \sum_{i=1}^{m} \hat{n}_{ci}. \tag{81}$$

Also $\hat{n}_{0i} = \hat{N}_{0i}, \hat{n}_{ci} = \hat{N}_{ci}$, and the system (1), (56) is asymptotically stable because

$$\hat{n}_0 - \hat{n}_c = \sum_{i=1}^{m} \hat{N}_{0i} - \sum_{i=1}^{m} \hat{N}_{ci} = p_0. \tag{82}$$

Moreover, the centre of the ith critical disc is at $\gamma_i = \frac{1}{2}(\beta_i + \alpha_i)$ and its radius is $\theta_i\delta_i = \frac{1}{2}\theta_i(\beta_i - \alpha_i)$. The conditions of the theorem verify (54) [resp (55)], and this is true for all D' having $R' \geqslant R$. All the conditions of Theorem 6 are satisfied and the proof is complete.

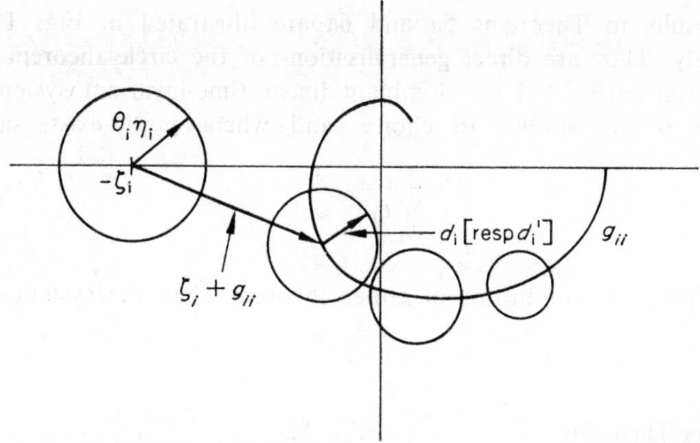

FIG. 1. The Gershgorin band is swept out by circles with radius $d_i(s)$ [resp $d_i'(s)$] as s goes round D. As usual, only the part corresponding to the positive imaginary segment of D is shown. The critical disc has centre at $-\zeta_i$ and radius $\theta_i\eta_i$. If the Gershgorin band does not have any point in common with the critical disc then (40) [resp (41)] is satisfied. Stability is then ensured by (71). A stronger result is obtained from Theorem 5b when the distance between the Gershgorin band and the critical disc is always greater than ε.

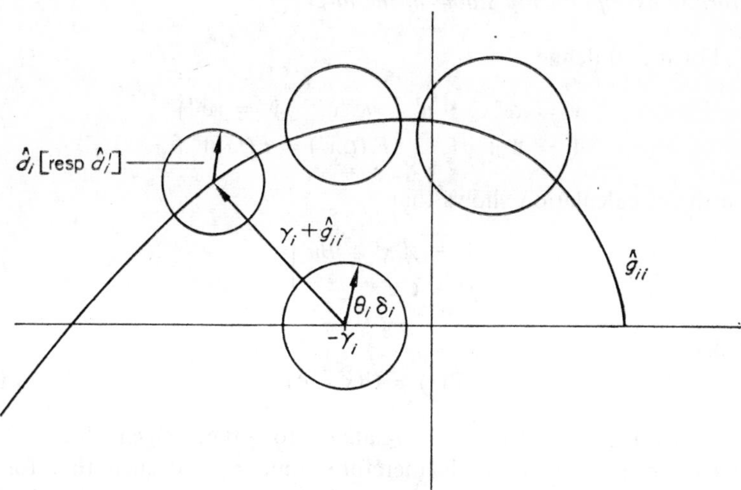

FIG. 2. The interpretation of Theorems 6a, 6b is similar to that of Theorems 5a, 5b. If the Gershgorin bands avoid the origin and the critical discs, and if (77) is satisfied, then Theorem 5a gives stability.

The results in Theorems 5a and 6a are illustrated in Figs 1 and 2 respectively. They are direct generalisations of the circle theorem and of results given earlier [7] for dominant linear time-invariant systems. The values of θ_i are subject to choice, and whenever θ_i exist, satisfying

$$\sum_{i=1}^{m} \theta_i^{-2} \leqslant 1$$

and verifying the conditions of either theorem, then the system (1), (3) is stable.

4. Stronger Theorems

The above theorems can now be strengthened to give asymptotic stability, and also to cover systems not having least order. An alternative procedure would be to seek a stronger form of Theorem 1, and obtain the following theorems directly in place of 5a and 6a.

THEOREM 5b. *Suppose that* (1) *does not necessarily have least order and let the least distance, between the Gershgorin row [resp column] discs centred on* g_{ii} *and the ith critical disc, be greater than* $\varepsilon > 0$, $i = 1, 2, \ldots, m$. *Let this be true for all* D' *having* $R' \geqslant R$. *If* (71) *is true, then the system* (1), (3) *is uniformly asymptotically stable in the large.*

Proof. For $\sigma > 0$ define

$$\left.\begin{aligned} x' &= xe^{\sigma t}, \quad & u' &= ue^{\sigma t}, \quad & y' &= ye^{\sigma t} \\ A' &= A + \sigma I, \quad & & & F'(t, y') &= F(t, y) \end{aligned}\right\} \tag{83}$$

Then a direct calculation shows that

$$\left.\begin{aligned} \dot{x}' &= A'x' + Bu' \\ y' &= Cx' \end{aligned}\right\} \tag{84}$$

which gives

$$G'(s) = G(s - \sigma). \tag{85}$$

The eigenvalues $\lambda_i(A')$ are related to the eigenvalues $\lambda_i(A)$ by $\lambda_i(A') = \sigma + \lambda_i(A)$. There is therefore some $\sigma_1 > 0$ such that for all $\sigma < \sigma_1$, A' has the same number p_0 of eigenvalues in the closed right half-plane as A. Further, as in the proof of Theorem 5a, no $g_{ij}(s)$ has a pole on any D' having $R' \geqslant R$, and so every $g_{ij}(s)$ is a continuous function of s in some neighbourhood of every point on such D'. Since $G(s) \to 0$

as $s \to \infty$, it follows that there is some $\sigma_2 > 0$ such that for all $\sigma < \sigma_2$, and all s on D' having $R' \geqslant R$, and for $i, j = 1, 2, \ldots, m$,

$$|g_{ij}(s) - g_{ij}(s - \sigma)| < \varepsilon/m \tag{86}$$

Choose $0 < \sigma < \min(\sigma_1, \sigma_2)$. Then A' has p_0 eigenvalues in the closed right half-plane, and for all s on D' having $R' \geqslant R$,

$$|\zeta_i + g_{ii}'(s)| - \sum_{j=1, j \neq i}^{m} |g_{ij}'(s)| > \theta_i \eta_i, \qquad i = 1, 2, \ldots, m \tag{87}$$

$$\left[\text{resp } |\zeta_i + g_{ii}'(s)| - \sum_{j=1, j \neq i}^{m} |g_{ji}'(s)| > \theta_i \eta_i, \qquad i = 1, 2, \ldots, m \right]. \tag{88}$$

Now carry out a canonical decomposition [8] of (84) to give

$$\left.\begin{aligned} \dot{x}_1 &= A_{11}x_1 \\ \dot{x}_2 &= A_{21}x_1 + A_{22}x_2 \\ \dot{x}_3 &= A_{31}x_1 \qquad\qquad\quad + A_{33}x_3 \qquad\qquad\quad + B_3u' \\ \dot{x}_4 &= A_{41}x_1 + A_{42}x_2 + A_{43}x_3 + A_{44}x_4 + B_4u' \\ y' &= C_1x_1 \qquad\qquad\quad + C_3x_3 \end{aligned}\right\} \tag{89}$$

where $(x')^T = (x_1^T, x_2^T, x_3^T, x_4^T)$. As in the proof of Theorem 5a, we can show that the closed-loop system defined by (1) and (42) is asymptotically stable, the proof being valid for systems not having least order*. But the decoupling zeros z_i of the system (1), (42) are those of (1), which consequently lie in the open left half-plane. By the choice of σ_1, the decoupling zeros $z_i - \sigma$ of (89) also lie in the open left half-plane: these decoupling zeros are the eigenvalues of A_{11}, A_{22}, A_{44}, taken all together.

Choose some arbitrary initial condition $x_i(0) = c_i$, $i = 1, 2, 3, 4$ for (89). Then there is some $\mu_1(c_1, c_2)$ and some $\sigma_4 > 0$ such that $\|x_1(t)\| + \|x_2(t)\| < \mu_1 e^{-\sigma_4 t}$ for all $t \geqslant 0$. The system

$$\left.\begin{aligned} \dot{x}_3 &= A_{33}x_3 + B_3u' \\ y' &= C_3x_3 \end{aligned}\right\} \tag{90}$$

has least order and gives $G'(s)$ satisfying (87) [resp (88)]. From (83) it follows that

$$0 < \alpha_i \leqslant f_i'(t, y') \leqslant \beta_i \tag{91}$$

and so the system defined by (9)) and

$$u' = -F'y' \tag{92}$$

* This point is considered in more detail in "Computer-aided Control System Design" by H. H. Rosenbrock (to be published).

obeys the conditions of Theorem 5, and is stable. A Liapunov function for this system is given in (122) of the Appendix [resp a similar result obtained from (126)]. The derivative of this Liapunov function, when $\dot{x}_3$ is given by (89), is

$$\frac{d}{dt} x_3{}^T P x_3 \leqslant 2x_3{}^T P(A_{31} - B_3 F' C_1)\, x_1 \tag{93}$$

If H is the positive definite square-root of P, (93) gives

$$\frac{d}{dt} \|Hx_3\|^2 \leqslant \|Hx_3\|\, \|2H(A_{31} - BF'C_1)\, x_1\| \tag{94}$$

$$\leqslant \|Hx_3\|\, \mu_2 e^{-\sigma_4 t} \tag{95}$$

by (91) and what was proved about $\|x_1\|$. Hence

$$\|x_3(t)\| \leqslant \mu_3 \tag{96}$$

for all $t \geqslant 0$.

Because the eigenvalues of A_{44} lie in the open left half-plane, there is a positive definite P_1 such that

$$A_{44}{}^T P_1 + P_1 A_{44} \tag{97}$$

is negative definite. Then when $\dot{x}_4$ is given by (89) we have

$$\frac{d}{dt} x_4{}^T P_1 x_4 \leqslant 2x_4{}^T P_1 [(A_{41} - B_4 F' C_1)\, x_1 + A_{42} x_2 + (A_{43} - B_4 F' C_3)\, x_3]. \tag{98}$$

If H_1 is the positive definite square root of P_1, we obtain from what has been proved above

$$\frac{d}{dt} \|H_1 x_4\|^2 \leqslant \|H_1 x_4\|\, (\mu_4 e^{-\sigma_4 t} + \mu_5) \tag{99}$$

$$\|x_4\| \leqslant \mu_6 + \mu_7 t \tag{100}$$

for all $t \geqslant 0$. Consequently, by (83) and what has been proved about $\|x_1\|$, $\|x_2\|$, $\|x_3\|$ and $\|x_4\|$,

$$\|x\| \leqslant (\mu_8 + \mu_9 t)\, e^{-\sigma t} \tag{101}$$

As the origin of time can be shifted without affecting the proof, this establishes the theorem.

THEOREM 6b. *Suppose that* (1) *does not necessarily have least order. Let the Gershgorin row [resp column] band based on $\hat{g}_{ii}(s)$ exclude the origin. Let the least distance, between the Gershgorin disc centred on $\hat{g}_{ii}(s)$ and the critical disc centred on γ_i, be not less than $\varepsilon|\gamma_i + \hat{g}_{ii}(s)|$. Let this be true for $i = 1, 2, \ldots, m$ and for all s on D' having $R' \geqslant R$. If* (77) *is true, then the system* (1), (3) *is uniformly asymptotically stable in the large.*

Proof. From (83), (84) and (85) we obtain

$$\hat{G}(s) = \hat{G}(s - \sigma) \tag{102}$$

By the definition of D, no $\hat{g}_{ii}(s)$ has a pole on D. The conditions of the theorem give, for row dominance,

$$|\gamma_i + \hat{g}_{ii}(s)| - \sum_{j=1, j\neq i}^{m} |\hat{g}_{ij}(s)| > \theta_i\eta_i + \varepsilon|\gamma_i + \hat{g}_{ii}(s)| \tag{103}$$

$$1 - \sum_{j=1, j\neq i}^{m} \left|\frac{\hat{g}_{ij}(s)}{\gamma_i + \hat{g}_{ii}(s)}\right| > \frac{\theta_i\eta_i}{|\gamma_i + \hat{g}_{ii}(s)|} + \varepsilon. \tag{104}$$

We may therefore write

$$\frac{\hat{g}_{ij}(s)}{\gamma_i + \hat{g}_{ii}(s)} = \mu_{ij} + \phi_{ij}(s) \tag{105}$$

where μ_{ij} is a constant and $\phi_{ij}(s)$ is a proper rational function. Then from (103), no $\hat{g}_{ij}(s)$ has a pole on D, and $\gamma_i + \hat{g}_{ii}(s)$ has no zero on D. Hence $\phi_{ij}(s)$ is a continuous function of s in some neighbourhood of every point on D, and moreover on every D' having $R' \geqslant R$. There is therefore some $\sigma_2 > 0$ such that for all $\sigma < \sigma_2$,

$$|[\mu_{ij} + \phi_{ij}(s)] - [\mu_{ij} + \phi_{ij}(s - \sigma)]| < \varepsilon/m, \quad i, j = 1,2, \ldots, m. \tag{106}$$

In a similar way there is a $\sigma_3 > 0$ such that for all $\sigma < \sigma_3$,

$$\left|\frac{\theta_i\eta_i}{\gamma_i + \hat{g}_{ii}(s)} - \frac{\theta_i\eta_i}{\gamma_i + \hat{g}_{ii}(s - \sigma)}\right| < \frac{\varepsilon}{m}, \quad i, j = 1,2, \ldots, m. \tag{107}$$

If we choose $\sigma < \min(\sigma_1, \sigma_2, \sigma_3)$, it follows that

$$1 - \sum_{j=1, j\neq i}^{m} \left|\frac{\hat{g}_{ij}(s - \sigma)}{\gamma_i + \hat{g}_{ii}(s - \sigma)}\right| > \frac{\theta_i\eta_i}{|\gamma_i + \hat{g}_{ii}(s - \sigma)|} \tag{108}$$

$$|\gamma_i + \hat{g}_{ii}(s - \sigma)| - \sum_{j=1, j\neq i}^{m} |\hat{g}_{ij}(s - \sigma)| > \theta_i\eta_i \tag{109}$$

and a similar result can be obtained for column dominance.

The remaining part of the proof follows similar lines to the proof of Theorem 5b, and is omitted.

5. Discussion

The results given have an obvious application to the design of multivariable systems, as will be clear from Figures 1 and 2. Mathematically, the chief interest probably lies in the demonstration that algebraic results of the type of Theorem 2 have direct applications to the analysis of stability. The passage from (17) to (19) involves two weak steps, and it is therefore likely that better results can be obtained. These would improve Theorems 5b and 6b by permitting values of θ_i not allowed by Theorem 2. In this connection it can be shown [11] that if (11) and (12) are true simultaneously, with $\theta_i = 1, i = 1, 2, \ldots, m$, then $\|Z\| = \|Z^*\| < 1$. Simultaneous achievement of (11) and (12), however, will usually be difficult to achieve, and for this reason Theorem 2 in the form given is preferred.

6. Acknowledgement

I am grateful to Professor R. W. Brockett for a number of helpful suggestions: also to Dr. P. A. Cook for a counterexample which exposed an error in a previous (incorrect) version of Theorem 2. This error was also pointed out by Professor B. D. O. Anderson.

References

1. Lur'e, A. I., "Some Non-linear Problems in the Theory of Automatic Control", H.M.S.O., London (1957) (translation of 1951 monograph).
2. Aizerman, M. A. and Gantmacher, F. R., "Absolute Stability of Regulator Systems". Holden-Day (1964).
3. Popov, V. M., Absolute stability of nonlinear control systems of automatic control. *Automation and Remote Control* **22** (1962), 857–875.
4. Willems, J. L., "Stability Theory of Dynamical Systems. Nelson (1970).
5. Newcomb, R. W., "Linear Multiport Synthesis". McGraw-Hill (1966).
6. Rosenbrock, H. H., "State-space and Multivariable Theory". Nelson (1970).
7. Rosenbrock, H. H., Progress in the design of multivariable control systems. *Measurement and Control* **4** (1971), 9–11.
8. Kalman, R. E., Canonical structure of linear dynamical systems. *Proc. Nat. Acad. Sci.* **48** (1962), 596–600.
9. Bodewig, E., "Matrix Calculus". Interscience (1956).
10. Anderson, B. D. O., A system theory criterion for positive real matrices. *SIAM J. Control* **5** (1967), 171–182.
11. Cook, P. A. Private Communication: also independently J. C. Willems.

7. Appendix

As in the conditions of Theorem 1, let the system (1) have least order. Associate with (1) the system

$$\left.\begin{aligned} \dot{x}_1 &= A_1 x_1 + B_1 u_1 = (A - B\beta C)\, x_1 - B(\beta\alpha^{-1} - I_m)\, u_1 \\ y_1 &= C_1 x_1 + D_1 u_1 = \beta C x_1 + \beta\alpha^{-1} u_1 \end{aligned}\right\} \tag{110}$$

Then we have

$$[sI_n - A + B\beta C, -B(\beta\alpha^{-1} - I_m)] = (sI_n - A, B)\begin{pmatrix} I_n & 0 \\ \beta C & I_m - \beta\alpha^{-1} \end{pmatrix} \tag{111}$$

and the last matrix in (111) is non-singular by (5), while

$$\begin{pmatrix} sI_n - A + B\beta C \\ -\beta C \end{pmatrix} = \begin{pmatrix} I_n & -B\beta \\ 0 & \beta \end{pmatrix}\begin{pmatrix} sI_n - A \\ -C \end{pmatrix} \tag{112}$$

and the first matrix on the right-hand side of (112) is again non-singular by (5). Hence the decoupling zeros of (110) are those of (1): that is (110) also has least order.

Now the transfer function corresponding to (110) is

$$G_1(s) = -\beta C(sI_n - A + B\beta C)^{-1} B(\beta\alpha^{-1} - I_m) + \beta\alpha^{-1} \tag{113}$$

$$= I_m + [I_m - \beta C(sI_n - A + B\beta C)^{-1} B]\,(\beta\alpha^{-1} - I_m) \tag{114}$$

$$= I_m + [I_m + \beta C(sI_n - A)^{-1} B]^{-1}\,(\beta\alpha^{-1} - I_m) \tag{115}$$

$$= [I_m + \beta G(s)]^{-1}\,[I_m + \beta G(s) + \beta\alpha^{-1} - I_m] \tag{116}$$

$$= [I_m + \beta G(s)]^{-1}\,[\beta\alpha^{-1} + \beta G(s)] \tag{117}$$

$$= [\beta^{-1} + G(s)]^{-1}\,[\alpha^{-1} + G(s)] \tag{118}$$

where $G(s) = C(sI_n - A)^{-1}B$ and (115) follows from (114) by a matrix identity attributed by Bodewig [9] to Frobenius. By assumption in Theorem 1, $G_1(s)$ is positive real, and Anderson's generalisation [10] of the Kalman–Yacubovich lemma shows that there exists a symmetric positive definite matrix P such that

$$\begin{pmatrix} A_1^T P + P A_1 & P B_1 - C_1^T \\ B_1^T P - C_1 & -2D_1 \end{pmatrix} = \begin{pmatrix} A_1^T P + P A_1 & P B_1 \\ B_1^T P & 0 \end{pmatrix} - \begin{pmatrix} 0 & C_1^T \\ C_1 & 2D_1 \end{pmatrix} \tag{119}$$

is negative semidefinite.

Equation (6) can be written

$$\dot{x} = \{A_1 + B_1[(\beta\alpha^{-1} - I_m)^{-1}(F - \beta)\beta^{-1}]\, C_1\}\, x \tag{120}$$

or

$$\left.\begin{aligned} \dot{x} &= A_1 x + B_1 u_2 \\ u_2 &= -(\beta\alpha^{-1} - I_m)^{-1}\beta^{-1}(\beta - F)\, C_1 x \triangleq -\mu(\beta - F)\, C_1 x \end{aligned}\right\} \tag{121}$$

on commuting diagonal matrices. Hence, using (119),

$$\frac{d}{dt} x^T P x = x^T(A_1{}^T P + P A_1)\, x + 2x^T P B_1 u_2 \tag{122}$$

$$\leqslant 2u_2{}^T C_1 x + 2u_2{}^T D_1 u_2 \tag{123}$$

$$= -2x^T C_1{}^T \mu(\beta - F)\, \{\mu^{-1} - \beta\alpha^{-1}(\beta - F)\}\, \mu C_1 x \tag{124}$$

$$= -2x^T C_1{}^T \mu(\beta - F)\, \{\beta\alpha^{-1}(F - \alpha)\}\, \mu C_1 x \tag{125}$$

and it is easy to see that this quadratic form is negative semi-definite. It follows that the system (6) is stable in the sense of Liapunov. This proves the first part of the theorem.

The second part is proved in a similar way by using instead of (110)

$$\left.\begin{aligned} \dot{x}_1 &= A_1 x_1 + B_1 u_1 = (A - B\beta C)\, x_1 - B\beta u_1 \\ y_1 &= C_1 x_1 + D_1 u_1 = (\beta\alpha^{-1} - I_m)\, C x_1 + \beta\alpha^{-1} u_1 \end{aligned}\right\} \tag{126}$$

giving as in (113) to (118)

$$G_1(s) = [\alpha^{-1} + G(s)]\, [\beta^{-1} + G(s)]^{-1} \tag{127}$$

Similarly the third part uses

$$\left.\begin{aligned} \dot{x}_1 &= A_1 x_1 + B_1 u_1 = (A - B\alpha C)\, x_1 + B(\beta - \alpha)\, u_1 \\ y_1 &= C_1 x_1 + D_1 u_1 = C x_1 + u_1 \end{aligned}\right\} \tag{128}$$

which gives

$$G_1(s) = C(sI_n - A + B\alpha C)^{-1}\, B(\beta - \alpha) + I_m \tag{129}$$

$$= \beta\alpha^{-1} + [I_m - C(sI_n - A + B\alpha C)^{-1}\beta\alpha]\, (I_m - \beta\alpha^{-1}) \tag{130}$$

$$= \beta\alpha^{-1} + [I_m + C(sI_n - A)^{-1}\beta\alpha]^{-1}\, (I_m - \beta\alpha^{-1}) \tag{131}$$

$$= \beta\alpha^{-1} + [I_m + G(s)\alpha]^{-1}\, (\alpha\beta^{-1} - I_m)\, \beta\alpha^{-1} \tag{132}$$

$$= \beta\alpha^{-1} + [I_m + G(s)\alpha]^{-1}\, [-I_m - G(s)\alpha + \alpha\beta^{-1} + G(s)\alpha]\, \beta\alpha^{-1} \tag{133}$$

$$= [I_m + G(s)\alpha]^{-1}\, [I_m + G(s)\beta] \tag{134}$$

Finally, the fourth part uses

$$\left.\begin{aligned} \dot{x}_1 &= A_1 x_1 + B_1 u_1 = (A - B\alpha C)\, x_1 + B u_1 \\ y_1 &= C_1 x_1 + D_1 u_1 = (\beta - \alpha)\, C x_1 + u_1 \end{aligned}\right\} \tag{135}$$

giving as in (129) to (134)

$$G_1(s) = [I_m + \beta G(s)]\,[I_m + \alpha G(s)]^{-1} \tag{136}$$

Discussion

COOK (*University of Loughborough*). Perhaps I can clarify the situation about this matrix theorem. The previous version of Theorem 2, the matrix theorem by Rosenbrock, is certainly wrong. What may be correct is the stability criterion which he attempted to base upon it without the θ_i. Several people have claimed that this can be proved although I'm not yet convinced.

BELL. Thank you for raising that. In fact, I believe Willems has hinted that he has a proof.†

COOK. Yes, he has but I haven't seen it.

MEES (University of Cambridge). You remember that Professor Lighthill and I presented a theorem for multivariable systems which was a generalization of the circle criterion.‡ This said that the eigenvalues, the spectrum if you like, of the linear part had to avoid a disc. This disc was much the same as your disc except that yours had a different α and β for each principal loop. This theorem is restricted to the case in which the operators are normal in the linear part. However, it turns out in this case that by applying a few transformations one can prove that Rosenbrock's conjectured stronger theorem, at least for the inverse Nyquist diagrams, reduces to sufficient conditions for this other generalized circle criterion. So presumably it gives sufficient conditions for stability. I don't think it's quite so easy doing it the right way up but it is certainly true for the inverse Nyquist diagrams.

† *Editor's Note.* Several attempts have been made to obtain a proof. The only new result to date is due to Dr. Cook and is given in the paper immediately following the above discussion.

‡ Figure 13 of Lighthill and Mees paper in this Volume. p. 14

25. Modified Multivariable Circle Theorems†

P. A. Cook

Department of Mathematics
Loughborough University of Technology,
Leicestershire, England

1. Introduction

In the above paper by Rosenbrock, (reference [1] of this paper), a generalisation of the circle criterion for stability is obtained which applies to diagonally-dominant multivariable linear systems with diagonal (nonlinear and/or time-varying) feedback. The generalised criterion requires that certain bands drawn around the Nyquist plots of diagonal elements of the open-loop transfer function matrix (or its inverse) should avoid certain discs in the complex plane. Unfortunately, these discs are not, as might be expected, those which would be associated with the corresponding feedback elements by the single loop circle criterion; although concentric with them, they have their radii expanded by certain factors (called θ_i by Rosenbrock) which, though to some extent adjustable, are always greater than unity. In this paper, we shall derive an alternative criterion which removes the θ_i-factors by employing a different definition of diagonal dominance.

We shall follow Rosenbrock's notation, and number our theorems consecutively with his paper, to avoid confusion. In Section 2, we prove a matrix theorem to be used in place of Rosenbrock's Theorem 2, and then use it to deduce counterparts of his Theorems 5 and 6. Finally, in Section 3, we state and discuss the graphical interpretation of these results.

2. Stability Criteria

We first prove the following theorem, which appears to be new, although related to known regularity theorems [2].

† December 1972. This work was supported by the Science Research Council under grant No. B/SR/8561.

THEOREM 7. *Let $\hat{Z}$ be a complex $m \times m$ matrix and let there be real numbers $\lambda_i > 0$ such that*

$$|\hat{z}_{ii}| - \sum_{j=1, j\neq i}^{m} \frac{\lambda_j}{\lambda_i}\left[\frac{|\hat{z}_{ij}| + |\hat{z}_{ji}|}{2}\right] > 1 \ (i = 1, 2, \ldots m). \qquad (1)$$

Then $\hat{Z}$ has an inverse Z with spectral norm

$$\|Z\| < 1.$$

Proof. Introduce the diagonal unitary matrix

$$\Phi = \mathrm{diag}(\hat{z}_{ii}/|\hat{z}_{ii}|)$$

and define the Hermitian matrix

$$X = \tfrac{1}{2}(\Phi^*\hat{Z} + \hat{Z}^*\Phi).$$

Then the diagonal elements of X are real and, by (1),

$$x_{ii} - \sum_{j=1, j\neq i}^{m} \frac{\lambda_j}{\lambda_i}|x_{ij}| > 1 \ (i = 1, 2, \ldots m).$$

Hence, by Gershgorin's Theorem, all the eigenvalues of X are greater than unity, and so, with v denoting a complex m-vector,

$$\inf_{v\neq 0}\frac{v^*Xv}{\|v\|^2} > 1. \qquad (2)$$

Also, using Schwartz's inequality,

$$\begin{aligned}\|\hat{Z}v\|\,\|v\| &= \|\hat{Z}v\|\,\|\Phi v\| \\ &\geqslant |v^*\Phi^*\hat{Z}v| \\ &\geqslant \mathrm{Re}(v^*\Phi^*\hat{Z}v) = v^*Xv\end{aligned}$$

and then (2) gives

$$\inf_{v\neq 0}\frac{\|\hat{Z}v\|}{\|v\|} > 1. \qquad (3)$$

Thus $\hat{Z}$ is nonsingular and the quantity on the L.H.S. of (3) is $\|Z\|^{-1}$, so that

$$\|Z\| < 1.$$

Now we use Theorem 7 to establish the next two results, which we state in terms of quantities defined in Rosenbrock's paper [1], assuming the conditions stated in his Section 1 to hold.

THEOREM 8. *Suppose that, for each s on D′ having R′ ⩾ R,*

$$|\zeta_i + g_{ii}(s)| - \sum_{j=1, j\neq i}^{m} \frac{|g_{ij}(s)| + |g_{ji}(s)|}{2} > \eta_i \quad (i = 1, 2, \ldots m) \tag{4}$$

Let (A, B, C) *be a least-order realization of* $G(s)$, *and let the linear time-invariant system*

$$\dot{x} = (A - B\beta C)x \tag{5}$$

be asymptotically stable.

Then the nonlinear time-dependent system

$$\dot{x} = (A - BFC)x \tag{6}$$

is stable in the sense of Liapunov.

Proof. By Theorem 7, identifying λ_i with $\sqrt{\eta_i}$, (4) gives

$$\|\sqrt{\eta}[\zeta + G(s)]^{-1}\sqrt{\eta}\| < 1 \tag{7}$$

where

$$\sqrt{\eta} \equiv \mathrm{diag}(+\sqrt{\eta_i})$$

and (7) can be written as

$$\|[\zeta + G_1(s)]^{-1}\eta\| < 1 \tag{8}$$

where

$$G_1(s) \equiv \sqrt{\eta}G(s)\sqrt{\eta}^{-1}$$

so that a least-order realization of $G_1(s)$ is given by

$$(A_1, B_1, C_1) = (A, B\sqrt{\eta}^{-1}, \sqrt{\eta}C).$$

But then, since (A, B, C) can be replaced in (5) and (6) by (A_1, B_1, C_1), the result follows from (8) exactly as in the proof of Rosenbrock's Theorem 5.

THEOREM 9. *Suppose that* $G(s)$ *has an inverse* $\hat{G}(s)$ *such that, for each s on D′ having R′ ⩾ R,*

$$|\gamma_i + \hat{g}_{ii}(s)| - \sum_{j=1, j\neq i}^{m} \frac{|\hat{g}_{ij}(s)| + |\hat{g}_{ji}(s)|}{2} > \delta_i \qquad (i = 1, 2, \ldots m). \tag{9}$$

Let (A, B, C) *be a least order-realization of* $G(s)$, *and let the linear time-invariant system*

$$\dot{x} = (A - B\alpha C)x \tag{10}$$

be asymptotically stable.

Then the nonlinear time-dependent system (6) *is stable in the sense of Liapunov.*

Proof. By Theorem 7, identifying λ_i with $\sqrt{\delta_i}$, (9) gives

$$\|\sqrt{\delta}[\gamma + \hat{G}(s)]^{-1} \sqrt{\delta}\| < 1 \tag{11}$$

where

$$\sqrt{\delta} \equiv \operatorname{diag}(+ \sqrt{\delta_i})$$

and we can rewrite (11) as

$$\|[\gamma + \hat{G}_2(s)]^{-1}\delta\| < 1 \tag{12}$$

where $\hat{G}_2(s)$ is the inverse of

$$G_2(s) \equiv \sqrt{\delta}\, G(s)\sqrt{\delta}^{-1}$$

which has the least-order realization

$$(A_2, B_2, C_2) = (A, B\sqrt{\delta}^{-1}, \sqrt{\delta}\, C).$$

Then, since (A, B, C) can be replaced by (A_2, B_2, C_2) in (6) and (10), the result follows from (12) as in the proof of Rosenbrock's Theorem 6.

3. Graphical Interpretation

We now consider the graphical interpretation of Theorems 8 and 9, stating the results in terms of quantities defined in Section 3 of [1]. We call the band swept out as s goes round D by the disc with centre $g_{ii}(s)$ and radius $[d_i(s) + d_i'(s)]/2$ the ith Gershgorin mean band, and the disc on $(-1/\alpha_i, -1/\beta_i)$ as diameter the ith critical disc.

THEOREM 8a. *Let* (A, B, C) *be a least-order realization of* $G(s)$, *with* A *having* p_0 *eigenvalues in the closed right half-plane. For* $i = 1, 2, \ldots m$, *let the* i*th Gershgorin mean band have no point in common with the* i*th critical disc and encircle it* N_{ci} *times clockwise as* s *goes once clockwise round* D, *these conditions remaining satisfied for all* D' *with* $R' \geqslant R$, *and suppose*

$$\sum_{i=1}^{m} N_{ci} = -p_0. \tag{13}$$

Then the system (6) *is stable in the sense of Liapunov.*

Proof. Since (4) is satisfied, we have, for each s on D',

$$\left|\frac{1}{\beta_i} + g_{ii}(s)\right| > \frac{d_i(s) + d'_i(s)}{2} \qquad (i = 1, 2, \ldots m)$$

and it follows, by arguments analogous to those in [3], that $|G(s) + \beta^{-1}|$ encircles the origin $\sum_{i=1}^{m} N_{ci}$ times clockwise as s goes once clockwise round D', and that (13) then guarantees that the system (5) is asymptotically stable. Thus the conditions of Theorem 8 are satisfied and the result follows.

Assuming next that $G(s)$ has an inverse $\hat{G}(s)$, we call the band swept out as s goes round D by the disc with centre $\hat{g}_{ii}(s)$ and radius $[\hat{d}_i(s) + \hat{d}'_i(s)]/2$ the ith inverse Gershgorin mean band and the disc on $(-\beta_i, -\alpha_i)$ as diameter the ith inverse critical disc.

THEOREM 9a. *Let (A, B, C) be a least-order realization of $G(s)$, with A having p_0 eigenvalues in the closed right half-plane. For $i = 1, 2, \ldots m$, let the ith inverse Gershgorin mean band have no point in common with the origin or with the ith inverse critical disc, let it encircle them $\hat{N}_{0i}$ and $\hat{N}_{ci}$ times respectively, clockwise as s goes once clockwise round D, these conditions remaining satisfied for all D' with $R' \geqslant R$, and suppose*

$$\sum_{i=1}^{m} (\hat{N}_{0i} - \hat{N}_{ci}) = p_0. \tag{14}$$

Then the system (6) *is stable in the sense of Liapunov.*

Proof. Since (9) is satisfied, we have, for each s on D',

$$|\alpha_i + \hat{g}_{ii}(s)| > \frac{\hat{d}_i(s) + \hat{d}'_i(s)}{2} \qquad (i = 1, 2, \ldots m)$$

and also, since each inverse Gerschgorin mean band avoids the origin,

$$|\hat{g}_{ii}(s)| > \frac{\hat{d}_i(s) + \hat{d}'_i(s)}{2} \qquad (i = 1, 2, \ldots m)$$

whence, by procedures analogous to those of [3], it follows that $|\hat{G}(s) + \alpha|$ and $|\hat{G}(s)|$ encircle the origin $\sum_{i=1}^{m} \hat{N}_{ci}$ and $\sum_{i=1}^{m} \hat{N}_{0i}$ times, respectively, clockwise, as s goes once clockwise round D', and that (14) then guarantees that the system (10) is asymptotically stable. Hence the conditions of Theorem 9 are satisfied and the result follows.

Theorems 8a and 9a are evidently counterparts of Theorems 5a and 6a of [1], their advantage being that the critical discs here are those of the single-loop circle criterion, while in [1] the ith disc has its radius magnified by a factor θ_i, for $i = 1, 2, \ldots m$, where the restriction

$$\sum_{i=1}^{m} \theta_i^{-2} \leqslant 1$$

must be satisfied. The Gershgorin discs in our case are also different, of course, their radii being the arithmetic means of those of the corresponding row and column discs used in [1].

By strengthening the assumptions, we can also derive counterparts of Theorems 5b and 6b of [1], which we state without proof, since they are obtained in essentially the same way as in [1].

THEOREM 8b. *Let the conditions of Theorem 8a be satisfied and let the least distance between the ith Gershgorin mean band and the ith critical disc be greater than* $\varepsilon > 0$, *for* $i = 1, 2, \ldots m$, *and for all* D' *with* $R' \geqslant R$.

Then the system (6) *is uniformly asymptotically stable in the large.*

THEOREM 9b. *Let the conditions of Theorem 9a be satisfied and let there be some* $\varepsilon > 0$ *such that*

$$(1 - \varepsilon)\,|\alpha_i + \hat{g}_{ii}(s)| > \frac{\hat{d}_i(s) + \hat{d}_i'(s)}{2} \qquad (i = 1, 2, \ldots m)$$

for each s on D' *with* $R' \geqslant R$.

Then the system (6) *is uniformly asymptotically stable in the large.*

Note. The assumption of least-order in the above theorems can be relaxed just as in [1].

Acknowledgement

The extent of the author's indebtedness to Professor H. H. Rosenbrock will be manifest to anyone who has read this paper.

References

1. Rosenbrock, H. H., Multivariable circle theorems. This Volume, pp. 345–365.
2. Marcus, M. and Minc, H., "A Survey of Matrix Theory and Matrix Inequalities". Allyn and Bacon (1964), pp 150–151.
3. Rosenbrock, H. H., "State-space and Multivariable Theory". Nelson (1970), Chapter 5, Section 6.

26. Numerical Comparison of Root-location Algorithms for Constant Linear Systems

I. S. PACE and S. BARNETT

School of Mathematics, University of Bradford, Yorkshire, England

1. Introduction

The problem of determining the roots of a polynomial has interested mathematicians for a considerable time. Since Galois showed that the roots of polynomials of degree five and higher could not be found explicitly, mathematicians have set their sights lower and aimed, for example, at determining how many roots were in a particular region of the complex plane. In many applications this is quite adequate since, for example, if all the roots of the characteristic polynomial

$$a_{n+1}\lambda^n + a_n\lambda^{n-1} + \ldots + a_2\lambda + a_1 = 0 \tag{1}$$

associated with the nth order linear differential equation

$$a_{n+1}\frac{d^{(n)}x}{dt^{(n)}} + a_n\frac{d^{(n-1)}x}{dt^{(n-1)}} + \ldots + a_2\frac{dx}{dt} + a_1x = 0 \tag{2}$$

have negative real parts then the solution $x(t)$ of the differential equation (2) tends to zero as t tends to infinity, and the differential equation is asymptotically stable. With the development of servomechanisms this problem became of great practical importance, an early and now famous study being by Maxwell [18] on the governing of steam engines. In a similar way the linear difference equation

$$a_{n+1}x(nT) + a_nx[(n-1)T] + \ldots + a_2x(T) + a_1x(0) = 0 \tag{3}$$

is asymptotically stable if all the roots of the associated characteristic polynomial (1) have modulus less than unity. These two cases form the basis of all the results presented here.

For many years the theorems of Routh, Hurwitz, Liénard–Chipart, Schur–Cohn and others provided the means by which these problems could be solved. Several survey articles and books have been published giving comprehensive details of these methods (see, for example, [1, 17, 21, 22]). Recently some of these classical theorems have been reformulated using companion form matrices [2, 4, 5]. When applied to the determinant-type methods, this work enables the orders of determinants to be reduced from n to $n/2$ or $(n-1)/2$ according as n is even or odd. This leads to a saving in computational effort making these methods more suitable for computer implementation than their rather bulky originals. It should be noted, however, that other theorems involving half-order determinants have been discovered and references are given in the Appendix.

An alternative way of dealing with (3) is to use the standard bilinear transformation to transform the interior of the unit circle into the left half plane, a convenient way of doing this being due to Duffin [9]. Also, a direct approach is to calculate the roots of (1) by a standard method or by applying the QR algorithm to the associated companion form matrix. Yet another method is to solve appropriate Liapunov matrix equations (see, for example, [7]).

Despite the variety of available techniques for determining root location there seems to be little published information as to their relative efficiency. The aim of this paper is to attempt to fill this gap in the literature by presenting the results of a critical numerical comparison of several algorithms.

In Section 2 we discuss the examples used, (consisting of polynomials up to 20th degree) and the relative merits of the methods. All problems were solved on the ICL 1909 computer at the University of Bradford.

In Section 3 we make some remarks on numerical difficulties that may occur, and give our conclusions. These are based on the computational results obtained, which agree closely with estimated times based on numbers of operations involved. It turns out that tabular methods (Routh, Jury, Miller), for both real and complex polynomials, are both simpler and faster than the corresponding determinantal methods. Direct computation of the roots is rather slower than other methods, but in compensation the precise information obtained can be utilised in further system analysis and design. Similarly, determination of the eigenvalues of the associated companion form matrix is slower, but the option to then calculate the corresponding eigenvectors with little extra effort can be useful. The Liapunov equation methods for determining stability are not competitive.

In an Appendix details of the methods considered are given, together with the steps to be taken when singular cases arise. Theoretical estimates are also given there of the numbers of arithmetic operations involved.

2. Examples and Results

Most of the polynomials used were chosen with random coefficients, but some published ones (taken from [12, 13]) were used which included singular cases.

The programs were run using double precision arithmetic (20–21 decimal digits). Because the computer is operated on a time sharing basis each example was run twenty times and an average obtained. For calculation of the determinants a standard Gaussian elimination routine was used.

We begin by giving the results for polynomials with randomly chosen coefficients, whose sizes ranged to $O(10^6)$ for the higher degree polynomials. Each polynomial was tested for roots in the left half plane and within the unit circle, the real polynomials being of degrees ranging from 4 to 20, and the complex polynomials of degrees 6 to 20. The respective results are presented in Tables 1, 2, 3 and 4. Only data for the two fastest methods in each case is presented. These are particularly interesting because the computing times agree closely with the times based upon the estimates of the number of operations involved with each method.

It should be noted that with higher degree polynomials the methods of Jury, Miller and Liénard–Chipart needed some modification. This was because machine overflow occured at some stage in the computation. This was easily overcome, however, by dividing by some appropriate positive number prior to the point at which overflow was going to occur.

A great advantage of tabular methods is their simplicity, especially when dealing with the singular types of polynomial. Compared to this, singular cases in the determinantal methods are rather cumbersome to handle. Whilst determination of stability is straightforward, the complete distribution can only be found by evaluating, for example, all the Hurwitz determinants.

Other methods of determining root distribution are more difficult, although they do have indirect advantages. Perhaps the most obvious approach would be to use some iterative technique to obtain the roots explicitly. The computing times for a real polynomial using Bairstow's method range from about 7 seconds for a polynomial of degree 6 to over a 100 seconds for one of degree 16. For determining root distribution only this method is not very efficient, but for system design purposes the numerical value of each root can be of value. A similar technique is to evaluate the

TABLE 1

Computing time in seconds and number of operations (divisions, multiplications, additions) for determination of number of roots of a real polynomial within the left half plane.

	Degree of polynomial						
	4	7	8	10	13	16	20
Routh array	0·05 (4, 4, 4)	0·10 (12, 12, 12)	0·10 (16, 16, 16)	0·20 (25, 25, 25)	0·10 (42, 42, 42)	0·35 (64, 64, 64)	0·85 (100, 100, 100)
Reduced Liénard–Chipart	0·05 (1, 4, 2)	0·05 (4, 18, 13)	0·15 (10, 38, 29)	0·50 (20, 80, 66)	0·96 (35, 156, 136)	2·00 (84, 420, 385)	4·75 (165, 960, 906)

TABLE 2

Computing time in seconds and number of operations (divisions, multiplications additions) for determination of presence of roots of a complex polynomial within the left half plane.

	Degree of polynomial					
	6	8	10	12	15	20
Routh array	0·10 (36, 36, 36)	0·45 (64, 64, 64)	0·80 (100, 100, 100)	0·85 (144, 144, 144)	1·60 (225, 225, 225)	2·50 (400, 400, 400)
Reduced Liénard–Chipart	1·65 (35, 150, 130)	2·20 (84, 420, 385)	5·60 (165, 960, 906)	8·30 (286, 1914, 1837)	13·10 (560, 4515, 4396)	50·75 (1330, 13870, 13661)

TABLE 3

Computing time in seconds and number of operations (divisions, multiplications, additions) for determination of number of roots of a real polynomial within the unit circle.

	Degree of polynomial						
	4	7	8	10	13	16	20
Jury array	0·10 (0, 20, 10)	0·25 (0, 56, 28)	0·25 (0, 72, 36)	0·65 (0, 110, 55)	0·45 (0, 182, 91)	0·70 (0, 272, 136)	1·30 (0, 420, 210)
Schur–Cohn	0·30 (10, 42, 33)	2·10 (56, 266, 239)	1·95 (84, 428, 393)	3·75 (165, 970, 916)	9·90 (364, 2613, 2523)	22·00 (680, 5816, 5681)	55·35 (1330, 13890, 13681)

TABLE 4

Computing time in seconds and number of operations (divisions, multiplications, additions) for determination of number of roots of a complex polynomial within the unit circle.

	Degree of polynomial					
	6	8	10	12	15	20
Miller	0·30 (2, 185, 53)	1·20 (2, 313, 87)	1·40 (2, 473, 129)	1·90 (2, 665, 179)	2·70 (2, 1013, 269)	4·50 (2, 1753, 459)
Schur–Cohn	2·90 (105, 729, 689)	7·40 (252, 1964, 1894)	15·60 (495, 4375, 4267)	29·65 (858, 8562, 8408)	65·75 (1680, 19800, 19562)	185·50 (3990, 59550, 59132)

eigenvalues of the associated companion form matrix. Using the QR algorithm the roots can be found in roughly the same time as taken by Bairstow's method. Additionally, the corresponding eigenvectors can be calculated and these can be used in certain situations, e.g. modal analysis [10].

Alternatively, we can solve one of the associated Liapunov matrix equations:

$$A^*P + PA = -Q$$

for the continuous case and

$$A^*PA - P = -Q$$

for the discrete-time case.

Special techniques are available for when the matrix A is in companion form (see [20] for a study of solution of Liapunov matrix equations). Typical computing times for the solution of the matrix equation are: $n = 4$, 4 secs; $n = 8$, 14 secs. To these figures must be added the time taken for determining whether or not the solution P is positive definite.

3. Discussion and Conclusions

One of the more noteworthy aspects of this work, from the computational point of view, was the problem of coefficient growth. Even for relatively low degree polynomials the Jury array, the complex alternative of Miller, and the companion-matrix version of the Liénard and Chipart method (although to a lesser extent) all suffered from this problem. For the tabular methods of Jury and Miller this can be easily overcome since we can divide any row by any positive quantity without affecting the rest of the computation. In fact we can make the Jury array more closely resemble the Routh array if, when evaluating the 2×2 determinants we divide by the modulus of the first column element. We can use the same procedure with Miller's algorithm, and the modification for the Liénard–Chipart algorithm is also straightforward. Here we get coefficient growth in two places, one in calculation of the matrix $f(G)$ or $g(F)$, and the other in evaluating the determinants. Clearly the two are related and since we are only interested in the sign of each determinant we can proceed as before. In this case during construction of the matrix $f(G)$ or $g(F)$ we must divide some row by a positive quantity, then all succeeding rows are divided by this same quantity. Suppose we divide row i by the positive number k, and that we are evaluating the jth principal minor ($j \geqslant i$), then the resulting determinant is less than the true determinant by a factor $k^{(j-i+1)}$. The choice of k and when to apply it can be argued as follows. Suppose we have a computer in which overflow is set if the result of any calculation is greater than 10^m, and at row i we have elements of the order 10^k. Then, generally, elements of row $i+1$ are going to be also of order 10^k at least, and may be as large as 10^{k+l}, where the coefficients of the polynomial are of the order 10^l. Thus, at this stage we may not have overflow, but in calculating the $(i+1)$th principal minor we certainly will if $k \geqslant m/2$. Therefore, we should perform a division when $k+l \geqslant m/2$. Clearly this division should be by a factor at least 10^{k+l}.

Whilst it is easy to see from the Tables that the tabular methods have a clear advantage in computing time, the determinant methods can be made faster, although with some loss of numerical accuracy, by calculating all the minors simultaneously. If we have the matrix $A = [a_{ij}], i, j = 1, 2, \ldots, n$ then we can calculate its principal minors A_i by pivoting down the diagonal. Clearly $A_1 = a_{11}$, and A_2 can be found by selecting a pivot from the second order leading principal submatrix and placing it in the (1, 1) position. By subtracting multiples of the first column from the other columns so that the first row of A becomes $[a_{11}', 0, \ldots, 0]$, the remaining elements being denoted by a_{ij}', we get $A_2 = \pm a_{11}' a_{22}'$, the sign being determined by the row and column interchanges that may have been performed. We next select a pivot from the 2×2 minor.

$$\begin{bmatrix} a_{22}' & a_{23}' \\ a_{32}' & a_{33}' \end{bmatrix}$$

and place it in the (2, 2) position and proceed as before. By this method all the principal minors can be determined in the same number of operations as required to evaluate the whole determinant. We can thus reduce the number of operations in the determinant methods from $O(n^4)$ to $O(n^3)$ but this still leaves them slower by a factor $O(n)$ than the tabular methods. This is a similar technique to that used by Jury when evaluating 'inners' [15].

When using hand computation the tabular methods have the advantage that the regular pattern of operations, in the absence of singular cases, involves only the simplest of computations. On a computer this is not so important except perhaps in the original programming. Additionally, the two types of singularity occurring are easily allowed for and do not affect the computing time significantly. Experience also shows that no difficulties are encountered with the tabular schemes if there are roots near the stability region boundary.

Alternatively the determinantal methods require a great deal of computation, even when using the companion matrix approach, although this can be useful in some cases when $n \leqslant 7$. To obtain the complete root distribution would necessitate evaluating all the Hurwitz determinants, thereby roughly doubling the computing times given here for the Liénard–Chipart criterion.

Other methods mentioned may have their own particular advantages when used in certain situations, but for the solution of the root distribution problem alone, none can compare with the speed and simplicity of the tabular methods. These are the Routh array for root distribution with respect to the imaginary axis, and the methods of Jury and Miller for real and complex polynomials, respectively, with respect to the unit disk.

Appendix: Summaries and Methods Used

Root location techniques fall into two basic groups. These are tabular methods and determinant methods. Each of these groups contains methods developed over a wide span of time. For example, for tabular methods when determining the number of roots in the left half plane we have the classical Routh array, and for determining the number of roots within the unit circle the more recent Jury array. The same applies for determinant methods where we have the classical Hurwitz determinants and the more recent developments of Barnett.

(i) *Tabular Methods*

(a) *Roots within the left half plane.* $\mathrm{Re}(\lambda) < 0$.

Given a real polynomial

$$a(\lambda) = a_1 + a_2\lambda + a_3\lambda^2 + \dots + a_n\lambda^{n-1} + a_{n+1}\lambda^n. \qquad \text{(A1)}$$

By constructing the Routh array

$$\begin{array}{llllllll}
\alpha_1 & \alpha_2 & \alpha_3 & \alpha_4 & . & . & . & . \\
\beta_1 & \beta_2 & \beta_3 & \beta_4 & . & . & . & . \\
\gamma_1 & \gamma_2 & \gamma_3 & \gamma_4 & . & . & . & . \\
\delta_1 & \delta_2 & \delta_3 & \delta_4 & . & . & . & . \\
\varepsilon_1 & \varepsilon_2 & . & . & . & . & . & . \\
\cdot & & & & & & & \\
\cdot & & & & & & & \\
\cdot & & & & & & & \\
\cdot & & & & & & &
\end{array}$$

where

$$\left.\begin{array}{l}\alpha_i = a_{n+3-2i} \\ \beta_i = a_{n+2-2i}\end{array}\right\}\ \begin{array}{ll} i = 1, 2, \dots, n/2 + 1, a_0 = 0 & n \text{ even} \\ i = 1, 2, \dots, (n+1)/2 & n \text{ odd},\end{array}$$

$$\gamma_i = -\begin{vmatrix} \alpha_1 & \alpha_{i+1} \\ \beta_1 & \beta_{i+1} \end{vmatrix} \bigg/ \beta_1$$

$$\delta_i = -\begin{vmatrix} \beta_1 & \beta_{i+1} \\ \gamma_1 & \gamma_{i+1} \end{vmatrix} \bigg/ \gamma_1 \quad \text{, etc.,}$$

then the number of sign changes in the sequence

$$(\alpha_1, \beta_1, \gamma_1, \delta_1, \ldots .) \tag{A2}$$

gives the number of roots with positive real parts, i.e. roots in the right half plane.

There are two types of singularity which can occur with the Routh array, both being easily dealt with. Suppose, for example, $\varepsilon_1 = 0$; then the two cases are:

(1) Some of the other ε_i are non-zero

(2) All the ε_i are zero.

For dealing with the first case Routh's algorithm states that we should replace ε_1 by a small quantity of some arbitrary sign and then proceed in the normal manner. On a computer this is easily done by assigning a positive sign to ε_1 and then setting the elements, ϕ_i, of the next row equal to $-\delta_1\varepsilon_{i+1}$. This is perfectly straightforward since

$$\phi_i = -\begin{vmatrix} \delta_1 & \delta_{i+1} \\ \varepsilon_1 & \varepsilon_{i+1} \end{vmatrix} \bigg/ \varepsilon_1 = (\varepsilon_1\delta_{i+1} - \delta_1\varepsilon_{i+1})/\varepsilon_1$$

and since multiplication by a positive quantity does not effect the signs in the sequence (A2) we can multiply by the 'small' positive quantity ε_1 giving $\phi_i = -\delta_1\varepsilon_{i+1}$ in the limit.

The second case is also dealt with in a simple manner. If the row $(\delta_1, \delta_2, \delta_3, \ldots)$ is the mth row of the Routh array, then it represents the polynomial

$$\delta_1\lambda^{n-m+1} + \delta_2\lambda^{n-m-1} + \ldots$$

and the elements of the next row should be set equal to the coefficients of the derivative of this polynomial,

i.e.

$$\varepsilon_i = (n - m + 3 - 2i)\delta_i.$$

The number of arithmetic operations involved with the Routh array, ignoring singular cases, are determined as follows:

n odd

Since $\gamma_1 = (\beta_1\alpha_2 - \alpha_1\beta_2)/\beta_1 = \alpha_2 - \alpha_1\beta_2/\beta_1$, to determine the third row requires $\frac{1}{2}(n + 1) - 1$ divisions, multiplications and additions.

Similarly the 4th row requires $\frac{1}{2}(n + 1) - 1$ divisions, multiplications and additions.

This is easily generalised to give the total operations as

$$2 \sum_{i=1}^{(n-1)/2} i = (n - 1)(n + 1)/4$$

divisions, multiplications and additions.

n even

Almost exactly the same procedure as above yields totals of

$$2 \sum_{i=1}^{n/2-1} i + n/2 = n^2/4$$

divisions, multiplications and additions.

For the complex polynomial

$$a(\lambda) = (a_1 + ib_1) + (a_2 + ib_2)\lambda + \dots + (a_{n+1} + ib_{n+1})\lambda^n \qquad \text{(A3)}$$

a method that reduces to the Routh array is the following:

If $a_{n+1} = 1$ and $b_{n+1} = 0$ then (A3) is a stable polynomial (i.e. has all its roots in the left half plane) if and only if the polynomial of degree $2n$

$$b(\lambda) = \lambda^{2n} + a_n\lambda^{2n-1} - b_n\lambda^{2n-2} - b_{n-1}\lambda^{2n-3} - a_{n-1}\lambda^{2n-4} - a_{n-2}\lambda^{2n-5}$$
$$+ b_{n-2}\lambda^{2n-6} + \dots \qquad \text{(A4)}$$

is a stable polynomial, i.e. for (A3) to be a stable polynomial there must be no changes of sign in the elements of the first column of the Routh array for $b(\lambda)$.

In this case the number of arithmetic operations involved is simply n^2 divisions, multiplications and additions.

(b) *Roots within the unit disc* $|\lambda| < 1$

For the polynomial (A1) a method analagous to the Routh array is the array of Jury [13].

With $a_{n+1} > 0$ the array is

$$\begin{array}{ccccc} a_1 & a_2 & a_3 & \ldots\ldots\ldots & a_{n-1}\; a_n\; a_{n+1} \\ b_1 & b_2 & b_3 & \ldots\ldots\ldots & b_{n-1}\; b_n \\ c_1 & c_2 & c_3 & \ldots\ldots\ldots & c_{n-1} \\ \cdot \\ \cdot \\ \cdot \\ \cdot \end{array}$$

where

$$b_i = \begin{vmatrix} a_1 & a_{n-i+2} \\ a_{n+1} & a_i \end{vmatrix}, \qquad c_i = \begin{vmatrix} b_1 & b_{n-i+1} \\ b_n & b_i \end{vmatrix}, \qquad \text{etc.}$$

The root distribution is then determined from the products $P_k = \delta_1\delta_2\ldots\delta_k$, where $\delta_1 = b_1, \delta_2 = c_2, \ldots$. The number of roots within the unit disc is equal to the number of products P_k with negative signs. Again, if say $l_1 = 0$, then two types of singularity exist:

(1) Some of the remaining l_i are non-zero

(2) All the l_i are zero.

The first case is dealt with as in the Routh array. The elements of the next row are to be set equal to $-l_r l_{r-i+1}$, where r is the number of elements in $\{l_i\}$. When calculating the signs of the products P_k then $l_1 = 0$ is taken to have a positive sign.

The second case of singularity is slightly more complicated. As with the Routh array, the rows of the Jury array represent polynomials. For example, if we have obtained

$$\begin{array}{cccccc} k_1 & k_2 & k_3 & \ldots\ldots & k_r & k_{r+1} \\ 0 & 0 & 0 & \ldots\ldots & 0 & \end{array}$$

then the k_i's represent the polynomial

$$k(\lambda) = k_1 + k_2\lambda + \ldots + k_{r+1}\lambda^r$$

and the row of zeros should be replaced by the coefficients of the derivative of this polynomial written in reverse order, i.e. replaced by the row

$$(rk_{r+1}, (r-1)k_r, \ldots, 2k_3, k_2)$$

representing the polynomial

$$l(\lambda) = rk_{r+1} + (r-1)k_r\lambda + \ldots + 2k_3\lambda^{r-2} + k_2\lambda^{r-1}$$

and $l(\lambda)$ has the same number of roots inside the unit circle as does $k(\lambda)$.

The determination of the root distribution does not now follow the normal pattern ([13] p. 114). In this case r of the roots of the original polynomial $a(\lambda)$ are distributed the same as the r roots of $k(\lambda)$, with an additional p roots inside the unit circle, where p is the number of products P_j with negative signs ($j = 1, \ldots, n-r$), and a further $n-r-p$ roots outside the unit circle. Also, the r roots of $k(\lambda)$ are either on the unit circle, or occur in pairs reciprocal to each other, or a combination of both. Because $k(\lambda)$ and $l(\lambda)$ have the same number of roots inside the unit circle, we can determine this number q from the number of negative products P_i in the Jury array for the polynomial $l(\lambda)$. From the special root distribution of $k(\lambda)$ we know then that $k(\lambda)$ has q roots inside and outside the unit circle and $r-2q$ roots on the unit circle, i.e. the full root distribution for $a(\lambda)$ is

$$\left.\begin{array}{ll} q+p & \text{inside the unit circle} \\ r-2q & \text{on the unit circle} \\ n-r-p+q & \text{outside the unit circle} \end{array}\right\}. \qquad \text{(A5)}$$

and

In this case the numbers of arithmetic operations are easily determined. To construct the Jury array requires a total of $2\sum_{i=1}^{n} i$ multiplications and $\sum_{i=1}^{n} i$ additions, i.e. $n(n+1)$ multiplications and $\frac{1}{2}n(n+1)$ additions. Determination of the products P_j is not strictly necessary since we need only count the negative elements of ($\delta_1 = b_1, \delta_2 = c_1, \ldots$).

For the complex polynomial (A3) [19] has recently given a generalisation of the Jury array. In this method a new polynomial $b(\lambda)$ is produced having degree one less than the degree of the original. This new polynomial is determined as follows:

$$b(\lambda) = \frac{a^*(0)a(\lambda) - a(0)a^*(\lambda)}{\lambda}$$

$$= \frac{(a_{n+1} - ib_{n+1})\,a(\lambda) - (a_1 + ib_1)\,a^*(\lambda)}{\lambda}$$

where $a^*(\lambda)$ is the polynomial whose coefficients are the complex conjugates of the coefficients of $a(\lambda)$ written in reverse order, i.e.

$$a^*(\lambda) = (a_{n+1} - ib_{n+1}) + (a_n - ib_n)\lambda + \dots + (a_2 - ib_2)\lambda^{n-1} + (a_1 - ib_1)\lambda^n.$$

Clearly, for real polynomials, the calculation of the coefficients of $b(\lambda)$ is the same as that for constructing another row of the Jury array, with a possible change of sign.

Provided that $|a_{n+1} - ib_{n+1}| \neq |a_1 + ib_1|$, Miller shows that if the polynomial $b(\lambda)$ has p_1 roots inside the unit circle, p_2 roots on the unit circle and p_3 roots outside the unit circle, such that $p_1 + p_2 + p_3 = n - 1$, then $a(\lambda)$ has

(i) $p_1 + 1$ roots inside the unit circle
p_2 roots on the unit circle
p_3 roots outside the unit circle,
if $|a_{n+1} - ib_{n+1}| > |a_1 + ib_1|$, or

(ii) p_3 roots inside the unit circle
p_2 roots on the unit circle
$p_1 + 1$ roots outside the unit circle,
if $|a_{n+1} - ib_{n+1}| < |a_1 + ib_1|$.

Two types of irregularity exist, the first being $|a_{n+1} - ib_{n+1}| = |a_1 + ib_1|$, the second being $b(\lambda) \equiv 0$ (in this case Miller calls $a(\lambda)$ a self-inversive polynomial.)

The first case can be treated in one of two ways: either by perturbing the coefficients of $a(\lambda)$ so that the overall root distribution is not affected, or by constructing another polynomial having the same number of roots inside the unit circle as $a(\lambda)$ (see [17], p. 204).

The second case is dealt with by replacing $b(\lambda)$ by the derivative of $a(\lambda)$. Miller has shown that if $a(\lambda)$ is self-inversive then it has p roots inside and outside the unit circle and $n - 2p$ roots on the unit circle for some $p \geqslant 0$. Now if $a'(\lambda)$ has $p + k - 1$ roots inside the unit circle, $n - 2p - k$ roots on the unit circle and p roots outside the unit circle, where k is the number of distinct roots of $a(\lambda)$ on the unit circle, then $a(\lambda)$ has the root distribution described in the previous sentence, i.e. if $a'(\lambda)$ has p_1 roots within the unit circle, p_2 roots on the unit circle, and p_3 roots outside the unit circle then $a(\lambda)$ has p_3 roots inside and outside the unit circle and $n - 2p_3$ roots on the unit circle.

To construct a polynomial of degree $i - 1$ from one of degree i would require totals of $8i$ multiplications and $2i$ additions. Therefore, to construct a sequence of polynomials of degrees $n - 1, n - 2, \ldots, 2, 1$ would require totals of

$$8\sum_{i=2}^{n} i = 4(n-1)(n+2) \text{ multiplications and}$$

$$2\sum_{i=2}^{n} i = (n-1)(n+2) \text{ additions.}$$

In addition to this the absolute values of the constant and leading terms of the polynomials of degrees 2 to n must be calculated, requiring a total of $4(n-1)$ multiplications and $2(n-1)$ additions, together with the determination of the location of the root of the linear polynomial, this taking a total of 2 divisions, 5 multiplications and 3 additions. Altogether this gives totals of

$$\begin{array}{ll} 4n^2 + 8n - 7 & \text{multiplications} \\ n^2 + 3n - 1 & \text{additions} \\ 2 & \text{divisions.} \end{array}$$

(ii) *Determinant Methods*

(a) *Roots within the left half plane*

The original theorem, due to Hurwitz, states that the polynomial (A1) is stable if all the a_i are positive and the ith order principal minors $H_i(i = 1, 2, \ldots, n)$ of the $(n \times n)$ Hurwitz matrix

$$\begin{bmatrix} a_n & a_{n-2} & a_{n-4} & \cdots & \begin{matrix} n \text{ odd} & n \text{ even} \\ a_1 & 0 \end{matrix} & \cdots & 0 \\ a_{n+1} & a_{n-1} & a_{n-3} & \cdots & \begin{matrix} a_2 & a_1 \end{matrix} & & \cdot \\ 0 & a_n & a_{n-2} & & \begin{matrix} a_3 & a_2 \end{matrix} & & \cdot \\ 0 & a_{n+1} & a_{n-1} & & \begin{matrix} \cdot & \cdot \end{matrix} & & \cdot \\ 0 & 0 & a_n & & \begin{matrix} \cdot & \cdot \end{matrix} & & \cdot \\ 0 & 0 & a_{n+1} & & \begin{matrix} \cdot & \cdot \end{matrix} & & \cdot \\ \cdot & \cdot & & & & & \\ \cdot & \cdot & & & & & \\ \cdot & \cdot & & & & & \\ 0 & 0 & \cdots & & & & \end{bmatrix}$$

are positive. The obvious disadvantage with this method is that it involves evaluation of minors up to order n. This in turn involves something of the order $n^4/12$ multiplications.

Fortunately, two other theorems enable the amount of work involved to be reduced considerably. The first was originally due to Liénard and Chipart [16] but was rediscovered more recently by Fuller [11]. This states (in one form) that necessary and sufficient conditions for the polynomial (A1) to be stable are that

$$a_1 > 0, a_n > 0, a_{n-2} > 0, \ldots$$

and
$$H_{n-1} > 0, H_{n-3} > 0, \ldots, \quad \begin{cases} H_3 > 0 & n \text{ even} \\ H_2 > 0 & n \text{ odd} \end{cases} \tag{A6}$$

This reduces the number of determinants to be evaluated by approximately half. However, other results by Liénard and Chipart [16], Cutteridge [8], Hermite (see [21]) and Barnett [4] enable the orders of determinants to be evaluated to be reduced by roughly a half. Here we present the form given by Barnett. With $a_{n+1} = 1$ there are two cases to consider:

n even

Define two polynomials as

$$\left.\begin{aligned} f(\lambda) &= a_1 + a_3\lambda + a_5\lambda^2 + \ldots + a_{n+1}\lambda^{n/2} \\ g(\lambda) &= a_2 + a_4\lambda + \ldots + a_n\lambda^{n/2-1} \end{aligned}\right\} \tag{A7}$$

where $a_{n+1} = 1$, and let the companion form matrix whose characteristic polynomial is $f(\lambda)$ be

$$F = \begin{bmatrix} 0 & 1 & 0 & \cdots & 0 \\ 0 & 0 & 1 & & \cdot \\ \cdot & & & & \cdot \\ 0 & & & & \cdot \\ \cdot & & & & \cdot \\ 0 & 0 & & & 1 \\ -a_1 & -a_3 & \cdots & \cdots & -a_{n-1} \end{bmatrix} \tag{A8}$$

Then the sequence (A6) is equivalent to the sequence $G_k, k = 2, 3, \ldots, n/2$, where G_k is the kth order principal minor of the matrix $g(F)J_{n/2}$.

n odd

In this case define two polynomials as

$$f^{(1)}(\lambda) = (a_1 - a_n a_2) + (a_3 - a_n a_4)\lambda + \ldots + (a_{n-2} - a_n a_{n-1})\,\lambda^{(n-3)/2}$$

$$g(\lambda) = a_2 + a_4\lambda + \ldots + a_{n+1}\lambda^{(n-1)/2}$$

o

where again $a_{n+1} = 1$. This particular form for $f^{(1)}(\lambda)$ is adopted because when n is odd, $f(\lambda)$ and $g(\lambda)$ in (A7) would have the same degree $(n-1)/2$. Use of the fact that $g(G) = 0$ then reduces $f(G)$ to the polynomial $f^{(1)}(G)$. The sequence (A6) is equivalent to the sequence F_k, $k = 1, 2, \ldots, (n-1)/2$, where F_k is the kth order principal minor of the matrix $f^{(1)}(G)J_{(n-1)/2}$.

Construction of the matrices $g(F)$ and $f^{(1)}(G)$ is easily performed by using a method of Barnett [3]. Briefly, this is as follows:
Consider any two polynomials

$$\alpha(\lambda) = \alpha_1 + \alpha_2\lambda + \ldots + \alpha_m\lambda^{m-1} + \lambda^m$$

$$\beta(\lambda) = \beta_1 + \beta_2\lambda + \ldots + \beta_m\lambda^{m-1}.$$

If the companion form matrix A of $\alpha(\lambda)$ has the form (A8) then

$$\beta(A) = [\boldsymbol{\beta}\,\boldsymbol{\beta}A\,\boldsymbol{\beta}A^2 \ldots \boldsymbol{\beta}A^{m-1}]^T \tag{A9}$$

where

$$\boldsymbol{\beta} = [\beta_1\,\beta_2 \ldots \beta_m].$$

When calculating the number of arithmetic operations involved with this method we use the fact that evaluation of a determinant of order i requires $i(i-1)/2$ divisions, $(i-1)(2i^2 - i + 6)/6$ multiplications and $i(i-1)(2i-1)/6$ additions, using standard Gaussian elimination.

n even

Calculation of the matrix $g(F)$ using (A9) and making use of the special form of the companion matrix requires

$$\tfrac{1}{2}n(\tfrac{1}{2}n - 1) \text{ multiplications}$$

and

$$(\tfrac{1}{2}n - 1)^2 \text{ additions.}$$

Additionally, to compute the determinants requires

$$\sum_{i=2}^{n/2} \{i(i-1)/2 \text{ divisions, } (i-1)(2i^2 - i + 6)/6 \text{ multiplications, and}$$

$$i(i-1)(2i-1)/6 \text{ additions}\}.$$

i.e. totals of

$$\frac{n}{48}(n-2)(n+2) \qquad \text{divisions,}$$

$$\frac{n}{192}(n-2)(n^2+2n+24) \quad \text{multiplications,}$$

and

$$\frac{n^2}{192}(n-2)(n+2) \quad \text{additions.}$$

Summing these figures gives grand totals of

$$\frac{n}{48}(n-2)(n+2) \quad \text{divisions,}$$

$$\frac{n}{192}(n-2)(n^2+2n+72) \quad \text{multiplications,}$$

and

$$\frac{n-2}{192}(n^3+2n^2+48n-96) \quad \text{additions.}$$

n odd

By similar techniques it is tedious but easy to show that the numbers of operations are

$$\frac{(n-1)}{48}(n+1)(n-3) \text{ divisions,}$$

$$\frac{192}{(n-1)}(n^3-3n^2+71n-117) \text{ multiplications}$$

and

$$\frac{1}{192}(n^4-4n^3+50n^2-188n+333) \text{ additions.}$$

As was seen in (i.a) the complex polynomial (A3) with $a_{n+1}=1$ and $b_{n+1}=0$ is stable if the polynomial (A4) of degree $2n$ is stable. Hence we can use the results of the previous paragraphs to deal with this case.

The numbers of arithmetic operations are easily deduced from previous results, replacing n by $2n$.

(b) *Roots within the unit circle*

The classical algorithm corresponding to that of Hurwitz is the Schur–Cohn theorem.

Briefly, for the polynomial

$$a(\lambda) = a_1 + a_2\lambda + \ldots + a_n\lambda^{n-1} + \lambda^n \tag{A10}$$

where the a_i may be real or complex, form the $n \times n$ Hermitian matrix $B = [b_{ij}]$, defined by

$$b_{ij} = \sum_{k=0}^{p} \{a_{n+k+2-i}\bar{a}_{n+k+2-j} - \bar{a}_{i-k}a_{j-k}\}$$

where $p = \min(i-1, j-1)$, $\quad i,j = 1, 2, \ldots, n+1, a_{n+1} = 1.$

Then the number of roots of $a(\lambda)$ outside the unit circle is equal to the number of changes in sign in the sequence $\{1, B_1, B_2, \ldots, B_n\}$, where the B_i are ith order leading principal minors of B, provided none of the B_i is zero.

When the a_i are real, Power [23] has shown the equivalence of the Schur–Cohn theorem with the Jury array.

Unfortunately, no reduction can be achieved in the orders of the determinants in theorems for roots within the unit circle. However, construction of the matrix B is simplified by use of another theorem [2] using the companion matrix approach. With $a(\lambda)$ as in (A10) define the polynomial $\hat{a}(\lambda)$ to be

$$\hat{a}(\lambda) = 1 + \bar{a}_n\lambda + \bar{a}_{n-1}\lambda^2 + \ldots + \bar{a}_2\lambda^{n-1} + \bar{a}_1\lambda^n.$$

It can be shown that the leading principal minors of the Schur–Cohn matrix are equal to the leading principal minors of the matrix $C = \hat{a}(A)$ where A is the companion form matrix of $a(\lambda)$ in the form (A8). Further, the first row of C is

$$\mathbf{c} = [1 - \bar{a}_1a_1, \bar{a}_n - \bar{a}_1a_2, \ldots, \bar{a}_2 - \bar{a}_1a_n] \tag{A11}$$

and the remaining rows are $\mathbf{c}A, \mathbf{c}A^2, \ldots, \mathbf{c}A^{n-1}$.

Use can be made of the left half plane results by mapping the interior of the unit circle $|w| < 1$ onto $\mathrm{Re}\,(\lambda) < 0$ by applying the bilinear transformation $w = (\lambda + 1)/(\lambda - 1)$. A convenient matrix method for carrying this out has been given by Duffin [9] (see also [6]).

The numbers of arithmetic operations can be calculated in a similar way to that used with the Liénard–Chipart criterion.

$a(\lambda)$ real

Calculation of the vector $\mathbf{c}$ (A11) requires n multiplications and additions. The remaining rows of the matrix C require a further $n(n-1)$ multiplications and $(n-1)^2$ additions; i.e. calculation of the matrix C requires a total of n^2 multiplications and $n^2 - n + 1$ additions.

Additionally, calculation of the determinants requires

$$\sum_{i=1}^{n} \{i(i-1)/2 \text{ divisions}, \quad (i-1)(2i^2-i+6)/6 \text{ multiplications},$$

$$i(i-1)(2i-1)/6 \text{ additions}\},$$

i.e.

$$n(n+1)(n-1)/6 \text{ divisions},$$

$$n(n-1)(n^2+n+6)/12 \text{ multiplications}$$

and

$$n^2(n+1)(n-1)/12 \text{ additions}.$$

Therefore the total number of operations is

$$n(n-1)(n+1)/6 \text{ divisions},$$

$$n(n^3+17n-6)/12 \text{ multiplications}$$

and

$$(n^4+11n^2-12n+12)/12 \text{ additions}.$$

$a(\lambda)$ complex

We can use the above results to calculate the number of operations in this case. Knowing that each complex division requires 3 divisions, 3 multiplications and 3 additions, each complex multiplication requires 4 multiplications and 2 additions, we can calculate the totals as

$$n(n-1)(n+1)/2 \text{ divisions},$$

$$n(2n^3+3n^2+34n-15)/6 \text{ multiplications}$$

and

$$(2n^4+3n^3+28n^2-21n+12)/6 \text{ additions}.$$

It is also worth mentioning here that Jury [14] has given a determinantal formulation which embraces the theorems for both the left half plane and unit circle problems, using what he terms 'inners'.

References

1. Anderson, B. D. O., *Int. J. Control* **5** (1967), 473–482.
2. Barnett, S., *Electron. Lett.* **6** (1970), 164–165.
3. Barnett, S., *Linear Algebra* **3** (1970), 7–9.
4. Barnett, S., *Proc. Camb. Phil. Soc.* **70** (1971), 269–274.
5. Barnett, S., *J. Inst. Maths. Applics.* **8** (1971), 240–250.
6. Barnett, S., "Matrices in Control Theory". Van Nostrand Reinhold, London (1971).

7. Barnett, S. and Storey, C., "Matrix Methods in Stability Theory". Nelson, London (1970).
8. Cutteridge, O. P. D., *Proc. Inst. elec. Engrs.* **106** (part C), (1959), 125–132.
9. Duffin, R. J., *SIAM Rev.* **11** (1969), 196–213.
10. Ellis, J. K. and White, G. W. T., *Control* **9** (1965), 193–197.
11. Fuller, A. T., *Proc. Camb. Phil. Soc. Math. Phys.* **53** (1957), 878–896.
12. Gantmacher, F. R., "Theory of Matrices". (Vol. 2) Chelsea, New York (1964).
13. Jury, E. I., "Theory and Application of the z-Transform Method". John Wiley, New York (1964).
14. Jury, E. I., *I.E.E.E. Trans. autom. Control* **AC-16** (1971), 233–240.
15. Jury, E. I. and Ahn, S. M., *I.E.E.E. Trans autom. Control* **AC-17** (1972), 541–543.
16. Liénard, A. and Chipart, M. H., *J. Math. Pures Appl.* **10** (1914), 291–346.
17. Marden, M., "Geometry of Polynomials". American Mathematical Society, Providence, Rhode Island (1966).
18. Maxwell, J. C., *Proc. Roy. Soc.* **16** (1868), 270–283.
19. Miller, J. J. H., *J. Inst. Maths. Applics.* **8** (1971), 397–406.
20. Pace, I. S. and Barnett, S. *Int. J. Control* **15** (1972), 907–915.
21. Parks, P. C., *Proc. Inst. Mech. Engrs.* **178** (1963) (part 3M), 7–17.
22. Porter, B., "Stability Criteria for Linear Dynamical Systems". Oliver and Boyd, Edinburgh (1967).
23. Power, H. M., *Electron. Lett.* **6** (1970), 39–40.

Discussion

PARKS (*University of Warwick*). In the case of the Hermite matrix, one can take advantage of the symmetry. Is this taken account of in your calculations?

PACE. No, I didn't take any account of that. Unfortunately, I didn't have time to write my own procedure for doing this.

PARKS. People have written on occasions and said because it is symmetric this is a great advantage computationally. Is this not true really?

PACE. I can see the case for inversion but for the calculation of determinants I cannot see all that advantage in it.

PARKS. I can see Dr. Barnett shaking his head vigorously so the answer is 'no'!

27. Insensitivity of Constant Linear Systems to Finite Variations in Parameters

BARBARA A. SHANE and S. BARNETT

School of Mathematics, University of Bradford, Yorkshire, England

1. Introduction

Consider the linear systems described by the differential and difference equations

$$\left.\begin{array}{c} \dot{x} = Ax \\ \text{or} \\ x_{k+1} = Ax_k \end{array}\right\} \tag{1}$$

respectively, where $A = [a_{ij}]$ is a constant $n \times n$ matrix. An important sensitivity problem associated with (1) is to determine the effect of variations in the a_{ij} on the characteristic roots of A (see Chapter 5 of [13]). If A is in companion form

$$\begin{bmatrix} 0 & 1 & 0 & & & 0 \\ 0 & 0 & 1 & 0 & & 0 \\ & & \cdot & \cdot & \cdot & \cdot \\ & & & \cdot & \cdot & \cdot \\ & & & \cdot & \cdot & \cdot \\ & & & & \cdot & 0 \\ 0 & 0 & \cdot & \cdot & 0 & 1 \\ -a_1 & -a_2 & \cdot & \cdot & -a_{n-1} & -a_n \end{bmatrix} \tag{2}$$

the parameters of the system are simply the coefficients a_i of the characteristic polynomial of A:

$$a(\lambda) = \lambda^n + a_n\lambda^{n-1} + \ldots + a_1. \tag{3}$$

One approach, [5, 7], is to obtain expressions for the variation in the roots for small changes in the a_i. If finite changes in the parameters are considered the problem is a difficult one ([12], p 125), and it is the aim of this paper to consider a more restricted problem in which the distribution of the roots of $a(\lambda)$ with respect to a given region Γ of the complex λ-plane is known. It is assumed throughout that no roots of $a(\lambda)$ occur on the boundary of Γ. It is then shown how expressions for finite changes in the a_{ij} and a_i can be found which do not affect this root distribution; in other words, A (or $a(\lambda)$) is 'insensitive' to these variations in parameters. This idea of insensitivity was discussed in detail for both linear and nonlinear systems in [1].

In Section 2 a unified summary is given, for comparison and reference purposes, of previous results [8, 9], which give expressions for the a_{ij} and a_i when Γ is the left half plane, a unit disc centred on the origin, or an arbitrary disc. This section also outlines the method of proof which relies on constructing special solutions of Liapunov-type matrix equations, these initially arising when a quadratic Liapunov function is associated with (1). For example, suppose Γ_1 is the left half plane: $\mathrm{Re}(\lambda) < 0$, and let A have r characteristic roots inside and $n-r$ roots outside this region (that is, A has *inertia* $(n-r, r, 0)$). Then from Liapunov theory (see [6] and Chapter 4 of [2]), the hermitian solution P of

$$A^*P + PA = -Q \tag{4}$$

has inertia $(r, n-r, 0)$, where Q is an arbitrary positive definite hermitian matrix. Expressions for finite changes in the a_{ij} and a_i are found by constructing a matrix $B = [b_{ij}]$ satisfying

$$(A + B)^*P + P(A + B) = -(Q + Q_0) \tag{5}$$

for some matrix Q_0 which makes $(Q + Q_0)$ positive definite. Similarly, if Γ_2 is the unit disc $|\lambda| < 1$ and A has r characteristic roots inside and $n-r$ outside Γ_2, the corresponding theorem in [11] states that the hermitian solution P of

$$A^*PA - P = -Q \tag{6}$$

has inertia $(r, n-r, 0)$. The matrix B is then constructed to satisfy

$$(A + B)^*P(A + B) - P = -(Q + Q_0) \tag{7}$$

for positive definite hermitian $(Q + Q_0)$.

In section 3 new results are presented for insensitivity with respect to Γ_1. Firstly it is shown that for real $a(\lambda)$ the a_i in (3) can be changed to $a_i + b_i$ if the b_i satisfy certain linear inequalities. The proof is accomplished by constructing a diagonal matrix Q to satisfy (4) with P having a given last row and column. Results can be generalised to complex a_i but the expressions involved become more complicated. Secondly, it is shown that any single coefficient a_k of $a(\lambda)$

can be changed by a finite amount b_k provided b_k lies inside a certain circle in the complex plane.

Throughout, the expressions derived provide only sufficient and not necessary conditions for insensitivity, but because of their relative simplicity and ease of application the results obtained seem potentially useful.

2. Some Insensitivity Theorems

The unifying feature in this section is that the matrix P in the solution of (4) or (6) is unaltered when B is constructed to satisfy (5) or (7) respectively.

Consider first the left half plane Γ_1:

THEOREM 1. (a) [4] *Given a matrix whose root distribution with respect to* Γ_1 *is known, then if*

$$B = P^{-1}(S - \tfrac{1}{2}Q_0) \tag{8}$$

where P *satisfies* (4), *the matrix* $(A + B)$ *has the same inertia as* A. *In* (8) Q_0 *is any hermitian matrix such that* $(Q + Q_0)$ *is positive definite and* S *is an arbitrary skew-hermitian matrix.*

(b) [8] *Given a polynomial* $a(\lambda)$ *with known inertia, let* $P = [p_{ij}]$ *be the solution of* (4) *with* A *in the form* (2). *The root distribution is unchanged on addition to* $a(\lambda)$ *of a polynomial*

$$b(\lambda) = b_n\lambda^{n-1} + b_{n-1}\lambda^{n-2} + \dots + b_1 \tag{9}$$

where

$$b_i = qp_{ni} \qquad i = 1, 2, \dots, n$$

and

$$-\lambda_{\min}(Q) \bigg/ \left[2 \sum_{i=1}^{n} p_{ni}^{2}\right] < q \tag{10}$$

$\lambda_{\min}(Q)$ *being the smallest characteristic root of* Q.

Proof. The proof of (a) follows immediately by substituting (8) into (5). The proof of part (b) is accomplished by constructing a matrix $B = [b_{ij}]$ to satisfy (5) where $b_{ij} = 0, i \neq n$ and $b_{nj} = -b_j, j = 1, \dots, n$, so that $A + B$ is also in the companion form (2), and ensuring that $(Q + Q_0)$ is positive definite. In this case

$$Q_0 = 2bb^*/q \tag{11}$$

where $b = [b_1\, b_2 \dots b_n]^T$ and a sufficient condition for $Q + Q_0$ to be positive definite is $\lambda_{\min}(Q) + \lambda_{\min}(Q_0) > 0$.

This condition when evaluated gives expression (10).

Consider the next case when Γ is the unit disc Γ_2:

THEOREM 2. [9] (a) *Given a matrix A with r characteristic roots inside Γ_2 and $n - r$ outside, then the addition to it of a matrix B does not affect this root distribution with respect to Γ_2 if*

$$B = (S - \tfrac{1}{2}P)^{-1}(SY + \tfrac{1}{2}PY + PA)$$

where Y is any hermitian solution of the Riccati-type algebraic equation

$$YPY + A^*PY + YPA + Q_0 = 0 \tag{12}$$

Q_0 being any hermitian matrix for which $(Q + Q_0)$ is positive definite, S is any skew-hermitian matrix and P any solution of (6).

(b) *Given a polynomial $a(\lambda)$ whose root distribution with respect to Γ_2 is known, the root distribution is unchanged on addition to it of a polynomial $b(\lambda)$ as defined in* (a) *with*

$$b_i = q\alpha_{in}/(2 + qp_{nn}) \qquad i = 1, \ldots, n \tag{13}$$

and either $q > 0$ or

$$-2q/(2 + qp_{nn})^2 < \lambda_{\min}(Q) \sum_{i=1}^{n} \alpha_{in}{}^2. \tag{14}$$

In (14), *$P = [p_{ij}]$ is a hermitian solution of* (6) *for a positive definite Q, the companion matrix* (2) *associated with $a(\lambda)$, and α_{in} denotes $(A^*P)_{in}$, the element in row i, column n of A^*P.*

The results of Theorem 2 can be extended to a general disc Γ_3 using a result of Howland [3]: if the boundary of the disc is given by

$$\phi(z, \bar{z}) + \bar{\phi}(z, \bar{z}) = 0$$

where

$$\phi(\mu, \nu) = c_0 + c_1\nu + c_2\mu + c_3\nu\mu$$

the c_i being constants, and A has r characteristic roots inside Γ_3 and $n - r$ outside, then the solution for the hermitian matrix P of the equation

$$(c_3 + \bar{c}_3)A^*PA + (c_0 + \bar{c}_0)P + (\bar{c}_1 + c_2)A^*P + (c_1 + \bar{c}_2)PA = -Q \tag{15}$$

where Q is an arbitrary positive-definite hermitian matrix, has inertia $(r, n - r, 0)$.

With the characteristic roots of A located as defined above the following results can be proved:

THEOREM 3. [9]. (a) *If*

$$B = (S - \tfrac{1}{2}P(c_3 + \bar{c}_3))^{-1}\{SY + P((c_1 + \bar{c}_2)I + (c_3 + \bar{c}_3)(Y + A))\}$$

where Y is the solution of the equation

$$\{(c_1 + \bar{c}_2)P + (c_3 + \bar{c}_3)A^*P\}\, Y + Y\{(\bar{c}_1 + c_2)P + (c_3 + \bar{c}_3)PA\} + (c_3 + \bar{c}_3)YPY + Q_0 = 0$$

for any hermitian matrix Q_0 for which $Q + Q_0$ is positive definite, then the roots of $(A + B)$ have the same distribution with respect to Γ_3 as A. The matrices P and Q are positive definite and satisfy (15) *and S is an arbitrary skew-hermitian matrix.*

(b) *If A is in companion form* (2) *with characteristic polynomial $a(\lambda)$ and if the coefficients of $b(\lambda)$ as defined in* (9) *are given by*

$$b_i = [q/\{1 + qp_{nn}(c_3 + \bar{c}_3)\}] \times [(c_1 + \bar{c}_2)P + (c_3 + \bar{c}_3)A^*P]_{in}$$

then the distribution of the roots of $a(\lambda) + b(\lambda)$ with respect to Γ_3 is the same as that for $a(\lambda)$, subject to the condition that either $q > 0$ or

$$\frac{-2q}{(qp_{nn}(c_3 + \bar{c}_3) + 1)^2} < \frac{\lambda_{min}(Q)}{\sum_{i=1}^{n}[(c_1 + \bar{c}_2)P + (c_3 + \bar{c}_3)A^*P]_{in}^{\;2}}.$$

3. Further Results for Polynomials

Some new results are now given for insensitivity of $a(\lambda)$ with respect to Γ_1. First suppose that the a_i in (3) are real:

THEOREM 4. *Let $a(\lambda)$ be a real polynomial whose roots all lie within Γ_1. If each coefficient a_i of $a(\lambda)$ is changed by the addition of qb_i, where the b_i satisfy the n inequalities*

$$q_{kk} = 2\sum_{i=\sigma}^{\theta}(-1)^{i+k}a_ib_j > 0 \quad k = 1, \ldots, n \tag{16}$$

with $j = 2k - i$,

$$\sigma = \begin{cases} 1 \quad , & 2k - 1 \leqslant n \\ 2k - n, & 2k - 1 > n \end{cases} \qquad \theta = \begin{cases} 2k - 1, & 2k - 1 \leqslant n \\ n + 1, & 2k - 1 > n \end{cases}$$

and $a_{n+1} = 1, b_{n+1} = 0, b_0 = 1$, then the distribution of the roots is unaltered provided

$$q > -\hat{q}/2\sum_{i=1}^{n} b_i^{\,2}, \quad where \quad \hat{q} = \min_k q_{kk}. \tag{17}$$

Proof. This differs from those of the preceding theorems in that Q in (4) is constructed to be diagonal, which imposes restrictions on the last row (and column) of P.

Let A be as in (2) and express $Q = [q_{ij}]$ in (4) in terms of $P = [p_{ij}]$ to give

$$q_{1i} = a_1 p_{in} + a_i p_{1n} - p_{1,i-1} \qquad i = 2,\ldots,n \tag{18}$$

$$q_{11} = 2a_1 p_{1n} \tag{19}$$

$$q_{ij} = (a_i p_{jn} + a_j p_{in} - p_{i,j-1} - p_{i-1,j}) \qquad i = 2, 3,\ldots,n \quad j = i, i+1, \ldots, n. \tag{20}$$

Since the aim is to make Q diagonal, take $q_{ij} = 0 \quad i \neq j$.

From (18) and (20)

$$a_1 p_{in} + a_i p_{1n} = p_{1,i-1} \qquad i = 2,\ldots,n \tag{21}$$

and

$$a_i p_{jn} + a_j p_{in} = p_{i,j-1} + p_{i-1,j} \qquad i = 2,\ldots,n \quad j = i+1,\ldots,n. \tag{22}$$

Expanding (22) gives

$$\left.\begin{aligned} p_{2,j-1} &= a_2 p_{jn} + a_j p_{2n} - p_{1j} && j = 3,\ldots,n \\ p_{3,j-1} &= a_3 p_{jn} + a_j p_{3n} - p_{2j} && j = 4,\ldots,n \\ &\;\vdots \\ p_{n-1,j-1} &= a_{n-1} p_{nn} + a_n p_{n-1,n} - p_{n-2,j}\,. \end{aligned}\right\} \tag{23}$$

Substituting for $p_{ij} (j \neq n)$ from the expressions in (23) gives with (21)

$$p_{kl} = \sum_{m=0}^{b} (-1)^m (a_{k-m} p_{l+m+1,n} + a_{l+m+1} p_{k-m,n}) \tag{24}$$

$$k = 1,\ldots,n-1 \qquad l = k,\ldots,n$$

$$b = \begin{cases} k-1, & k+l \leqslant n \\ n-l, & k+l > n \end{cases} \quad \text{with} \quad p_{n+1,n} = 0, \quad a_{n+1} = 1, \quad p_{0n} = 1.$$

Consequently all the elements of P can be expressed in terms of the last row of P, and because of (19) and (20) the diagonal elements of Q can be similarly expressed, namely

$$q_{kk} = 2(a_k p_{kn} - p_{k-1,k}), \; k = 1,\ldots,n \quad \text{and} \quad p_{01} = 0. \tag{25}$$

Substituting from (24) into (25) gives

$$q_{kk} = 2\left\{a_k p_{kn} - \sum_{m=0}^{b} (-1)^m (a_{k-m-1} p_{k+m+1,n} + a_{k+m+1} p_{k-m-1,n})\right\}$$

$$q = 2\left\{\left(\sum_{m=b}^{0} (-1)^{m+1} a_{k-m-1} p_{k+m+1,n}\right) + a_k p_{kn}\right.$$

$$\left. + \left(\sum_{m=0}^{b} (-1)^{m+1} a_{k+m+1} p_{k-m-1,n}\right)\right\}$$

$$= 2\left\{\sum_{i=\sigma}^{k-1} (-1)^{k-i} a_i p_{jn} + a_k p_{kn} + \sum_{i=k+1}^{\theta} (-1)^{i-k} a_i p_{jn}\right\}$$

with

$$\sigma = \begin{cases} 1, & 2k-1 \leqslant n \\ 2k-n+1, & 2k-1 > n \end{cases}, \quad \theta = \begin{cases} 2k-1, & 2k-1 \leqslant n \\ n+1, & 2k-1 > n \end{cases}$$

$$\text{and} \quad j = 2k - i$$

or

$$q_{kk} = 2\left\{\sum_{i=\sigma}^{\theta} (-1)^{|k-i|} a_i p_{jn}\right\} \qquad k = 1, \ldots, n. \tag{26}$$

Since Q is chosen to be positive definite we have

$$q_{kk} > 0 \qquad k = 1, \ldots, n. \tag{27}$$

From Theorem 1b the root distribution of $a(\lambda)$ is unchanged by the addition of $qb(\lambda)$ with $b(\lambda)$ as in (9) and

$$b_i = q p_{ni}/q = p_{ni}.$$

Substituting for p_{in} in (26) and (27) gives inequality (17). Finally condition (10) of Theorem 1b gives inequality (17).

Example 1. Consider as a very simple example

$$a(\lambda) = \lambda^2 + 2\lambda + 1$$

which clearly has both roots within Γ_1. From (16) the polynomial

$$\hat{a}(\lambda) = a(\lambda) + q(b_1\lambda + b_2) \tag{28}$$

has roots with negative real parts if

$$q_{11} = 2 \sum_{i=1}^{1} (-1)^{|1-i|} a_i b_{2-i} > 0$$

and

$$q_{22} = 2 \sum_{i=2}^{3} (-1)^{|2-i|} a_i b_{4-i} > 0$$

or

$$\left.\begin{aligned} 2a_1 b_1 > 0 \\ 2(a_2 b_2 - b_1) > 0 \end{aligned}\right\} \tag{29}$$

The region of the b_1, b_2 plane which satisfies (29) is shown in Fig. 1.

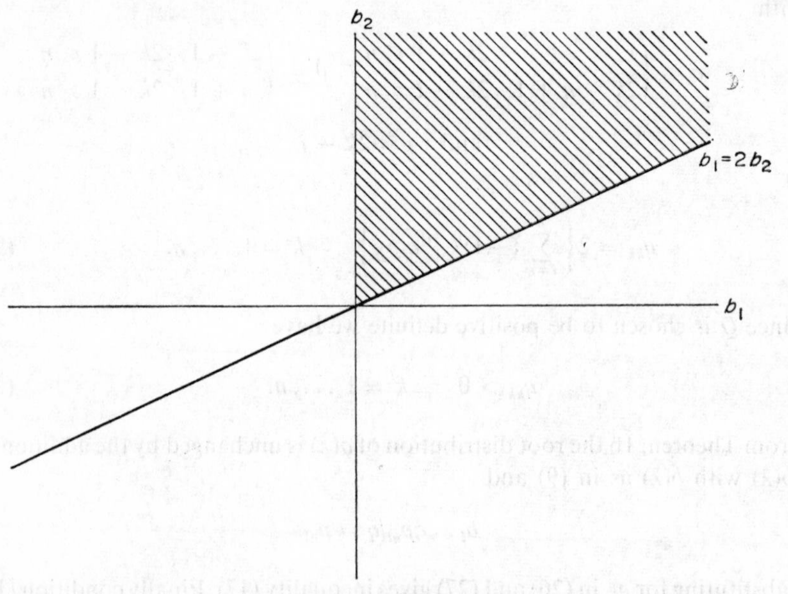

FIG. 1. Region for Example 1.

Suppose $b_1 = 1$ and $b_2 = 1$; then (28) gives

$$\hat{a}(\lambda) = \lambda^2 + (2+q)\lambda + (1+q).$$

Condition (17) gives the sufficient condition $q > -\frac{1}{2}$ for $\hat{a}(\lambda)$ to have roots within Γ_1. In fact using the Routh–Hurwitz criterion the necessary and sufficient condition is easily seen to be $q > -1$.

Unlike all the preceding theorems we now give a result which enables just one of the coefficients a_i in (3) to be changed:

THEOREM 5. *Let $a(\lambda)$ be a complex polynomial whose roots lie inside Γ_1. If a particular coefficient a_k is changed by the addition of b_k then the distribution of the roots of the resulting polynomial is unaltered provided that b_k is contained in the circle centre $(\lambda_{\min}(Q)/s^2)(h_{kn}, -g_{kn})$, radius*

$$\left.\begin{aligned} &\lambda_{\min}(Q)\sqrt{\Sigma|p_{kn}|^2/s^2} \\ \text{where} \quad & s^2 = \sum_{\substack{i=1\\ i\neq k}}^{n} |p_{in}|^2 \end{aligned}\right\} \tag{30}$$

and $P = [p_{ij}]$ and Q satisfy (4) *with* $\mathrm{Re}(p_{in}) = h_{in}$, $\mathrm{Im}(p_{in}) = g_{in}$.

Proof. Consider the companion matrix A as in (2) and let $(A + B)$ be the companion matrix associated with $a(\lambda) + b_k \lambda^{k-1}$. Then $B = [b_{ij}]$ is given by $b_{nk} = -b_k, b_{ij} = 0$ otherwise. Substituting for A from (4) into (5) gives $Q_0 = [q_{ij}^0]$ where

$$\left.\begin{aligned} \bar{q}_{ki}^0 = q_{ik}^0 &= \bar{p}_{in} b_k \qquad i = 1,\ldots,k-1,k+1,\ldots,n \\ q_{kk}^0 &= \bar{p}_{kn} b_k + p_{kn} b_k \end{aligned}\right\} \tag{31}$$

and $Q + Q_0$ is positive definite if

$$x^*(Q + Q_0)x/x^*x > 0. \tag{32}$$

Since $x^*Q_0x/x^*x \geqslant \lambda_{\min}(Q_0)$ and $x^*Qx/x^*x \geqslant \lambda_{\min}(Q)$, (32) is satisfied if

$$\lambda_{\min}(Q_0) + \lambda_{\min}(Q) > 0. \tag{33}$$

The roots of Q_0 are easily evaluated since Q_0 has characteristic equation

$$\lambda^{n-2}(\lambda^2 + (\bar{b}_k p_{kn} + b_k \bar{p}_{kn})\lambda - b_k \bar{b}_k s^2) = 0$$

and the minimum root of Q_0 is thus

$$\lambda_{\min}(Q_0) = -(h_{kn} b_k' + g_{kn} b_k'') - |\sqrt{b_k \bar{b}_k s^2 + (h_{kn} b_k' + g_{kn} b_k'')^2}|$$

defining $b_k = b_k' + ib_k''$ $(i = \sqrt{-1})$. Condition (33) is thus satisfied if

$$\lambda_{\min}(Q) - (h_{kn} b_k' + g_{kn} b_k'') > |\sqrt{b_k \bar{b}_k s^2 + (h_{kn} b_k' + g_{kn} b_k'')^2}|$$

or, after squaring both sides,

$$\left(b_k' - \frac{h_{kn}\lambda_{\min}(Q)}{s^2}\right)^2 + \left(b_k'' + \frac{g_{kn}\lambda_{\min}(Q)}{s^2}\right)^2 - \frac{\lambda_{\min}^2(Q)}{s^4}\left(\sum_{i=1}^{n} p_{in}\bar{p}_{in}\right) > 0$$

giving the interior of the circle (30).

Example 2. Consider

$$a(\lambda) = \lambda^2 + (3 - i)\lambda + 4.$$

The solution of (4) with $Q = 24I_2$ and A as in (2) is

$$P = \begin{bmatrix} 30 & 3+i \\ 3-i & 5 \end{bmatrix}.$$

Considering changes in a_1, (30) shows that b_1 lies within the circle centre (2·88, −0·96), radius ≃ 5·68. In comparison, a result given in [10] states that $a(\lambda) + b_1$ is a stable polynomial if

$$|b_1| < \Pi\,|\mathrm{Re}(\alpha_i)| \tag{34}$$

where α_i are the roots of $a(\lambda) = 0$. In our example

$$a(\lambda) = (\lambda + 1 + i)(\lambda + 2 - 2i)$$

and (34) gives $|b_1| < 2$.

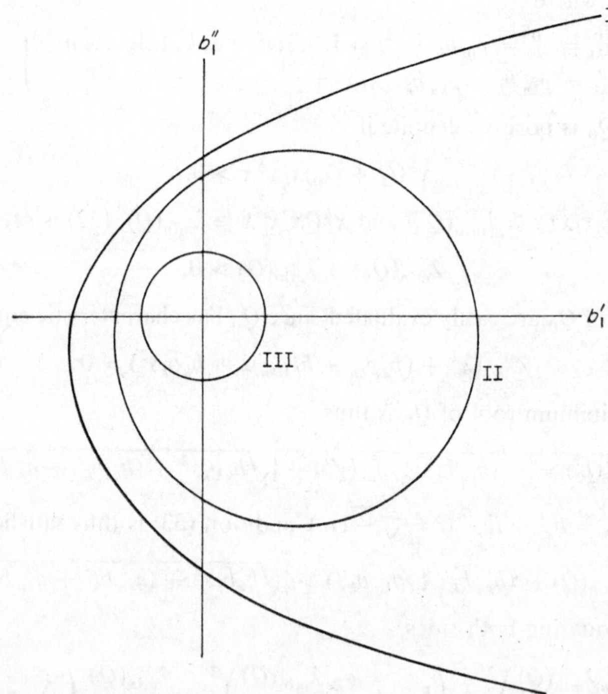

FIG. 2. Regions for Example 2. (I) Parabola $(b_1'' + 1{\cdot}5)^2 < 9(b_1' + 4{\cdot}25)$ obtained from Bilharz determinant; (II) Circle centre (2·88, −0·96), radius 5·68, obtained from Theorem 5; (III) Circle $|b_1| < 2$ obtained from a result of Tallis and Gordon [10].

The necessary and sufficient condition for $a(\lambda) + b_1$ to be stable is given by the Bilharz determinant

$$\begin{vmatrix} 1 & -1 & -b_1'-4 & 0 \\ 0 & 3 & b_1'' & 0 \\ 0 & 1 & -1 & -b_1'-4 \\ 0 & 0 & 3 & b_1'' \end{vmatrix} > 0$$

or

$$\tfrac{1}{9}(b_1'' + 1{\cdot}5)^2 < b_1' + 4{\cdot}25$$

This represents the interior of the parabola which is shown, together with the regions defined by the two sufficient conditions given above, in Fig. 2.

4. Concluding Remarks

To obtain necessary and sufficient conditions for insensitivity of $a(\lambda)$ to variations b_i in the coefficients a_i classical theorems, such as that of Routh–Hurwitz for Γ_1, can be applied. This method is sometimes suggested in standard texts on linear control systems, but the resulting inequalities involving the b_i are nonlinear and it is generally not possible to find explicit regions containing the b_i unless most of the b_i are taken zero. Furthermore, if A is not in companion form then interpretation in terms of the elements a_{ij} is difficult. It has been shown in this paper that the use of linear matrix equations enables sufficient conditions for insensitivity to be obtained with relative ease. The expressions obtained for polynomials are especially straightforward and hence potentially useful. Further developments are currently being studied (for example, the use of partitioning to determine insensitivity of A to variations in only certain of its elements) and these will be presented elsewhere.

Acknowledgement

The first-named author carried out this work whilst holding a United Kingdom Science Research Council Studentship.

References

1. Barnett, S., *Int. J. Control* **10** (1969), 665–675.
2. Barnett, S., "Matrices in Control Theory". Van Nostrand Reinhold, London (1971).
3. Howland, J. L., *J. Math. Anal. Applics.* **33** (1971), 683–691.
4. Joyce, G. T. and Barnett, S., *Lin. Algebra and Applications.* **3** (1970), 1–5.
5. Morgan, B. S., *Electron Lett.* **2**, (1966) 197–198.
6. Ostrowski, A. and Schneider, H., *J. Math. Anal. and Applics* **4** (1962), 72–84.

7. Rosenbrock, H. H., *Electron Lett.* **1** (1965), 1–4.
8. Shane, B. A. and Barnett, S., *I.E.E.E. Trans autom. control* **AC-17** (1972), 148–150.
9. Shane, B. A. and Barnett, S., *J. Math. Anal. Applics.* (1973) (To appear).
10. Tallis, G. M. and Gordon, G., *SIAM J. Appl. Maths.* **21** (1971), 186–190.
11. Taussky, O., *J. Math. Anal. and Appl.* **2** (1961), 105.
12. Taussky, O., "Recent Advances in Matrix Theory". University of Wisconsin Press, Wisconsin (1964).
13. Tomovic, R. and Vukobratovic, M., "General Sensitivity Theory". Elsevier, New York (1972).

Discussion

FULLER (*University of Cambridge*). I wonder if you could make this approach the starting point of some iterative scheme to find the exact values that would give you instability? So you would find the variation which you would calculate with your method first of all and then, starting from this new set of coefficients, start all over again with a new derivation and so on. Would that be possible?

SHANE. The trouble with this is you don't know what you're going to end up with until you've actually done it. Say you worked out a new polynomial and then chose another Q and solved it all again. You'd just get more bounds and there is no reason why you should get nearer to the polynomial which is almost unstable. You'll only get sufficient conditions—that's the trouble.

28. Extension of Existing Knowledge of the Theory of Multivariable Systems in the Z Domain by Using the Theory of Multivariable Infinite Matrices

M. A. MAKHLOUF*

Texas Instruments Incorporated, Petroleum Exploration Division, Dallas, Texas, U.S.A.

1. Introduction

Consider the multivariable feedback discrete system shown in Fig. (1)

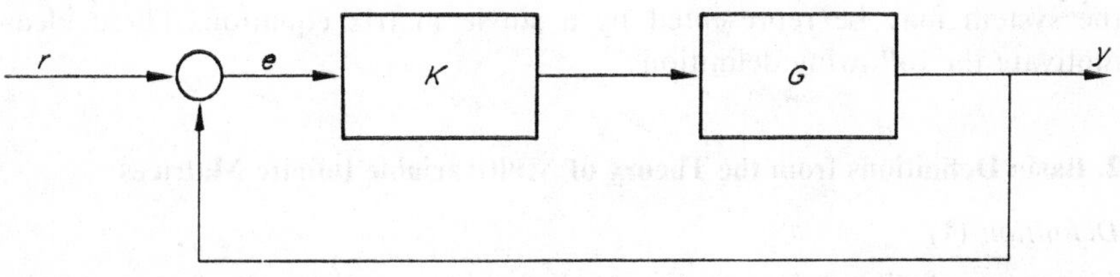

FIG. 1.

In the present approach to the analysis of the above system in the Z domain different components of the system are specified as follows:

(a). The input, the error and the output are specified each as a sequence of n-dimensional vectors $r = \{r_i\}_0^\infty, e = \{e_i\}_0^\infty$ and $y = \{y_i\}_0^\infty$. Thus for example r is the sequence in the space R^n defined by $r = \{r_i\}_0^\infty : N \to R^n$ where N is the set of positive integers and R^n is the set of n-dimensional vectors with real elements. The Z transform of the sequence r is the function of the complex variable z defined by $\tilde{r}(z) = \sum_{i=0}^\infty r_i z^{-i}$.

(b). The linear time invariant dynamical plant G is specified as a sequence of $n \times n$ matrices given by $G = \{G_i\}_0^\infty : N \to R^{n\times n}$ where $R^{n\times n}$ is the set of $n \times n$ matrices with real elements.

* Research in this paper was carried out when Dr. Makhlouf was attached to The City University, London.

(c). The memoryless time varying gain K is specified as a sequence of $n \times n$ matrices $K = \{K_i\}_0^\infty$.

By using the above definitions, the best available results on multivariable systems in the Z domain are given by four theorems in [1]. These theroems are limited to the analysis of time invariant systems and no investigation was made of the nature of the transformation relating the input and the output sequences in the sense of investigating the structural properties of this transformation and giving it a concrete representation.

Such an investigation can be carried out for both the time invariant and the time varying cases by formulating the problem in terms of the theory of multivariable infinite matrices. Instead of treating the n components of the output, for example, at the successive sample instants as a sequence of n-dimensional vectors, the whole sequence of sample values for the ith output $(i = 1, 2, \ldots, n)$ is taken as an infinite vector, and n such infinite vectors form the components of the output vector of dimension $n\infty$. Such a vector is called a multivariable infinite vector and may be thought of as a point in a product space $\prod_{i=1}^n X_i$, in which X_i is the infinite dimensional space of sample values of the ith component. The input vector is treated in the same way and the system may be represented by a single matrix equation. These ideas motivate the following definitions.

2. Basic Definitions from the Theory of Multivariable Infinite Matrices

Definition (1)

The space of all product spaces of dimension n (where the factors of the product space are separable infinite dimensional spaces and may be different) is called an infinite dimensional multivariable space of dimension $n\infty$. Such a space is defined by the set of all n-dimensional column vectors ${}^s x_n$ whose n elements are infinite column vectors, i.e.

$$
{}^s x_n = \begin{bmatrix} {}^s x_{(1)} \\ {}^s x_{(2)} \\ \vdots \\ {}^s x_{(i)} \\ \vdots \\ {}^s x_{(n)} \end{bmatrix}
$$

where for all i, ${}^s x_{(i)}$ is an element of an infinite dimensional space. ${}^s x_n$ is called an infinite dimensional multivariable vector.

If each factor of the product space $\prod_{i=1}^n X_i$ is of dimension m then the multivariable space is called finite dimensional multivariable space of dimension nm. An element of this space is denoted by $x_n^{(m)}$.

The infinite output vector ${}^s y_n$ will be regarded as limits

$$ {}^s y_n = \lim_{r \to \infty} y_n^{(r)} $$

of vectors $y_n^{(r)}$ in the finite dimensional multivariable space of dimension nr representing the values of the output at the first r sampling instants.

Definition (2)

(a). A multivariable operator g_{nm} is defined as the mapping

$$ g_{nm} : \prod_{i=1}^{m} X_i \to \prod_{i=1}^{n} Y_i. $$

If the order of the product spaces of both the domain and range of the operator g_{nm} is equal to one, the operator is a single variable operator.

(b). Let the multivariable operator

$$ g_{nm} : \prod_{i=1}^{m} X_i \to \prod_{i=1}^{n} Y_i $$

be given, where the factors of the product spaces are separable infinite dimensional spaces. Since ${}^s y_{(i)} \in Y_i$ for all i, there exist n row vectors ${}^s G_{(i)}$ each of which is dimension $1 \times m$, i.e.

$$ {}^s G_{(i)} = [{}^s G_{(i1)}\, {}^s G_{(i2)} \ldots {}^s G_{(ij)} \ldots {}^s G_{(im)}] $$

where the m elements ${}^s G_{(ij)}$ are single variable infinite matrices (the matrix representation of single variable operators).

Hence the ith projection of the range of the operator g_{nm} can be defined by the infinite column vector ${}^s y_{(i)}$ where

$$ {}^s y_{(i)} = \sum_{j=1}^{m} {}^s G_{(ij)}\, {}^s x_{(j)}. \tag{1} $$

Therefore g_{nm} is represented by a matrix ${}^s G_{nm}$ which has an $n \times m$ block structure and where elements may be characterized by four indices indicating the block and the position within the block, thus g_{ijlk} is in the ij block and occupies the position lk within this block.

Each block such as ${}^s G_{(ij)}$ in (1) is in effect the transfer matrix of the system G relating the jth input to the ith output.

Consequently the mapping

$$ g_{nm} : \prod_{i=1}^{m} X_i \to \prod_{i=1}^{n} Y_i $$

can be represented by the matrix transformation

$$^{s}y_n = {}^{s}G_{nm}\,{}^{s}x_m$$

where

$$^{s}G_{nm} = [{}^{s}G_{(ij)}],\ i = 1, 2, \ldots, n;\ j = 1, 2, \ldots, m.$$

$^{s}G_{nm}$ is called a multivariable infinite matrix of dimension $n \times m \times \infty \times \infty$.
If

$$g_{nm} : \prod_{i=1}^{m} X_i \to \prod_{i=1}^{n} Y_i$$

is an operator mapping a finite dimensional multivariable space of dimension qm into another of dimension pn, then g_{nm} can be represented by a matrix of $n \times m$ block structure where each block is a finite matrix of dimension $p \times q$. This matrix representation of g_{nm} is denoted by $G_{nm}^{(pq)}$ and is called a multivariable finite matrix of dimension $np \times mq$.

Definition (3)

Let a multivariable infinite matrix $^{s}G_{nm}$ be given. $^{s}G_{nm}$ is called lower semi-infinite (l.s.i.) if and only if all the ij block matrices $^{s}G_{(ij)}$ are lower semi-infinite matrices, i.e. for all i and j

$$^{s}G_{(ij)} = [g_{ijlk} : g_{ijlk} = 0\,\forall\, k > l].$$

A multivariable l.s.i. matrix $^{s}G_{nm}$ is called a stationary matrix if for each $^{s}G_{(ij)}$, the elements along any parallel to the principal diagonal are constant (the values along different parallels being in general different), i.e. $^{s}G_{nm}$ can be represented by

$$^{s}G_{nm} = [{}^{s}G_{(ij)}];\ i = 1, 2, \ldots, n;\ j = 1, 2, \ldots, m.$$

where

$$^{s}G_{(ij)} = [g_{ijl}];\ i, j \ \text{ fixed } \ ;\ l = 1, 2, \ldots, \infty$$

Otherwise $^{s}G_{nm}$ is nonstationary.

In the multivariable infinite matrix formulation of the control system, the plant G is defined by the multivariable l.s.i. matrix $^{s}G_{nn}$ which consists of $n \times n$ blocks each of which is an infinite matrix in the manner described in the last definition. This contrasts with the previous approach of defining G as a sequence of $n \times n$ matrices, $G = \{G_i\}_0^{\infty} : N \to R^{n \times n}$. Furthermore $^{s}G_{nn}$ will be regarded as $^{s}G_{nn} = \lim\limits_{r \to \infty} G_{nn}^{(rr)}$ where $G_{nn}^{(rr)}$ is a multivariable finite matrix of dimension $nr \times nr$.

3. Representation of Systems by Infinite Matrices

Consider the single input—single output causal time invariant plant in Fig. 2.

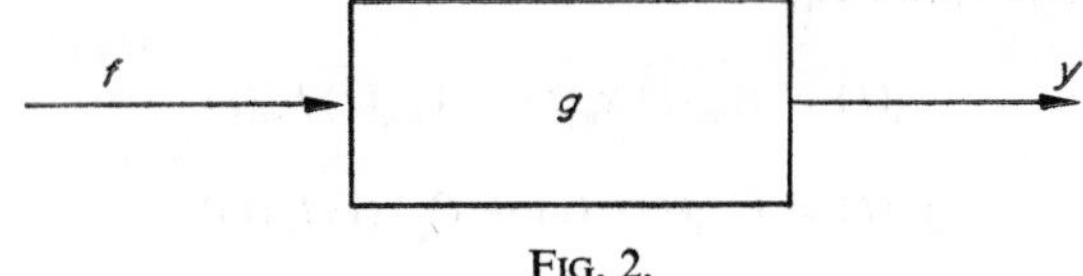

FIG. 2.

The output y in the Z domain is given by

$$y(z) = \sum_{n=0}^{\infty} z^{-n} \sum_{l=0}^{\infty} g(nT - lT)f(lT)$$

where T is the sampling interval.

The above equation corresponds to the infinite matrix equation ${}^s y = {}^s G {}^s f$ which in expanded form is

$$\begin{bmatrix} y(0) \\ y(T) \\ \vdots \\ y(iT) \\ \vdots \end{bmatrix} \begin{bmatrix} g(0) & 0 & \ldots 0 \ldots \\ g(T) & g(0)\,. & \vdots \\ \vdots & \vdots \quad \ddots & \vdots \\ g(iT) & g((i-1)T)\ldots & g(0)\ldots \\ \vdots & \vdots & \vdots \end{bmatrix} \begin{bmatrix} f(0) \\ f(T) \\ \vdots \\ f(iT) \\ \vdots \end{bmatrix}.$$

Hence the system g is defined by a stationary single variable l.s.i. matrix. If g is time varying then it can be represented by a nonstationary l.s.i. matrix. Similarly if G is a time varying multivariable plant with m inputs and m outputs, G can be represented by a nonstationary multivariable l.s.i. matrix ${}^sG_{mm}$ [2]. In subsequent investigations it is important to define ${}^sG_{mm}$ as

$${}^sG_{mm} = \lim_{r \to \infty} G_{mm}^{(rr)}$$

where $G_{mm}^{(rr)}$ is the mapping of a finite dimensional multivariable space of dimension mr into itself. The role played by the sequence of matrices $\{G_i\}_0^{\infty} : N \to R^{n \times n}$ will prove to be of secondary importance in establishing subsequent results. Similarly the controller K can be defined either as a time invariant or time varying system.

4. The Stability Problem

4.1 *Stability assessment by the first approach:*

In the first approach stability is considered as an internal property of the system where the system is excited by initial conditions. Boundedness or convergence in some sense of the state for future time is taken as the basic requirement for stability.

If the time varying multivariable plant G has the internal structure shown in Fig. 3, the state space equations are

$$x_n(t) = A_{nn}(t)\,x_n(t) + V_{nm}(t)f_m(t)$$

$$y_p(t) = C_{pn}(t)x_n(t) + Q_{pm}(t)f_m(t)$$

where the subscripts denote the dimensions of the respective vectors and matrices.

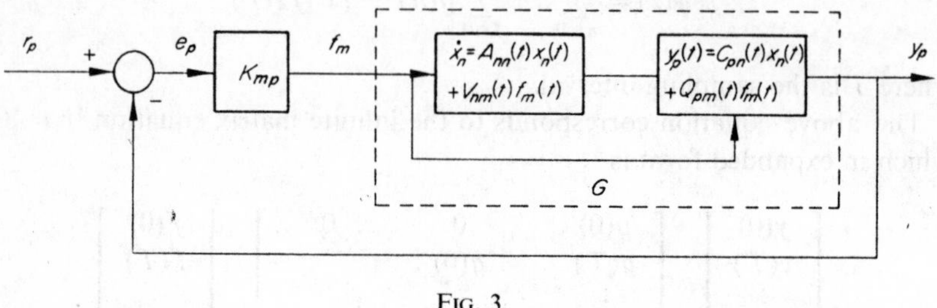

FIG. 3.

The multivariable infinite matrix representation of the state space equations is

$${}^s\dot{x}_n = {}^s\hat{A}_{nn}\,{}^s x_n + {}^s\hat{V}_{nm}\,{}^s f_m$$

$${}^s y_p = {}^s\hat{C}_{pn}\,{}^s x_n + {}^s\hat{Q}_{pm}\,{}^s f_m$$

where ${}^s\hat{A}_{nn}$ is given by the multivariable l.s.i. matrix

$${}^s\hat{A}_{nn} = [{}^s\hat{A}_{(ij)}]; \quad i, j = 1, 2, \ldots, n.$$

The ij block being

$${}^s\hat{A}_{(ij)} = \begin{bmatrix} a_{ij}(T) & 0 & 0 & \cdots \\ 0 & a_{ij}(2T) & 0 & \\ 0 & 0 & a_{ij}(3T) & \\ & & & \ddots \end{bmatrix}$$

$\hat{V}_{nm}$, ${}^s\hat{C}_{pn}$, and ${}^s\hat{Q}_{pm}$ are similarly expressed. Note that in the last equation the product ${}^s\hat{A}_{nn}\,{}^s x_n$ is not a convolution product.

Let

$${}^s R_{pp} = {}^s\hat{Q}_{pm}\,{}^s K_{mp}$$

where K is considered as a time varying multivariable system. Then with the feedback loop, the output is given by

$$^{s}y_p = (^{s}I_{pp} + {^{s}R_{pp}})^{-1}(^{s}\hat{C}_{pn}\, ^{s}x_n + {^{s}R_{pp}}\, ^{s}r_p)$$

and the state space equation becomes

$$^{s}\dot{x}_n = {^{s}\mathscr{A}_{nn}}\, ^{s}x_n + {^{s}\mathscr{V}_{np}}\, ^{s}r_p$$

where

$$\begin{aligned} ^{s}\mathscr{A}_{nn} &= {^{s}\hat{A}_{nn}} - {^{s}\hat{V}_{nm}}\, ^{s}K_{mp}\, (^{s}I_{pp} + {^{s}R_{pp}})^{-1}\, ^{s}\hat{C}_{pn} \\ ^{s}\mathscr{V}_{np} &= {^{s}\hat{V}_{nm}}\, ^{s}K_{mp}\, (^{s}I_{pp} - (^{s}I_{pp} + {^{s}R_{pp}})^{-1}\, ^{s}R_{pp}). \end{aligned} \tag{2}$$

To assess the stability of the feedback system the following theorems are required for evaluation of the nonstationary multivariable l.s.i. matrix $^{s}\mathscr{A}_{nn}$ and its associated eigenvalues.

Definition (4)

(a). Given the nonstationary multivariable lower finite matrix

$$G_{nn}^{(rr)} = [G_{(ij)}^{(rr)}];\ i, j = 1, 2, \ldots, n$$

where each ij block matrix $G_{(ij)}^{(rr)}$ is given by

$$G_{(ij)}^{(rr)} = [g_{ijlk} : g_{ijlk} = 0\ \forall\, k > l];\ l, k = 1, 2, \ldots, r$$

define

$$G_{nn}(m, e) = [g_{ijme}];\ i, j = 1, 2, \ldots, n.$$

Thus $G_{nn}(m, e)$ is the submatrix formed by taking the m, e element from each block i.e. formed by the elements common to rows $m + ir$; $i = 0, 1, 2, \ldots, (n - 1)$ and columns $e + jr$; $j = 0, 1, 2, \ldots, n$.

(b).

$$D_n(m) = |G_{nn}(m, m)|.$$

Definition (5)

The determinant of a nonstationary multivariable l.s.i. matrix $^{s}G_{nn}$ will be defined by

$$|^{s}G_{nn}| = \lim_{i \to \infty} |G^{(ii)}|.$$

THEOREM 1. *The determinant of the nonstationary multivariable l.s.i. matrix* $^{s}G_{nn}$ *is given by*

$$|^{s}G_{nn}| = \lim_{i \to \infty} \prod_{r=1}^{i} D_n(r) \tag{3}$$

provided the limit exists.

If $\sum_{r=1}^{\infty} \ln D_n(r)$ converges to a finite constant c then the infinite product of Eqn (3) converges to a non-zero limit and the determinant is given by

$$|{}^sG_{nn}| = \exp(c)$$

Proof. By suitable interchanges of rows and identical interchanges of columns, $|G_{nn}^{(rr)}|$ may be rearranged in the form

$$|G_{nn}^{(rr)}| = \begin{vmatrix} G_{nn}(1,1) & \theta_{nn} & \cdots & \theta_{nn} \\ G_{nn}(2,1) & G_{nn}(2,2) & & \vdots \\ \vdots & \vdots & \ddots & \vdots \\ G_{nn}(r,1) & G_{nn}(r,2) & \cdots & G_{nn}(r,r) \end{vmatrix}$$

where θ_{nn} is the $n \times n$ zero matrix. It follows that

$$|G_{nn}^{(rr)}| = |G_{nn}(1,1)|\,|G_{nn}(2,2)|\ldots|G_{nn}(r,r)|$$

$$|G_{nn}^{(rr)}| = D_n(1)\,D_n(2)\ldots D_n(r)$$

Finally, if the infinite product $\prod_{i=1}^{\infty} D_n(i)$ converges to a limit other than zero then

$$\lim_{r\to\infty} |G_{nn}^{(rr)}| = |{}^sG_{nn}|.$$

But the infinite product $\prod_{i=1}^{\infty} D_n(i)$ converges to a limit other than zero if and only if $\sum_{i=1}^{\infty} \ln D_n(i)$ converges [3]; if the last summation converges with sum c then

$$|{}^sG_{nn}| = \exp(c).$$

COROLLARY 1. *If ${}^sG_{nn}$ is a stationary multivariable l.s.i. matrix then*

$$|{}^sG_{nn}| = \prod_{i=1}^{\infty} D_n(i)$$

where $D_n(i)$ equals $D_n(1)$ for every i.

COROLLARY 2: *The determinant of the nonstationary multivariable lower finite matrix $G_{nn}^{(rr)}$ is given by*

$$|G_{nn}^{(rr)}| = \prod_{i=1}^{r} G_{nn}(i,i)$$

If $G_{nn}^{(rr)}$ is stationary then all the determinants $|G_{nn}(i,i)|$ in the above equation are equal to $|G_{nn}(1,1)|$.

COROLLARY 3. *The determinant of a single variable l.s.i. matrix is given by*

$$|{}^sG| = \lim_{j\to\infty} \prod_{i=1}^{j} g_{ii}$$

provided the limit exists.

THEOREM 2. *The inverse of a nonstationary multivariable l.s.i. matrix ${}^sF_{nn} = {}^sG_{nn}^{-1}$ is defined by the recursion formula:*
for

$$m = 1, 2, \ldots, \infty$$

$$F_{nn}(m, m) = G_{nn}^{-1}(m, m);$$

for

$$k = m - 1, \ldots, 1$$

$$F_{nn}(m, k) = -F_{nn}(m, m) \sum_{r=k}^{m-1} G_{nn}(m, r)\, F_{nn}(r, k).$$

COROLLARY 1. *If ${}^sG_{nn}$ is a stationary l.s.i. matrix then ${}^sF_{nn} = {}^sG_{nn}^{-1}$*
for

$$m = 1, 2, \ldots, \infty$$

as follows. If

$$m = 1 \quad \textit{then} \quad F_{nn}(1) = G_{nn}^{-1}(1)$$

and if $m > 1$ then

$$F_{nn}(m) = -F_{nn}(1) \sum_{k=2}^{k=m} G_{nn}(k)\, F_{nn}(r)$$

where

$$r = m + 1 - k \qquad \textit{and} \qquad G_{nn}(k) = [g_{ijk}];\, i, j = 1, 2, \ldots, n.$$

COROLLARY 2. *The inverse of a nonstationary single variable l.s.i. matrix ${}^sF = {}^sG^{-1}$ is defined by the recursion formula:*
for

$$m = 1, 2, \ldots, \infty$$

$$f_{mm} = 1/g_{mm}.$$

For

$$k = m - 1, \ldots, 1$$

$$f_{mk} = -f_{mm} \sum_{r=k}^{m-1} g_{mr} f_{rk}$$

If ${}^sG$ is stationary then ${}^sG^{-1} = {}^sF$ is defined by the recursion formula: for $m = 1, 2, \ldots, \infty$ as follows:
if

$$m = 1, \quad f_1 = 1/g_1$$

and if

$$m > 1, \quad f_m = -f_1 \sum_{k=2}^{m} g_k f_r$$

where

$$r = m + 1 - k.$$

The above theorem is useful for evaluation of the matrix of Eqn (2)

$$({}^sI_{pp} + {}^sR_{pp})^{-1}.$$

Multiplication of two nonstationary multivariable l.s.i. matrices is defined by the recursion formula:
if

$${}^sF_{nn} = {}^sG_{nn}\,{}^sW_{nn}$$

then

$$f_{ijrl} = \sum_{k=1}^{n} \sum_{d=l}^{r} g_{ikrd} w_{kjdl}$$

THEOREM 3. *The characteristic function $f_n^{(\infty)}(\lambda)$ of the nonstationary multivariable l.s.i. matrix ${}^sG_{nn}$ is given by*

$$f_n^{(\infty)}(\lambda) = \prod_{i=1}^{\infty} D_n^{(\lambda)}(i)$$

where

$$D_n^{(\lambda)}(i) = |\lambda I_{nn} - G_{nn}(i, i)|$$

and I_{nn} is the $n \times n$ identity matrix.

If $n = 2$ then for $r = 1, 2, \ldots$ the eigenvalues of ${}^sG_{nn}$ are given by

$$\lambda_r = \tfrac{1}{2}(T_2(r) \pm (T_2{}^2(r) - 4D_2(r))^{\frac{1}{2}})$$

where $T_2(r)$ is the trace of $G_{22}(r, r)$.

COROLLARY 1. *If ${}^sG_{nn}$ is a stationary multivariable l.s.i. matrix then*

$$f_n{}^{(\infty)}(\lambda) = \prod_{i=1}^{\infty} D_n{}^{(\lambda)}(i)$$

where $D_n(i) = D_n{}^{(\lambda)}(1)$ for all i.

If $n = 2$ then the eigenvalues of ${}^sG_{nn}$ are a repetition of the following two eigenvalues

$$\lambda = \tfrac{1}{2}(T_2(1) \pm (T_2{}^2(1) - 4\,D_2(1))^{\frac{1}{2}}).$$

Theorem (3) will prove useful for the design of nonstationary multivariable control systems.

COROLLARY 2. *The characteristic function $f_n{}^{(r)}(\lambda)$ of the nonstationary multivariable lower finite matrix $G_{nn}^{(rr)}$ is given by*

$$f_n{}^{(r)}(\lambda) = \prod_{i=1}^{r} |\lambda I_{nn} - G_{nn}(i, i)|.$$

If $G_{nn}^{(rr)}$ is stationary then

$$|\lambda I_{nn} - G_{nn}(i, i)| = |\lambda I_{nn} - G_{nn}(1, 1)|$$

for every i.

LEMMA 1. *The set of eigenvalues of a diagonal multivariable l.s.i. matrix ${}^sG_{nn}$ is given by the totality of the eigenvalues of the separate block matrices where the eigenvalues of a block matrix are its principal diagonal elements.*

Formulation of the problem in terms of the theory of multivariable infinite matrices is useful not only for evaluation of the determinant, the inverse and the eigenvalues of the system but also for defining further mathematical entities fot the system under study as shown by the following theorems.

THEOREM 4. *The exponential function exists for every multivariable l.s.i. matrix, all its values are l.s.i. and it misses all the nilpotent values in the field.*

Definition 6.

In the field of diagonal multivariable l.s.i. matrices define

(a).

$$({}^s\hat{G}_{nn})_r = {}^sG_{nn}\, G_{nn}^{-1}(r, r)\,. \tag{4}$$

Since $G_{nn}(r, r)$ is a diagonal matrix, the elements of $G_{nn}^{-1}(r, r)$ are merely $1/g_{iirr}$ for $i = 1, 2, \ldots, n$. The effect of $G_{nn}^{-1}(r, r)$ on ${}^sG_{nn}$ in Eqn (4) is therefore to multiply the ith diagonal block by the scalar g_{iirr}^{-1} $(i = 1, 2, \ldots, n)$. Hence in $({}^s\hat{G}_{nn})_r$ the diagonal element in the $iirr$ position equals one for every block matrix ${}^sG_{(ii)}$

(b).

$$({}^sG_{nn})_k = \prod_{r=1}^{\infty}{}^{(k)}({}^sI_{nn} - ({}^s\hat{G}_{nn})_r)$$

where (k) signifies that the factor where $r = k$ is omitted and ${}^sI_{nn}$ is the identity element in the field of multivariable infinite matrices which is defined by the diagonal matrix all of whose block matrices are equal to ${}^sI$ (the identity element in the field of single variable l.s.i. matrices).

(c).

$$(C_{nn})_k = \prod_{r=1}^{\infty}{}^{(k)}({}^sI_{nn} - ({}^s\hat{G}_{nn})_r)$$

THEOREM 5. *In the field of diagonal multivariable l.s.i. matrices if*

$$\sum_{k=1}^{\infty} \|G_{nn}^{-1}(k, k)\|$$

is convergent then

(a). *Infinite products of the form*

$$\prod_{r=1}^{\infty}({}^sI_{nn} + ({}^s\hat{G}_{nn})_r) \qquad \textit{exist}$$

(b).

$$\prod_{r=1}^{\infty}({}^sI_{nn} - ({}^s\hat{G}_{nn})_r) = {}^s\theta_{nn}$$

(c). *For all k both* $({}^sG_{nn})_k$ *and* $(C_{nn})_k$ *exist and*

$$({}^sG_{nn})_k\, ({}^sG_{nn})_r = {}^s\theta_{nn}\ \forall\, k \neq r.$$

THEOREM 6.*In the field of diagonal multivariable l.s.i. matrices if the eigenvalues of each block matrix are distinct (they can be repeated among different block matrices) and if*

$$\sum_{k=1}^{\infty} \|G_{nn}^{-1}(k,k)\|$$

is convergent then

(a).

$$\sum_{k=1}^{\infty} ({}^sG_{nn})_k\,(C_{nn}^{-1})_k = {}^sI_{nn}$$

and

$$({}^sG_{nn})_k\,(C_{nn}^{-1})_k \qquad \textit{is idempotent}$$

(b).

$${}^sG_{nn} = \sum_{k=1}^{\infty} G_{nn}(k,k)\,({}^sG_{nn})_k\,(C_{nn}^{-1})_k$$

(c). *If $p(x)$ is any function containing only positive and negative powers of x then*

$$p({}^sG_{nn}) = \sum_{k=1}^{\infty} p\big(G_{nn}(k,k)\big)\,({}^sG_{nn})_k\,(C_{nn}^{-1})_k.$$

Definition 7.

$$\{f_n^{(\infty)}(\lambda)\}_i = \prod_{k=1}^{\infty}{}^{(i)}\left(\frac{\lambda-\lambda_k}{\lambda_i-\lambda_k}\right)$$

where (i) signifies that the factor where $k = i$ is omitted.

THEOREM 7. *If the eigenvalues of a multivariable l.s.i. matrix ${}^sG_{nn}$ are such that $(\lambda-\lambda_i)\sum_{k=1}^{\infty} 1/(\lambda-\lambda_k)$ is convergent i.e. all the eigenvalues are distinct then the set $\{f_n^{(\infty)}({}^sG_{nn})\}$ forms a linearly orthonormal set of matrices.*

THEOREM 8. *If $p(x)$ is any scalar rational function whose denominator has no common factor with the reduced characteristic function of the multivariable l.s.i. matrix ${}^sG_{nn}$ such that*

$$(\lambda-\lambda_i)\sum_{k=1}^{\infty}\frac{1}{\lambda-\lambda_k}$$

is convergent then

$$p({}^sG_{nn}) = \sum_{i=1}^{\infty} p(\lambda_i)\,\{f_n^{(\infty)}({}^sG_{nn})\}_i\,.$$

For proofs of Theorems 4, 5, 6, 7 *and* 8 *see* [2].

4.2 *Stability assessment by the second approach*

In the second approach stability is considered as an input–output property of the system where the system is regarded as mapping of a normed space into another normed space and boundedness of the map is taken as the basic requirement for stability.

If G is a time varying multivariable plant with p inputs and m outputs, G can be represented by a nonstationary multivariable l.s.i. matrix ${}^sG_{pm}$. For generation of the infinite matrix representation of nonlinear, stochastic or linear systems cascaded with a partial order hold circuit of any order see [2].

Matrix equations of the control system of Fig. 1 are given by

$${}^se_p = {}^sr_p - {}^sy_p, \; {}^sf_m = {}^sK_{mp}\,{}^se_p, \; {}^sy_p = {}^sG_{pm}\,{}^sf_m.$$

Hence the system response at the sampling instants is given by

$${}^sy_{pp} = {}^sL_{pp}\,{}^sr_p$$

where

$${}^sL_{pp} = ({}^sI_{pp} + {}^sG_{pm}\,{}^sK_{mp})^{-1}\,{}^sG_{pm}\,{}^sK_{mp}.$$

${}^sL_{pp}$ is the nonstationary multivariable l.s.i. matrix representing the overall closed loop system.

Assessment of the stability of the system as an input–output property can be carried out by means of the following results which are an extension of known results in the theory of single variable infinite matrices. They clearly show that by formulating the problem in terms of the theory of multivariable infinite matrices, the multivariable case can be treated in a manner which in principle is identical to the single variable case.

Definition 8

The space $(l_p)_n$ is defined to be the set of all n-dimensional column vectors ${}^sx_n$ whose elements are infinite column vectors which belong to l_p where l_p is the space of pth summable sequences.

Similarly the multivariable null space $(c_0)_n$ is defined as the set of all vectors ${}^sx_n$ whose components are the null sequences i.e. sequences which converge to zero. The space $(c)_n$ is the multivariable space of convergent sequences. It can be readily shown that the space $(l_p)_n$ is a Banach space and $(l_2)_n$ forms a Hilbert space.

Definition 9

For a multivariable infinite matrix ${}^sG_{nm}$ two bounds denoted by $(K_r)_{nm}$ and $(K_c)_{nm}$ are defined as follows:

$$(K_r(l))_{nm} = \sum_{i=1}^{n} \sum_{j=1}^{m} \sum_{k=1}^{\infty} |g_{ijlk}|; \qquad (K_r)_{nm} = \sup_l (K_r(l))_{nm}$$

$$(K_c(k))_{nm} = \sum_{i=1}^{n} \sum_{j=1}^{m} \sum_{l=1}^{\infty} |g_{ijlk}|; \qquad (K_c)_{nm} = \sup_k (K_c(k))_{nm}.$$

If ${}^sG_{nm}$ is a l.s.i. matrix it is clear that

$$(K_r(l))_{nm} = \sum_{i=1}^{n} \sum_{j=1}^{m} \sum_{k=1}^{l} |g_{ijlk}|$$

$$(K_c(k))_{nm} = \sum_{i=1}^{n} \sum_{j=1}^{m} \sum_{l=k}^{\infty} |g_{ijlk}|.$$

THEOREM 9. *The norm of a multivariable infinite matrix defined by the* $(K_r)_{nm}$ *and the* $(K_c)_{nm}$ *bounds satisfies the axioms of a normed linear space. The* $(K_r)_{nn}$ *and* $(K_c)_{nn}$ *bounds satisfy the axioms of a normed algebra, i.e. in addition to satisfying the axioms of a normed linear space they satisfy the product axiom of a normed algebra.*

THEOREM 10. *If for a given multivariable infinite matrix* ${}^sG_{nm}$, $\lim_{l\to\infty} g_{ijlk} = 0$ *for each fixed i, j and k and if the bound* $(K_r)_{nm}$ *is finite then* ${}^sG_{nm}$ *defines a bounded operator on* $(c_0)_m$ *into* $(c_0)_n$.

For proofs of Theorems 9 and 10 see [2].

Given a multivariable infinite matrix it is possible to define the nature of the operator which the matrix represents by evaluating the $(K_r)_{nm}$ and the $(K_c)_{nm}$ bounds then checking if the sufficient and necessary conditions given in the last theorems and the next one are satisfied.

THEOREM 11. *Given a multivariable infinite matrix* ${}^sG_{nm}$.
(1) *If the* $(K_r)_{nm}$ *bound in finite and if for each r*

$$\lim_{l\to\infty} \sum_{i=1}^{n} \sum_{j=1}^{m} \sum_{k=r}^{\infty} g_{ijlk}$$

is finite then ${}^sG_{nm}$ *defines an operator mapping* $(c)_m$ *into* $(c)_n$ *where c is the set of all convergent sequences. If* $n = m = 1$ *this result is the Kojima–Schurr theorem.*

(2). *If for* $1 \leqslant p < \infty$

$$\sup_k \sum_{i=1}^{n} \sum_{j=1}^{m} \sum_{l=1}^{\infty} |g_{ijlk}|^p$$

is finite then ${}^sG_{nm}$ *defines an operator mapping* $(l_1)_m$ *into* $(l_p)_n$.

Since $(l_p)_n$, $(c_0)_n$ and $(c)_n$ are Banach spaces then it can be shown that the set of matrices mapping $(l_1)_n$ into $(l_p)_n$ or $(c_0)_n$ into $(c_0)_n$ or $(c)_n$ into $(c)_n$ form Banach algebras.

Acknowledgement

The author wishes to thank Dr. A. W. Gillies, a Reader in the Mathematics Department, The City University for helpful discussions on writing this paper.

References

1. Desoer, C. A. and Wu, M. Y., *J. Franklin Inst.*, **290** (1) (1970).
2. Makhlouf, M. A., "The Algebras, Structural Properties and Operator Calculi of Multivariable Operators and their Martix Representation". Ph.D. Thesis, The City University, London (1972).
3. Titchmarsh, E. C., "Theory of Functions". Oxford University Press (1932).

Discussion

FULLER (*University of Cambridge*). Would you say that this was a rather numerical approach or is the importance from the algebraist's aspect?

MAKHLOUF. The theory presented in this talk provides a link between the theory of functional analysis and numerical mathematics. In fact, I have used this for the development of an operator calculus which makes use of the theory of direct operational methods in the theory of generalised functions. The calculus avoids some of the troubles met on solving time-varying differential equations.† Solution of differential equations by the above methods requires the inversion of a determinant in terms of the operator D and to do it numerically one has to use an algorithm by Chen and this can be quite tedious†. Whereas, by using the inverse operation defined in Theorem 2 of this talk the solution is easier to implement.

† Liverman, T. P. Generalized Functions and Direct Operational Methods, Prentice Hall Inc., 1964.

29. A Geometric Approach to the Decoupling of Linear Multivariable Systems by Output Feedback

M. J. DENHAM

Department of Computing and Control,
Imperial College of Science and Technology,
London, England

Notation

Capital letters denote matrices, bold face capital letters denote linear vector spaces. The same letter is used to denote a matrix and its map. A^T signifies the transpose of A, and $A^{-1}\mathbf{X} \equiv \{y\colon Ay \in \mathbf{X}\}$. $\{B\}$ or $\mathbf{B}$ denotes the range of B, and $\mathbf{N}(C)$ the null space of C. The empty set is denoted by $\varnothing$, the zero space by 0. $\mathbf{X}^{\perp}$ signifies the orthogonal complement space of $\mathbf{X}$. For k a fixed positive integer, $k \equiv \{1, 2, \ldots, k\}$. Dimension of X is denoted by $d(\mathbf{X})$.

1. Introduction

The problem of decoupling a linear multivariable system

$$\dot{x} = Ax + Bu \tag{1a}$$

$$y = Cx \tag{1b}$$

by means of feedback control from the output y, is considered here in geometric terms. A general partitioning of the output vector into k subvectors is assumed

$$\begin{bmatrix} y_1 \\ \cdot \\ \cdot \\ \cdot \\ y_k \end{bmatrix} = \begin{bmatrix} C_1 \\ \cdot \\ \cdot \\ \cdot \\ C_k \end{bmatrix} x. \tag{2}$$

The decoupling problem is to construct a closed loop system with a set of input vectors v_i, $i \in \mathbf{k}$, such that v_i completely controls y_i, $i \in \mathbf{k}$, without affecting y_j, $j \neq i, j \in \mathbf{k}$. In geometric terms, the problem can be interpreted as that of constructing a set of controllability subspaces, $\{\mathbf{R}_i\}_{\mathbf{k}}$, where $\mathbf{R}_i \subset \mathbf{X} \equiv$ system state space, $i \in \mathbf{k}$, which satisfy

$$\mathbf{R}_i + \mathbf{N}_i = \mathbf{X} \qquad i \in \mathbf{k} \tag{3a}$$

$$\mathbf{R}_i \subset \bigcap_{j \neq i,\, j \in \mathbf{k}} \mathbf{N}_j \qquad i \in \mathbf{k} \tag{3b}$$

where $\mathbf{N}_i \equiv \mathbf{N}(C_i)$, and $\mathbf{R}_i$ is a controllability subspace of (A, B) if, for some F_i,

$$\mathbf{R}_i = \{A + BF_i \mid \mathbf{B} \cap \mathbf{R}_i\} \equiv \mathbf{B} \cap \mathbf{R}_i + (A + BF_i)(\mathbf{B} \cap \mathbf{R}_i)$$
$$+ \ldots + (A + BF_i)^{n-1}(\mathbf{B} \cap \mathbf{R}_i) \tag{4}$$

Writing $F(\mathbf{R}_i)$ for the set of all F_i such that (4) is true, in addition to the ability to construct $\{\mathbf{R}_i\}$ satisfying (3a, b) the existence of a solution to the state feedback decoupling problem requires that

$$\bigcap_{i \in \mathbf{k}} F(\mathbf{R}_i) \neq \varnothing \tag{5}$$

In order to define the concept of a controllability subspace, the following theorem is stated without proof [1]:

THEOREM 1. *Given A, B, $\mathbf{R} \subset \mathbf{X}$ fixed, $\mathbf{R}$ is a controllability subspace of (A, B) if and only if*

$$A\mathbf{R} \subset \mathbf{R} + \mathbf{B} \tag{6}$$

and

$$\mathbf{R} = \lim \mathbf{S}^j, \qquad j = 0, 1, \ldots \tag{7}$$

where

$$\mathbf{S}^0 \equiv 0$$

$$\mathbf{S}^r = (A\mathbf{S}^{r-1} + \mathbf{B}) \cap \mathbf{R}, \qquad r = 1, 2, \ldots$$

If a matrix K exists such that, for

$$F \in \bigcap_{i \in \mathbf{k}} F(\mathbf{R}_i), \qquad F = KC,$$

then a solution to the output feedback decoupling problem has been found, since feedback is now from the output y, which we assume to be measurable. In the following, we assume that no such K exists. In this case, it is necessary to introduce additional dynamics into the system, in order to find a solution to the problem for the higher order system.

2. State Space Extension

Additional dynamics can be introduced by adjoining $\tilde{n}$ integrators, $\tilde{n}$ to be determined, i.e.

$$\dot{\tilde{x}} = I\tilde{u} \tag{8}$$

where $\tilde{x} \in \tilde{\mathbf{X}}$, an extension of $\mathbf{X}$. The extended state space is denoted by $\hat{\mathbf{X}} = \mathbf{X} \oplus \tilde{\mathbf{X}}$. The system can now be represented in $\hat{\mathbf{X}}$ by

$$\dot{\hat{x}} = \hat{A}\hat{x} + \hat{B}\hat{u} \tag{9a}$$

$$\hat{y} = \hat{C}\hat{x} \tag{9b}$$

where

$$\hat{x} = \begin{bmatrix} x \\ \tilde{x} \end{bmatrix}, \quad \hat{y} = \begin{bmatrix} y \\ \tilde{x} \end{bmatrix}, \quad \hat{u} = \begin{bmatrix} u \\ \tilde{u} \end{bmatrix}$$

$$\hat{A} = \begin{bmatrix} A & 0 \\ 0 & 0 \end{bmatrix}, \qquad \hat{B} = \begin{bmatrix} B & 0 \\ 0 & I \end{bmatrix}, \quad \hat{C} = \begin{bmatrix} C & 0 \\ 0 & I \end{bmatrix}.$$

The outputs to be decoupled remain unaltered, and given by

$$y_i = [C_i\, 0]\hat{x} \equiv \overline{C}_i\hat{x}, \qquad i \in \mathbf{k} \tag{10}$$

The state feedback decoupling problem in the extended state space is that of constructing a set of controllability subspaces $\{\hat{\mathbf{R}}_i\}_{\mathbf{k}}$ of $(\hat{A}, \hat{B})$ where $\hat{\mathbf{R}}_i \subset \hat{\mathbf{X}}$, $i \in \mathbf{k}$ such that

$$\hat{\mathbf{R}}_i + \overline{\mathbf{N}}_i = \hat{\mathbf{X}}, \qquad i \in \mathbf{k} \tag{11a}$$

$$\hat{\mathbf{R}}_i \subset \bigcap_{j \neq i,\ j \in k} \overline{\mathbf{N}}_j, \qquad i \in \mathbf{k} \tag{11b}$$

where $\overline{\mathbf{N}}_i$ denotes $\mathbf{N}(\overline{C}_i)$, with the further requirement that

$$\bigcap_{i \in \mathbf{k}} \hat{F}(\hat{\mathbf{R}}_i) \neq \varnothing \tag{12}$$

The main result can now be stated.

THEOREM 2. *Given A, B and $\mathbf{N}_i$, $i \in \mathbf{k}$, there exists a set of controllability subspaces $\{\hat{\mathbf{R}}_i\}_\mathbf{k}$ of $(\hat{A}, \hat{B})$, where $\hat{\mathbf{R}}_i \subset \hat{\mathbf{X}} = \mathbf{X} \oplus \tilde{\mathbf{X}}$, such that* (11a, b) *and* (12) *are satisfied, and* (i) *there exists a $\hat{K}$ such that*

$$\hat{K}\hat{C} \in \bigcap_{i \in \mathbf{k}} \hat{F}(\hat{\mathbf{R}}_i)$$

(ii)
$$\bigoplus_{i \in \mathbf{k}} \hat{\mathbf{R}}_i = \hat{\mathbf{X}}$$

if and only if

$$\sum_{i \in \mathbf{k}} \mathbf{R}_i^m = \mathbf{X} \quad \text{and} \quad \mathbf{R}_i^m + \mathbf{N}_i = \mathbf{X}, \qquad i \in \mathbf{k} \tag{13}$$

where $\mathbf{R}_i^m \equiv$ maximal controllability subspace contained in

$$\bigcap_{j \neq i} \mathbf{N}_j.$$

The theorem will be proved by construction of $\{\hat{\mathbf{R}}_i\}_\mathbf{k}$ and using the following lemmas, the first two of which are proved elsewhere [2].

LEMMA 1. *A necessary condition for the existence of a solution to the state feedback decoupling problem in the extended state space $\hat{\mathbf{X}}$ is that*

$$\mathbf{R}_i^m + \mathbf{N}_i = \mathbf{X} \qquad i \in \mathbf{k}.$$

LEMMA 2. *Given $\mathbf{R} \subset \mathbf{X}$, a controllability subspace of (A, B), and $\tilde{\mathbf{X}}$, an extension of $\mathbf{X}$ such that $\hat{\mathbf{X}} = \mathbf{X} \oplus \tilde{\mathbf{X}}$, let P be the projection of $\hat{\mathbf{X}}$ onto $\mathbf{X}$ along $\tilde{\mathbf{X}}$. Then*

$$\hat{\mathbf{R}} = (P + M)\mathbf{R} \subset \hat{\mathbf{X}}$$

where $\{M\} \subset \tilde{\mathbf{X}}$, is a controllability subspace of $(\hat{A}, \hat{B})$, where

$$\hat{A} = \begin{bmatrix} A & 0 \\ 0 & 0 \end{bmatrix}, \qquad \hat{B} = \begin{bmatrix} B & 0 \\ 0 & I \end{bmatrix}.$$

LEMMA 3. *Given $\bar{\mathbf{R}} \subset \hat{\mathbf{X}}$, a controllability subspace of $(\hat{A}, \hat{B})$, where $\hat{\mathbf{X}}$, $\hat{A}$, and $\hat{B}$ are defined in Lemma 2, then*

$$\hat{\mathbf{R}} = \bar{\mathbf{R}} + \mathbf{U}$$

where $\mathbf{U} \subset \tilde{\mathbf{X}}$, is also a controllability subspace of $(\hat{A}, \hat{B})$.

Proof. *Since $\bar{\mathbf{R}}$ is a controllability subspace of $(\hat{A}, \hat{B})$,*

$$\bar{\mathbf{R}} = \lim \bar{\mathbf{S}}^m \qquad m = 0, 1, \ldots$$

where $\overline{\mathbf{S}}^0 = 0$, $\overline{\mathbf{S}}^m = (\hat{A}\overline{\mathbf{S}}^{m-1} + \hat{\mathbf{B}}) \cap \overline{\mathbf{R}}$.

Let

$$\hat{\mathbf{S}}^0 = 0, \qquad \hat{\mathbf{S}}^m = (\hat{A}\hat{\mathbf{S}}^{m-1} + \hat{\mathbf{B}}) \cap \hat{\mathbf{R}}.$$

Then

$$\hat{\mathbf{S}}^1 = \hat{\mathbf{B}} \cap \hat{\mathbf{R}} = (\hat{\mathbf{B}} + \bar{\mathbf{X}}) \cap (\overline{\mathbf{R}} + \mathbf{U}) = (\hat{\mathbf{B}} \cap \overline{\mathbf{R}}) + \mathbf{U} = \overline{\mathbf{S}}^1 + \mathbf{U}$$

If

$$\hat{\mathbf{S}}^m = \overline{\mathbf{S}}^m + \mathbf{U},$$

$$\begin{aligned} \hat{\mathbf{S}}^{m+1} &= (\hat{A}\hat{\mathbf{S}}^m + \hat{\mathbf{B}}) \cap \hat{\mathbf{R}} \\ &= [(\hat{A}\hat{\mathbf{S}}^m + \hat{\mathbf{B}}) \cap \overline{\mathbf{R}}] + \mathbf{U} \\ &= [(\hat{A}\overline{\mathbf{S}}^m + \hat{\mathbf{B}}) \cap \overline{\mathbf{R}}] + \mathbf{U} \\ &= \overline{\mathbf{S}}^{m+1} + \mathbf{U}. \end{aligned}$$

By induction,

$$\hat{\mathbf{R}} = \overline{\mathbf{R}} + \mathbf{U} = \lim \overline{\mathbf{S}}^m + \mathbf{U} = \lim \hat{\mathbf{S}}^m.$$

Since

$$\hat{A}\hat{\mathbf{R}} = \hat{A}\overline{\mathbf{R}} \subset \overline{\mathbf{R}} + \hat{\mathbf{B}} \subset \hat{\mathbf{R}} + \hat{\mathbf{B}}$$

the result of Theorem 1 completes the proof.

LEMMA 4. *Given a set of controllability subspaces* $\{\hat{\mathbf{R}}_i\}_{\mathbf{k}}$ *of* $(\hat{A}, \hat{B})$, *such that* $\bigoplus_{i\in\mathbf{k}} \hat{\mathbf{R}}_i = \hat{\mathbf{X}}$, *let* $\hat{F}(\hat{\mathbf{R}}_i) \neq \varnothing$ *be the set of* $\hat{F}$ *such that*

$$\{\hat{A} + \hat{B}\hat{F} \mid \hat{\mathbf{R}}_i \cap \hat{\mathbf{B}}\} = \hat{\mathbf{R}}_i.$$

If there exists $\hat{F}_i \in \hat{F}(\hat{\mathbf{R}}_i)$, $i \in \mathbf{k}$, *such that*

$$\hat{F}_i\hat{R}_i = \hat{K}_i\hat{C}\hat{R}_i$$

for some $\hat{K}_i$, *then there exists*

$$\hat{F} \in \bigcap_{i\in\mathbf{k}} \hat{F}(\hat{\mathbf{R}}_i),$$

such that $\hat{F} = \hat{K}\hat{C}$ *for some* $\hat{K}$.

Proof: Since the $\hat{\mathbf{R}}_i$ are disjoint, it can be shown [1] that

$$\hat{F} = \sum_{i\in\mathbf{k}} \hat{F}_i P_i \in \bigcap_{i\in\mathbf{k}} \hat{F}(\hat{\mathbf{R}}_i)$$

where P_i is the projection of $\hat{\mathbf{X}}$ onto $\hat{\mathbf{R}}_i$ along

$$\bigoplus_{j\neq i} \hat{\mathbf{R}}_j.$$

Thus

$$\hat{F}^T = \sum_{i\in\mathbf{k}} P_i^T \hat{F}_i^T$$

But $\hat{F}_i\hat{R}_i = \hat{K}_i\hat{C}\hat{R}_i$ implies that

$$\hat{R}_i^T \mathbf{F}_i^T \subset \hat{R}_i^T \hat{\mathbf{C}}^T$$

and since $\sum_{i\in\mathbf{k}} P_i^T = I$

$$\hat{\mathbf{F}}^T \subset \sum_{i\in k} P_i^T \hat{\mathbf{C}}^T = \hat{\mathbf{C}}^T$$

and there exists $\hat{K}$ such that

$$\hat{F} = \hat{K}\hat{C}.$$

LEMMA 5. *Given* $\mathbf{R} \subset \mathbf{X}$, *a controllability subspace of* (A, B), *a sufficient condition for the existence of a matrix* K *such that*

$$\{A + BKC \mid \mathbf{R} \cap \mathbf{B}\} = \mathbf{R}$$

is that

$$\mathbf{N}(CR) = 0$$

where R *is a basis matrix for* $\mathbf{R}$.

Proof: Since $\mathbf{R}$ is a controllability subspace of (A, B),

$$A\mathbf{R} \subset \mathbf{R} + \mathbf{B}.$$

Let $\mathbf{X}_1$ be such that $\mathbf{X} = \mathbf{X}_1 \oplus \mathbf{R}$, and define Q as the projection of $\mathbf{X}$ on $\mathbf{X}_1$ along $\mathbf{R}$. Then

$$QA\mathbf{R} \subset Q\mathbf{B}$$

or

$$QAR = QBZ$$

for some Z, where R is a basis matrix for $\mathbf{R}$.

If $\mathbf{N}(CR) = 0$, there exists a solution K to

$$\mathbf{Z} = -KCR.$$

Thus

$$Q(A + BKC)R = 0$$

or

$$(A + BKC)\mathbf{R} \subset \mathbf{R}.$$

But [1],

$$\{A + BF \mid \mathbf{B} \cap \mathbf{R}\} = \mathbf{R}$$

for all F such that $(A + BF)\mathbf{R} \subset \mathbf{R}$. Thus

$$\{A + BKC \mid \mathbf{B} \cap \mathbf{R}\} = \mathbf{R}.$$

3. Construction of $\{\hat{\mathbf{R}}_i\}_{\mathbf{k}}$

For $\mathbf{R}_i \subset \mathbf{X}, i \in \mathbf{k}$, define

$$\mathbf{R}^* = \bigcap_{i \in \mathbf{k}} \mathbf{R}_i^*$$

where

$$\mathbf{R}_i^* = \sum_{j \neq i} \mathbf{R}_j.$$

Given that the set of controllability subspaces $\{\mathbf{R}_i^m\}_{\mathbf{k}}$ has been constructed [1], which satisfies the condition of Lemma 1, and that

$$\sum_{i \in k} \mathbf{R}_i^m = \mathbf{X},$$

the construction proceeds as follows.

Writing $\mathbf{R}_i$ for $\mathbf{R}_i^m$ in the following, define $\mathbf{V} \subset \mathbf{X}$ such that

$$\mathbf{N}(C) = \mathbf{V} \oplus \mathbf{R}^*. \tag{14}$$

Let R_i, V be basis matrices for $\mathbf{R}_i$, $\mathbf{V}$, $i \in \mathbf{k}$. Set

$$R = [R_1 \ldots R_k\, V]$$

and let

$$\{M^T\} = \mathbf{N}(R)$$

where

$$M = [M_1 \ldots M_{k+1}].$$

Set

$$\bar{R} = \begin{bmatrix} R \\ M \end{bmatrix} = \begin{bmatrix} R_1 \ldots R_k\, V \\ M_1 \ldots M_k M_{k+1} \end{bmatrix}$$

and define

$$\bar{\mathbf{R}}_i = \left\{ \begin{bmatrix} R_i \\ M_i \end{bmatrix} \right\} \quad i \in \mathbf{k} \quad \text{and } \bar{\mathbf{V}} = \left\{ \begin{bmatrix} V \\ M_{k+1} \end{bmatrix} \right\}$$

If $\mathbf{V} = 0$ set $\hat{\mathbf{R}}_i = \bar{\mathbf{R}}_i \quad i \in \mathbf{k}$.

If $\mathbf{V} \neq 0$, find linearly independent subspaces $\tilde{\mathbf{R}}_i \subset \mathbf{M}_{k+1}$, for $i \in \tilde{\mathbf{k}} \subset \mathbf{k}$, such that

$$N(C) \cap (\bar{\mathbf{R}}_i + \tilde{\mathbf{R}}_i) = 0 \tag{15}$$

and

$$\bigoplus_{i\in\mathbf{k}} \tilde{\mathbf{R}}_i = \mathbf{M}_{k+1}. \tag{16}$$

For $i \in \tilde{\mathbf{k}}$, set $\hat{\mathbf{R}}_i = \overline{\mathbf{R}}_i + \tilde{\mathbf{R}}_i$ and, for $i \in \mathbf{k} - \tilde{\mathbf{k}}$, set $\hat{\mathbf{R}}_i = \overline{\mathbf{R}}_i$.

4. Proof of Theorem 2.

The proof will show that the construction procedure above yields $\{\hat{\mathbf{R}}_i\}_{\mathbf{k}}$ which satisfy the requirements of Theorem 2.

For $i \in \mathbf{k}$, by construction

$$\mathbf{M}_i^T = R_i^{-1}\left(\sum_{j\neq i}\mathbf{R}_j + \mathbf{V}\right).$$

Now,

$$R_i^{-1}\left(\sum_{j\neq i}\mathbf{R}_j\right) + R_i^{-1}\mathbf{V} \subset R_i^{-1}\left(\sum_{j\neq i}\mathbf{R}_j + \mathbf{V}\right) = \mathbf{M}_i^T$$

and

$$\begin{aligned} R_i^T\left(\sum_{j\neq i}\mathbf{R}_j\right)^{\perp} &= R_i^T\left(\bigcap_{j\neq i}\mathbf{R}_j^{\perp}\right) \\ &= R_i^T\sum_{i\in\mathbf{k}}\left(\bigcap_{j\neq i}\mathbf{R}_j^{\perp}\right) \\ &= R_i^T(\mathbf{R}^*)^{\perp}. \end{aligned}$$

Thus

$$\begin{aligned} R_i^{-1}\mathbf{R}^* + R_i^{-1}\mathbf{V} &= R_i^{-1}(\mathbf{R}^* \oplus \mathbf{V}) \\ &= R_i^{-1}\mathbf{N}(C) \\ &\subset \mathbf{M}_i^T. \end{aligned}$$

Now,

$$\mathbf{N}(\hat{C}\overline{R}_i) = \mathbf{N}\begin{bmatrix} CR_i \\ M_i \end{bmatrix} \neq 0$$

implies the existence of $\mathbf{L}_i \neq 0$, such that $\mathbf{L}_i \subset R_i^{-1}\mathbf{N}(C)$ and $\mathbf{L}_i \subset \mathbf{N}(M_i) = (\mathbf{M}_i^T)^{\perp}$. But $\mathbf{L}_i \subset R_i^{-1}\mathbf{N}(C) \subset \mathbf{M}_i^T$ implies $\mathbf{L}_i = 0$. Therefore,

$$\mathbf{N}(\hat{C}\overline{R}_i) = 0. \tag{17}$$

By Lemma 2, the $\{\overline{\mathbf{R}}_i\}_{\mathbf{k}}$ are controllability subspaces of $(\hat{A}, \hat{B})$, and by Lemmas 4 and 5, there exists a matrix $\hat{K}$ such that

$$\{\hat{A} + \hat{B}\hat{K}\hat{C} \mid \overline{\mathbf{R}}_i \cap \hat{\mathbf{B}}\} = \overline{\mathbf{R}}_i \qquad i \in \mathbf{k}.$$

Satisfaction of (11a, b) by $\{\overline{\mathbf{R}}_i\}_{\mathbf{k}}$ follows simply, assuming (13).

If $\mathbf{V} = 0$, therefore

$$\bigoplus_{i\in \mathbf{k}} \overline{\mathbf{R}}_i = \hat{\mathbf{X}} \tag{18}$$

and the $\{\overline{\mathbf{R}}_i\}_{\mathbf{k}}$ satisfy the conditions of Theorem 2.

Consider now when $\mathbf{V} \neq 0$. Since

$$M_{k+1}^T = V^{-1} \sum_{i\in k} \mathbf{R}_i = V^{-1}\mathbf{X}$$

it follows that

$$\mathbf{N}(M_{k+1}) = 0.$$

Thus, independent subspaces $\tilde{\mathbf{R}}_i \subset \mathbf{M}_{k+1}$, $i \in \tilde{\mathbf{k}} \subset \mathbf{k}$, can be chosen such that

$$\bigoplus_{i\in \mathbf{k}} \tilde{\mathbf{R}}_i = \mathbf{M}_{k+1}. \tag{19}$$

Write

$$\begin{bmatrix} V \\ M_{k+1} \end{bmatrix} = \begin{bmatrix} V_1 \ldots V_s \\ \tilde{M}_1 \ldots \tilde{M}_s \end{bmatrix}$$

such that, for each $j \in \mathbf{s}$, there exists $\mathbf{R}_i$, $i \in \mathbf{k}$ for which

$$\mathbf{V}_j \cap \mathbf{R}_i = 0.$$

This partitioning is always possible, since for arbitrary V_j, let

$$\mathbf{Z}_i = \mathbf{V}_j \cap \mathbf{R}_i \qquad i \in \mathbf{k}.$$

If $\mathbf{Z}_i \neq 0$, for all $i \in \mathbf{k}$, and the $\mathbf{Z}_i$ are independent, $\mathbf{V}_j$ can be further partitioned into the set of $\mathbf{Z}_i$.

Then, for $L \neq i, i \in \mathbf{k}$

$$\begin{aligned} \mathbf{Z}_i \cap \mathbf{R}_L &= \mathbf{V}_j \cap \mathbf{R}_i \cap \mathbf{R}_L \\ &= (\mathbf{V}_j \cap \mathbf{R}_i) \cap (\mathbf{V}_j \cap \mathbf{R}_L) \\ &= \mathbf{Z}_i \cap \mathbf{Z}_L \\ &= 0. \end{aligned}$$

If the $\mathbf{Z}_i$, $i \in \mathbf{k}$ are not independent

$$0 \neq \mathbf{Z}^* \subset \mathbf{R}^* \cap \mathbf{V}$$

which is false by construction of $\mathbf{V}$.

Now, set $\tilde{\mathbf{R}}_i = \tilde{\mathbf{M}}_j$ and

$$\hat{\mathbf{R}}_i = \bar{\mathbf{R}}_i + \tilde{\mathbf{R}}_i \qquad i \in \tilde{\mathbf{k}}$$

$$\hat{\mathbf{R}}_i = \bar{\mathbf{R}}_i \qquad i \in \mathbf{k} - \tilde{\mathbf{k}}.$$

For $\hat{\mathbf{R}}_i = \bar{\mathbf{R}}_i$, it has been shown that $\mathbf{N}(\hat{C}\hat{R}_i) = 0$.

For $\hat{\mathbf{R}}_i = \bar{\mathbf{R}}_i + \tilde{\mathbf{R}}_i$, write

$$\hat{C}\hat{R}_i = \begin{bmatrix} CR_i & 0 \\ M_i & \tilde{M}_j \end{bmatrix} \qquad i \in \tilde{\mathbf{k}}.$$

Now, $\mathbf{N}(\hat{C}\hat{R}_i) \neq 0$, implies the existence of a subspace $\mathbf{T}_i \neq 0, i \in \tilde{\mathbf{k}}$, such that

$$\mathbf{T}_i \subset [R_i\, 0]^{-1}\, \mathbf{N}(C)$$

and

$$\mathbf{T}_i \subset \mathbf{N}[M_i\, \tilde{M}_j] = \left\{ \begin{bmatrix} M_i^{\,T} \\ \tilde{M}_j^{\,T} \end{bmatrix} \right\}^{\perp}.$$

Also,

$$[R_i V_j]^{-1} \sum_{m \neq i,\, m \in \mathbf{k}} \mathbf{R}_m + [R_i V_j]^{-1} \sum_{L \neq j,\, L \in \mathbf{s}} \mathbf{V}_L$$

$$\subset [R_i V_j]^{-1} \left(\sum_{m \neq i,\, m \in \mathbf{k}} \mathbf{R}_m + \sum_{L \neq j,\, L \in \mathbf{s}} \mathbf{V}_L \right) = \left\{ \begin{bmatrix} M_i^{\,T} \\ \tilde{M}_j^{\,T} \end{bmatrix} \right\}.$$

But

$$\begin{aligned} [R_i V_j]^{-1} \sum_{m \neq i,\, m \in \mathbf{k}} \mathbf{R}_m &= R_i^{-1} \sum_{m \neq i,\, m \in \mathbf{k}} \mathbf{R}_m + V_j^{-1} \sum_{m \neq i,\, m \in \mathbf{k}} \mathbf{R}_m \\ &= R_i^{-1}\, \mathbf{R}^* + V_j^{-1} \sum_{m \in \mathbf{k}} \mathbf{R}_m \\ &= R_i^{-1}\, \mathbf{R}^* + \tilde{\mathbf{X}} \end{aligned}$$

since $V_j^{-1}\, \mathbf{R}_i = 0$ by construction. Also

$$\begin{aligned} [R_i V_j]^{-1} \sum_{L \neq j,\, L \in \mathbf{s}} \mathbf{V}_L &= R_i^{-1} \sum_{L \neq j,\, L \in \mathbf{s}} \mathbf{V}_L + V_j^{-1} \sum_{L \neq j,\, L \in \mathbf{s}} \mathbf{V}_L \\ &= R_i^{-1}\, \mathbf{V} \end{aligned}$$

since

$$\mathbf{V}_L \cap \mathbf{V}_j = 0, \qquad L \in \mathbf{s}$$

and

$$R_i^{-1}\, \mathbf{V}_j = 0,$$

by construction.

Thus,

$$[R_i\,0]^{-1}\,\mathbf{N}(C) = [R_i\,0]^{-1}(\mathbf{R}^* + \mathbf{V})$$
$$\subset \left\{ \begin{bmatrix} M_i^{\,T} \\ \tilde{M}_j^{\,T} \end{bmatrix} \right\}$$

which implies that $\mathbf{T}_i = 0, \ i \in \tilde{\mathbf{k}}$.

By Lemma 3, the $\hat{\mathbf{R}}_i$, $i \in \tilde{\mathbf{k}}$, are controllability subspaces and clearly satisfy (11a, b) assuming (13). Furthermore, since

$$\mathbf{N}(\hat{R}_i) \subset \mathbf{N}(\hat{C}\hat{R}_i) = 0 \qquad i \in \mathbf{k} \tag{20}$$

and

$$\bigoplus_{i \in \tilde{\mathbf{k}}} \tilde{\mathbf{R}}_i = \mathbf{M}_{k+1}$$

it follows that

$$\bigoplus_{i \in \mathbf{k}} \hat{\mathbf{R}}_i = \hat{\mathbf{X}} \tag{21}$$

and $\{\hat{\mathbf{R}}_i\}_{\mathbf{k}}$ satisfies the conditions of Theorem 2.

Necessity of (13), i.e.

$$\mathbf{R}_i + \mathbf{N}_i = \mathbf{X} \qquad i \in \mathbf{k}$$

is proved by Lemma 1, since output feedback is a special case of state feedback.

5. Construction of the Feedback Control

The output feedback decoupling control is given by

$$\hat{u} = \hat{K}\hat{y} + \hat{G}v. \tag{22}$$

Using the construction of Lemma 5, given $\{\hat{\mathbf{R}}_i\}_{\mathbf{k}}$, $\hat{K}_i$, $i \in \mathbf{k}$, can be determined such that

$$(\hat{A} + \hat{B}\hat{K}_i\hat{C})\hat{\mathbf{R}}_i \subset \hat{\mathbf{R}}_i \qquad i \in \mathbf{k}$$

and Lemma 4 guarantees the existence of $\hat{K}$ such that

$$(\hat{A} + \hat{B}\hat{K}\hat{C})\,\hat{R}_i \subset \hat{R}_i \qquad i \in \mathbf{k}.$$

Write

$$\hat{G} = [\hat{G}_1 \ldots \hat{G}_k].$$

Then, $\hat{G}_i$ is chosen such that

$$\hat{B}\hat{\mathbf{G}}_i = \hat{\mathbf{B}} \cap \overline{\mathbf{R}}_i \qquad i \in \mathbf{k}. \tag{23}$$

6. Example

$$A = \begin{bmatrix} 0 & 1 & 1 & 0 & 0 \\ 0 & -1 & 0 & 0 & 0 \\ 0 & 1 & -2 & 1 & 1 \\ 0 & 1 & 0 & -2 & 1 \\ 1 & 0 & 1 & 0 & -1 \end{bmatrix}, \quad B = \begin{bmatrix} 0 & 0 & 0 \\ 1 & 0 & 1 \\ 0 & 0 & 1 \\ 0 & 0 & 0 \\ 0 & 1 & 0 \end{bmatrix}$$

$$C_1 = \begin{bmatrix} 1 & 0 & 0 & 0 & 0 \\ 0 & 1 & 1 & 0 & 0 \end{bmatrix}, \quad C_2 = [0 \ \ 0 \ \ 0 \ \ 1 \ \ 0]$$

$\{\mathbf{R}_i^M\}$ can be constructed as

$$\mathbf{R}_1^M = \left\{ \begin{bmatrix} 1 \\ 0 \\ 0 \\ 0 \\ 0 \end{bmatrix}, \begin{bmatrix} 0 \\ 1 \\ 0 \\ 0 \\ -1 \end{bmatrix}, \begin{bmatrix} 0 \\ 0 \\ 1 \\ 0 \\ 0 \end{bmatrix} \right\}, \quad \mathbf{R}_2^M = \left\{ \begin{bmatrix} 0 \\ 1 \\ -1 \\ 0 \\ 0 \end{bmatrix}, \begin{bmatrix} 0 \\ 0 \\ 0 \\ 1 \\ 0 \end{bmatrix}, \begin{bmatrix} 0 \\ 0 \\ 0 \\ 0 \\ 1 \end{bmatrix} \right\}$$

Thus,

$$\mathbf{R}_i^M + \mathbf{N}_i = \mathbf{X}, \qquad i = 1, 2$$

and

$$\mathbf{R}_i^M + \mathbf{R}_2^M = \mathbf{X}.$$

Then, for

$$\mathbf{N}(C) = \mathbf{V} \oplus \mathbf{R}^*$$

let

$$\mathbf{V} = \{[0 \ \ 0 \ \ 0 \ \ 0 \ \ 1]^T\}.$$

Thus

$$\overline{\mathbf{R}} = \left[\begin{array}{ccc:ccc:c} 1 & 0 & 0 & 0 & 0 & 0 & 0 \\ 0 & 1 & 0 & 1 & 0 & 0 & 0 \\ 0 & 0 & 1 & -1 & 0 & 0 & 0 \\ 0 & 0 & 0 & 0 & 1 & 0 & 0 \\ 0 & -1 & 0 & 0 & 0 & 1 & 1 \\ \hdashline 0 & 1 & -1 & -1 & 0 & 1 & 0 \\ 0 & 0 & 0 & 0 & 0 & 1 & -1 \end{array}\right] = \left[\begin{array}{c:c:c} R_1 & R_2 & V \\ \hdashline M_1 & M_2 & M_3 \end{array}\right].$$

Since $\mathbf{V} \cap \mathbf{R}_1 = 0$, $\hat{\mathbf{R}}_1$ and $\hat{\mathbf{R}}_2$ are chosen as follows.

$$\hat{\mathbf{R}}_1 = \left\{ \begin{bmatrix} 1 \\ 0 \\ 0 \\ 0 \\ 0 \\ 0 \\ 0 \end{bmatrix}, \begin{bmatrix} 0 \\ 1 \\ 0 \\ 0 \\ -1 \\ 1 \\ 0 \end{bmatrix}, \begin{bmatrix} 0 \\ 0 \\ 1 \\ 0 \\ 0 \\ -1 \\ 0 \end{bmatrix}, \begin{bmatrix} 0 \\ 0 \\ 0 \\ 0 \\ 0 \\ 0 \\ -1 \end{bmatrix} \right\} \hat{\mathbf{R}}_2 = \left\{ \begin{bmatrix} 0 \\ 1 \\ -1 \\ 0 \\ 0 \\ -1 \\ 0 \end{bmatrix}, \begin{bmatrix} 0 \\ 0 \\ 0 \\ 1 \\ 0 \\ 0 \\ 0 \end{bmatrix}, \begin{bmatrix} 0 \\ 0 \\ 0 \\ 0 \\ 1 \\ 1 \\ 1 \end{bmatrix} \right\}$$

The feedback matrix $\hat{K}$ then has the form

$$\hat{K} = \begin{bmatrix} k_{11} & k_{12} & k_{13} & 0 & 1 \\ k_{21} & k_{22} & k_{23} & -\frac{1}{2} & 1 \\ -1-k_{11}-k_{21} & -\frac{1}{2}-k_{12}-k_{22} & -\frac{1}{2}-\frac{1}{2}k_{13} & 1 & -2 \\ k_{11} & \frac{1}{2}+k_{12} & \frac{1}{2}+k_{23}-\frac{1}{2}k_{13} & -3/2 & 1 \\ k_{51} & k_{52} & k_{23} & \frac{1}{2} & -1 \end{bmatrix}$$

Writing $\tilde{A} = \hat{A} + \hat{B}\hat{K}\hat{C}$, $\hat{A}$ can be transformed [1] into block diagonal form

$$\tilde{A}_1 = M\tilde{A}M^{-1}$$

$$= \left[\begin{array}{cccc|ccc} 0 & 1 & 0 & 0 & 0 & 0 & 0 \\ -2-2k_{21}-k_{11} & -1-k_{12}-2k_{22} & -2 & -3 & 0 & 0 & 0 \\ -2k_{11} & -2k_{12} & -2 & -2 & 0 & 0 & 0 \\ -1-k_{21}+k_{51} & k_{52}-k_{22} & -1 & -2 & 0 & 0 & 0 \\ \hline 0 & 0 & 0 & 0 & -3/2 & -\frac{1}{2}+k_{23}+\frac{1}{2}k_{13} & \frac{1}{2} \\ 0 & 0 & 0 & 0 & 1 & -2 & 0 \\ 0 & 0 & 0 & 0 & 1 & 1+2k_{23}-k_{13} & -1 \end{array}\right]$$

Similarly,

$$\bar{C}M^{-1} = \left[\begin{array}{cccc|ccc} 1 & 0 & 0 & 0 & 0 & 0 & 0 \\ 0 & 1 & 0 & 0 & 0 & 0 & 0 \\ \hline 0 & 0 & 0 & 0 & 0 & 1 & 0 \end{array}\right]$$

and for $\hat{G}$ determined as in Section 5.

$$M\hat{B}\hat{G} = \left[\begin{array}{cc:cc} 0 & 0 & 0 & 0 \\ 2 & 1 & 0 & 0 \\ 0 & 2 & 0 & 0 \\ 1 & 0 & 0 & 0 \\ \hdashline 0 & 0 & 2 & -1 \\ 0 & 0 & 0 & 0 \\ 0 & 0 & 0 & 2 \end{array}\right]$$

The partitioning indicates the decoupled nature of the system.

The dimension of the extension in this case is the dimension of $\mathbf{N}(C)$. In general, $\tilde{n} = n_0 + v$, where

$$n_0 = \dim\left(\sum_{i \in \mathbf{k}} \mathbf{R}_i^m\right) - \sum_{i \in \mathbf{k}} \dim \mathbf{R}_i^m$$

$v = \dim(\mathbf{V})$, using the construction of Section 3.

7. Conclusion

A solution of the output feedback decoupling problem, leading to a controllable and observable system, assuming controllability and observability of the original system, has been obtained. It should be noted that the existence of a state feedback decoupling control for the original system (1), (2), is not a necessary condition for the existence of this solution.

The geometric approach used here does not appear to offer a means of determining the minimal order of such an extension. However, the concept of invariant subspaces has yielded some insight into the structure of multivariable systems, and has also found application in the fields of disturbance rejection [3], multivariable tracking problems [4], and observer theory [5].

References

1. Morse, A. S. and Wonham, W. M., Status of non-interacting control. *I.E.E.E. Trans. autom. Control* **AC-16** (1971), 568–581.
2. Morse, A. S. and Wonham, W. M., Decoupling and pole assignment by dynamic compensation. *SIAM J. Control* **8** (1970), 317–337.

3. Wonham, W. M. and Morse, A. S., Decoupling and pole assignment in linear multivariable systems: A geometric approach. *SIAM J. Control* **8** (1970), 1–18.
4. Bhattacharyya, S. P. and Pearson, J. B., On error systems and the servo-mechanism problem. *Int. J. Control* **15** (1972), 1041–1062.
5. Wonham, W. M., Dynamic observers—geometric theory. *I.E.E.E. Trans. autom. Control* **AC-15** (1970), 258–259.

Discussion

MURDOCH (*Brunel University*). Can one alter the eigenvalues of the decoupled system, and what freedom do you have?

DENHAM. One's freedom in choosing the closed loop eigenvalues is limited by the fact that feedback is restricted to be from the output, and the extension of the state space proposed in my paper is not sufficient in general to allow arbitrary assignment of all the eigenvalues.

FULLER (*University of Cambridge*). It occurs to me to wonder how practical this problem is. Do you know any problems where you have to have decoupled outputs?

DENHAM. The decoupled structure is desirable in several practical situations, including aircraft engine control, nuclear reactors, helicopters and VTOL aircraft.

Author Index

Numbers in *italics* refer to the pages where references are listed in full.

W

Y

Z

Subject Index

(Numbers in *italics* indicate that the subject is referred to in Discussion)

R

S

T

V

W

Z